The

SOUND

of the

SEA

The

SOUND

of the

SEA

Seashells and the
Fate of the Oceans

Cynthia Barnett

W. W. NORTON & COMPANY
Independent Publishers Since 1923

For information about permission to reproduce selections from this book, write to
Permissions, W. W. Norton & Company, Inc., 500 Fifth Avenue, New York, NY 10110

For information about special discounts for bulk purchases, please contact
W. W. Norton Special Sales at specialsales@wwnorton.com or 800-233-4830

Manufacturing by LSC Communications, Harrisonburg
Production manager: Beth Steidle

ISBN 978-0-393-65144-7

W. W. Norton & Company, Inc., 500 Fifth Avenue, New York, N.Y. 10110
www.wwnorton.com

W. W. Norton & Company Ltd., 15 Carlisle Street, London W1D 3BS

1 2 3 4 5 6 7 8 9 0

To my Mom, Gerry,
for a lifetime of seashells.

We can have a surfeit of treasures—an excess of shells,
where one or two would be significant.

—Anne Morrow Lindbergh, *Gift from the Sea*

CONTENTS

PART III: ORACLE

The

SOUND

of the

SEA

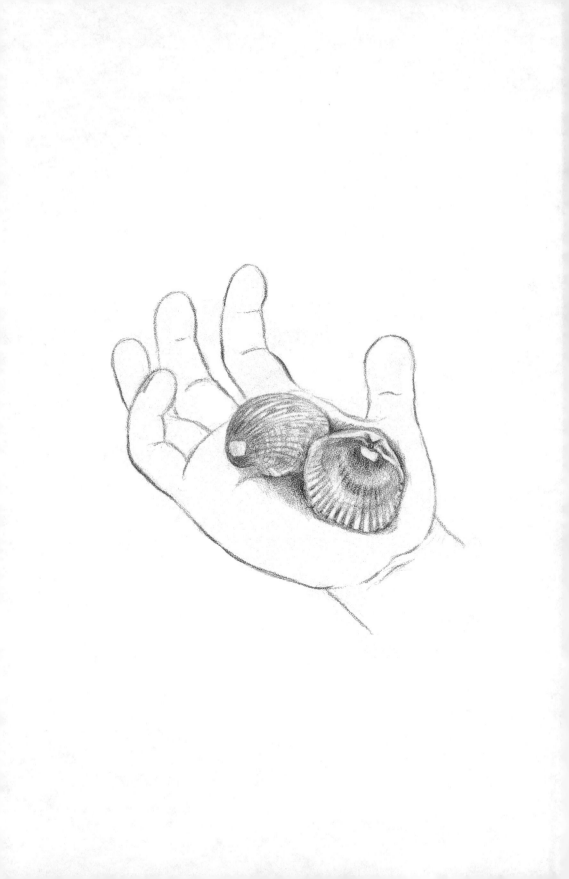

COCKLES

One hundred thousand years ago, a human cousin walked a rock-ribbed beach along the Mediterranean Sea, her head lowered and her large eyes scanning the shoreline. Now and again she stopped, bent her strong body, and picked up a seashell.

Among the polished whorls and sturdy half-shells washed ashore a couple miles from her cave, the Neanderthal girl knew precisely what she was looking for: cockle shells of a certain size and shape—about an inch across, perfectly round, and with a natural hole in the top.

She was picky about the hole, too. She collected those shells with eyelets she deemed best for threading. Her appreciation for seashells beyond food, and her imagination to string them together for a necklace or some other intention, would help scientists overturn nearly two centuries of assumptions and poorly conceived science that Neanderthals were dim-witted brutes.

The cockle shells gathered in Neanderthal times were discovered fused into the maw of a sea cave overlooking Spain's Cartagena Harbor. Several other shells found in the cave from the same era had been harvested live, for eating. Archaeologists could tell from their unblemished contours that they'd never bumped along the rocky shore.

The cockles had tumbled onto the beach empty. Someone collected them intentionally, but not for food. One keeper seashell, from a bittersweet clam, had been painted red. Another, from a thorny oyster, had a long second life as a cosmetic case. It still held a reddish pigment hand-ground from bits of hematite, pyrite, and other minerals, none found naturally in the cave.

These eons later, the powder still sparkles. And the girl's human cousins are still picking up seashells.

WHEN I READ about the Neanderthal shell cache, I wondered whether the collector could have been a child. I imagined a young girl about five. That was my daughter's age when, during a beach weekend on the east coast of Florida, she became obsessed with collecting only those shells with perfect holes in the top, for stringing necklaces and driftwood mobiles.

Those were the Bead Years in our house. In precisely ordered tackle boxes, she amassed colored beads and clear beads, owl beads and Scottie-dog beads, alphabet beads to spell out her friends' names and I♥u. Now, as we slowly walked a bank of shell and seaweed sculpted by the high tide, that same collector's gene was switched on at the Atlantic Ocean. Her fixated silence amplified the breaking waves beside us, the scold of gulls above, and the clink, clank, clatter of shells into her purple sand bucket. She skipped the shining olive shells, shark's eyes, and other coiled prizes pressed into the wet sand. Like our ancestor, Ilana chose rounded half-shells: orange Atlantic cockles, purple-striped calico scallops, and scads of pendant-sized surf clams with hard-candy stripes and colors, all with a little round hole in the top.

When she'd chosen all she wanted, she wrote her name in big letters in the sand, along with the name of our town a couple hours inland, as if signing a seashell invoice from King Neptune.

Ten years later, those shells are still stashed here in our landlocked town, in a heavy little bag pushed to the back of a cabinet in my study. They've been tucked there since I rescued them from a pile of household detritus my husband was about to toss for spring cleaning. The shell necklaces and mobiles we strung on fishing line are long broken and thrown away. But I couldn't bring myself to toss seashells so carefully chosen by the hands of a kindergartner, especially now that she's a teenager keeping us at arm's length.

I know of lots of other shell stashes, and inherited one. My mother-in-law once gave me a hand-painted porcelain cup from her late mother's china cabinet. The delicate legacy was not the porcelain, but the two dozen jingle shells inside, shimmering pale yellow and orange. My husband's grandmother had collected the translucent "mermaid's toenails" with her young daughters along the beach at Long Island's Peconic Bay. She stashed the memory in the small cup. Seventy years later, the shells still tinkle when I sift my fingers through their diaphanous forms. Finer than the porcelain, they are an order of magnitude stronger.

I wonder how many small but weighty bags and boxes of seashells are similarly hoarded in cabinets and closets from Muskegon to Mumbai; near the sea and many miles from it. Amid all the natural wonders cumulated at Thomas Jefferson's Monticello, a small Money Cowrie, a *Monetaria moneta* not acquired by the founding father, is for me the most compelling. The humble cowrie was discovered in a subfloor pit beneath a slave house. A hole in the back of the shell and two grooves rubbed by a thread that had once run through it are part of the evidence that an enslaved African likely brought it to Virginia. The shell might have been attached to clothing, or somehow survived as a necklace.

It might have been a person's secret stash; a connection to home.

A SHELL TOO is a home, and the life's work of the animal that secretes it layer by layer with minerals from the surrounding environment. Consider the mollusks; soft animals far better known for the shells they build than for the lives within. The second-largest group of animals behind the arthropods that include insects, mollusks are everywhere—from the hundreds of snail species high in the Himalayas to the bone-white clams clustered at Earth's greatest depths, filtering hydrothermal vents at Mariana Trench in the western Pacific Ocean.

Seashells are the work of marine mollusks, the most diverse group of animals in the oceans. They inhabit worlds tiny—spiraled *Ammonicera* washing up on beaches around the globe with exquisite stripes too small to admire; and worlds vast—*Tridacna gigas*, or giant clams, weighing hundreds of pounds and glowing with millions of microalgae.

Marine mollusks settle on reefs and rocks and seagrasses and sandy beaches and mudflats and countless places above and below. The violet sea snail *Janthina janthina* lives only on tropical surface waters, a molluscan Huck Finn floating on its own bubble raft. If anything happens to its homemade boat, the purple-shelled Huck will sink and die. Thin-spindled *Tibia fusus* hunkers deep in the sand thanks to its siphon that draws water for respiration through a long, thin shell canal like a hypodermic needle from a vial. The carrier snails, *Xenophoridae*, cement other shells, bits of coral, and even little pebbles onto their own shells in elaborate camouflage.

Marine mollusks are vegetarians and cannibals, fish hunters and filter feeders, algae distillers, and carrion eaters. They are sedentary blobs that leap and swim. Shy beings that create the showiest architecture of all time. Squishy invertebrates that make some of the hardest building materials known. Vulnerable species with the longest evolutionary history of any living today.

From the shell cults of prehistory to the impressive number of mollusk-inspired Pokémon characters, no creatures have stirred human admiration for as long or as intimately. Yet even in our time of Extinction Rebellion in the streets, and images of endangered species projected 1,250 feet up the side of the Empire State Building, the mollusks remain almost wholly anonymous artists.

Earth's great shell middens—mounds of oyster, whelk, and other shells piled high along the world's coastlines—testify to their significance as food since at least the early Stone Age. Raw or roasted, mollusks have often satisfied our appetites. Their iron, zinc, and other brain-boosting nutrients may have helped evolve the bigger brains that made us human.

But it is their shells that have captured our imaginations. Seashells were money before coin, jewelry before gems, art before canvas. Fossilized mussel shells found on the banks of the Solo River in Java, Indonesia, site of "Java Man," bear geometric zigzags engraved half a million years ago by a purposeful hand. The decorated shells represent cognition among our human predecessors *Homo erectus*, and some of the world's oldest-known art.

Seashells are the earliest-known keepsakes tucked into graves. A small cone shell, *Conus ebraeus*, holds its rose-colored tint after 75,000 years interred. The stubby cone was unearthed from the grave of a four-to-six-month-old infant in a large rock shelter in South Africa known as Border Cave. It had been notched by hand, strung onto a pendant, and worn for many years before being placed with the Stone Age baby.

Shells are the most-collected naturalia along with rocks; easier to amass than butterflies and more affordable than gemstones. They are collected by children and by kings. A shell collection was unearthed from the ruins of Pompeii. Their devoted aficionados, known as conchologists, admit to a certain madness. But the polished forms captivate even casual admirers strolling the beach or museum displays: The perfect symmetry of a Chambered Nautilus. The pink-glossed lip of a Queen Conch. The pearlescent inlay of an abalone. The exceptional spines of a murex—raptor talons in some species, delicate doll combs in others. The distant roil in a trumpet shell, held to the ear for wisdom from the sea.

WE'VE ALWAYS TRIED to listen to shells. It's striking how often they've led us to clear truths in murky times. Shells of unfamiliar species like ammonites provided evidence of evolution and extinction in an era of loyal belief that God made all creatures at once in everlasting perfection. Seashells on mountaintops told a story of shifting continents and rising and falling seas, articulating an Earth history much older than the six thousand years in the Bible. Layered in canyon walls and cliffsides and strata belowground, marine shells recorded a fossil diary for half a billion years, leaving one of Earth's most complete archives of past life and global change.

Just as they hold Earth's memory in mountains, or a mother's memory in a small cup, seashells are more accurate recorders of human history than the humans who got to write it down. Shell middens once rose in North America like temples in the ancient world. Some early scientists and historians considered them mere garbage heaps of nomadic people. But the shells—contoured by long-ago hands to gird homes, sanctuaries, and public buildings, or buried in ancient cemeteries and shellwork factories—establish major pre-Columbian cities on U.S. soil. The "great cities of shell" make clear that the New World was hardly new, much less settled by bearded men from sailing ships. Around the world, shells are correcting history, fact-checking vanquishers.

The Portuguese archaeologist João Zilhão has spent his career tunneling deep into rock shelters and caves to understand how Neanderthals lived. Interpreting their marine shells from caves along Spain's Iberian Peninsula has helped him shed light on Neanderthals' intelligence, no less their humanity. The cockles join several tantalizing shell discoveries in proving that notions of symbolism and beauty long predate anatomically modern humans.

As early people interacted with greater numbers of outsiders, a cockle-shell necklace or other shell pendant might have been a way to burnish individual identity or declare allegiance to a social group. Coastal dwellers naturally adorned themselves with marine animals. Farther inland, the adornments were eagle claws or mammal teeth. Once trade networks took off, the transcendent appeal of shells carried them far from their ocean homes. Different species of the spiny oyster or *Spondylus*, the wild-spiked, blood-red bivalve that held the sparkly Neanderthal powder, are found in Neolithic burial sites across Europe as well as the rituals and jewelry of pre-Columbian cultures in South America, where they made their way from the depths of the Pacific Ocean to the tops of the Andes.

I asked Zilhão whether the cockle shells hidden in the Neanderthal sea cave could have been collected by a child. He didn't hesitate. "Children and adolescents are more open to discovery," he said. "One might guess that this use for an important social purpose was originally prompted by child's play; a collection that begins while a child is helping out fishing or shell fishing on the coast, and running along and picking up these beautiful objects.

"There is something fundamental about shells' aesthetics that pleases the brain, that must be very powerful. This is not just symbolic thinking. It's this very modern sense of what is beautiful."

ONE MUGGY JUNE night at a ballroom seashell auction in Captiva, Florida, the barrier island where Anne Morrow Lindbergh wrote *Gift from the Sea*, her beloved 1955 book of shell wisdom, I watched two collectors try to outbid one another for a scarlet *Spondylus crassisquama*. Its two halves still attached at the hinge, the shell was big and round as a baseball and covered with at least a hundred curved spines, jutting short and long like a protozoan pincushion.

I appreciated those two modern women hungry for a species once revered by the Indigenous people of the Andes as the food of Pachamama, a fertility goddess who was also considered Earth's mother. Bidding started at $50 and went up in discreet $25 waves of cardboard auction paddles to a final price of $250.

It wasn't close to the highest price of the night for a single shell. A man named Donald Dan paid $2,000 for a rare slit shell, its conical steppes built by a secretive deep-sea mollusk named *Entemnotrochus adansonianus bermudensis*. Dan, a well-known shell dealer in Florida, grew up in the Philippines, where his boyhood acumen for seashells got him invited to shell club meetings at the presidential palace in Manila. Dan has helped the police solve the theft of rare shells from the American Museum of Natural History. He has helped scientists identify numerous species. They have named at least eight new species in his honor.

Among perils facing the sea, shell collecters' harm to mollusks might compare to personal car trips versus the global fossil fuels industry in accruing the carbon emissions warming Earth. How you drive your car matters because transportation is the biggest source of U.S. emissions; our individual actions reflect the larger ethos that could help us live within the planet's ecological limits. Yet one family's way of living means little if we don't change the larger industrial systems now breaking those limits.

The disappearance of mollusks and their shells from bays, beaches, and estuaries is most often linked to destruction of their habitats, including pollution. Mollusks famously clean up the water around them; scientists sometimes call them "the liver of our rivers." Like livers, their soft bodies can take only so much. The digestive glands of marine mollusks living near human shores brim with dozens of contaminants such as PCBs and pesticides including DDT banned in the United States in 1972, revealing how everything we put out into the world comes back to us. Plastic is spreading yet farther. Tropical islands where no humans live are stifled in grocery-bag sludge thick as seaweed. Mollusks hunkered in the most remote arctic and deepest seas are ingesting microplastic fibers shed from our yoga pants.

Meanwhile the shell-makers most beloved to humans for their beauty—like Queen Conchs and Chambered Nautilus—are the ones we're killing off for their beauty. Other threatened species are not listed as such, or studied as much, because mollusks don't draw the attention or research dollars of animals like sea turtles and pandas whose eyes are big, soulful, and not mounted on tentacles. The Red List of the International Union for Conservation of Nature—the official gauge for the staggering decline of animals now underway around the world—severely underestimates loss of invertebrates, which make up an estimated 97 percent of all creatures.

The pages of history fill in some of what's been lost. Early American coastlines were dense in oysters, scallops, clams—add abalone on the west coast—before we dredged or buried them alive to make way for waterfront development. When Henry Hudson sailed his ship *Half Moon* into New York Harbor in 1609, he had to navigate 350 square miles of oyster reefs. Within three centuries, oysters no longer colonized the harbor.

Colorful giant clams grew so abundantly in the shallow coasts of the Indo-Pacific that the nineteenth-century British conchologist Hugh Cuming described drifting over a solid mile of them on a collecting trip. Today the largest species are locally extinct in China, Taiwan, Singapore, and numerous smaller islands where they were overharvested for their adductor muscle—a sashimi delicacy—and their shells.

Ambitious restoration projects are underway in New York and other historic oyster bays around the world, and in giant-clam nurseries—some at top-secret locations to evade poachers—in the Pacific. Bringing back these seminal creatures could help restore the seas we share and establish the clean ocean farming we need to feed people and save wild fish. Yet their vulnerability to the warming, acidifying seas makes success far from certain.

The carbon dioxide we send into the atmosphere by burning coal and oil; manufacturing cement and plastics; and leveling the world's great forests is warming the Earth unevenly. The sea and its life are taking a far greater blow than those of us on land. The oceans have silently absorbed 90 percent of the additional heat—already, some places have become too warm for mollusks. Oceans have also taken up a third of the carbon dioxide, which has made seawater 30 percent more acidic than at the start of the industrial era.

This chemical change, known as ocean acidification, has begun to limit the carbonate that mollusks use to make their shells. Acidic waters are also boring into some shells, pitting or eroding them. One of the world's tiniest shelled creatures, the sea butterflies—a source of food for other sea life including shorebirds and whales—have thin, hard shells especially sensitive to the changing ocean chemistry. Scientists around the world have found these pteropod shells thinning, or corroding in their delicate outer layers.

The luminous fairies may be signaling what could happen to other shelled life as the seas turn more acidic. In the Pacific Northwest, young oysters have died off en masse, unable to build shell in lower-pH seawater. In California, scientists detect radical changes in the way mussels build their sleek black shells, trying to adapt. In the lab, ubiquitous periwinkles—wee bivalves common to rocky shores and soup bowls—build a weaker shell when subjected to seawater just a bit more acidic than today's. In experiments that replicate the acidity forecast for a century from now, conch shells deteriorate. Scallops and clams build thinner homes. Scientists find that triton shells living near seeps with predicted future levels of CO_2 grow thinner—and a third smaller—than those in normal conditions. The large spirals blown by the Greek god Triton to calm the seas or raise the waves are sending us a signal.

IN WILLIAM WORDSWORTH'S autobiographical poem *The Prelude*, the narrator falls asleep near the sea and starts to dream. He holds a seashell to his ear and hears

> *A loud prophetic blast of harmony;*
> *An Ode, in passion uttered, which foretold*
> *Destruction to the children of the earth*
> *By deluge, now at hand.*

Seashells do not really echo their native ocean, as people have believed for centuries. Nor do they foretell coming storms, as old superstition had it. Contrary to a more modern theory still found in some kids' fact books, they don't magnify the sound of blood through our veins.

No, the poet got much closer to the science when his narrator held the shell to his ear and heard his own mind's fear. A large spiraled shell like a conch, whelk, or India's sacred chank is simply the perfect resonating chamber. Like a hand cupped to the ear, it picks up ambient noise in the environment—amplifying exactly what's going on around us.

The modern signs from seashells are as clear as those that showed earlier generations the age of Earth or the rise and fall of ancient seas. They are also pointing to the considerable solutions that lie beneath the ocean's waves. The mollusks, and the seagrass meadows where many of them begin life, sequester tons of carbon. They build some of the world's most efficient homes, and the best storm barriers known. They tap sunlight and algae for fuel.

Their ranks include the longest-lived animal known—the burrowing Ocean Quahog, *Arctica islandica*, that can live beyond four hundred years—and the longest-surviving. The fabled nautilus has lived through warming, acidic oceans before. They do hold wisdom from the sea.

THIS BOOK WAS born in a record warm and rainy winter (records now shattered) on Sanibel Island in southwest Florida, where every street is named for the seashells that wash ashore on the southern beaches. The marine biologist José H. Leal had invited me to give a book talk at Sanibel's Bailey-Matthews National Shell Museum, which is devoted entirely to shells and their makers. Leal, who grew up near the beach in Rio de Janeiro, has the lithe build, leather bracelets, and serene bearing of a lifelong surfer. He is an expert in the biodiversity of mollusks and in their constantly changing scientific nomenclature. Fluent in four languages, he reads another two. He has worked in the world's great shell collections, from the Smithsonian in Washington D.C. to the Muséum National d'Histoire Naturelle in Paris, and edits *The Nautilus*, one of the oldest scientific journals of mollusks. Yet he found what he considers his most vital role in a place that hosts shell-craft lessons where visitors glue googly eyes onto nature's masterworks. For Leal, and a number of marine scientists I've met in the years since, helping people understand what's happening to the world and its life has become even more important than their research. (I once asked Leal for his opinion on shell craft; he would say only that some of his best friends glue googly eyes on shells.)

Ten years before I met Leal, the shell museum had surveyed its visitors, many of them tourists and their children visiting Florida, to find out how much they already knew about seashells. The survey revealed that *90 percent* of the visitors had no idea that a shell is made by a living animal. Most people thought they were stones.

As much as the modern crisis of truth is a conceit of politics, it is also a consequence of that severing from nature. When Pokémon characters are more familiar to children than the snails that helped inspire them, when drifts of plastic are far more common than seashells on many beaches of the world, natural history and life's struggle to survive are hard to know.

The chapters that follow take the form of a coiled shell, circling out from an apex in Sanibel, spiritual home of every mad sheller. The apex, the pointy top of the spire, is where the mollusk begins its life and work of shell-building. I, too, was born in this part of the world, in the county that is also home to Sanibel Island. As a child I loved collecting shells and visiting the cheesy, self-proclaimed "largest shell shop in the world" on the mainland. Only as an adult did I learn about the great cities of shell built by Indigenous coastal people living here more than a thousand years ago. The Calusa left the most extensive shellworks known in the United States: cultural and natural archives almost entirely flattened by twentieth-century road graders.

From Florida, the narrative winds outward to tell stories of some of humanity's most iconic seashells, the animals that make them, the people intertwined with them, and the changing seas we share. Modern shell madness rose in Europe with the first mega-corporations: the Dutch and British East India companies, whose ships carried home tropical seashells and other high-demand goods from the East and lit the flame of global consumerism that has become an inferno.

The era also saw the rise of the first worldwide currency, a small white shell with an outsized imprint on history and culture. Harvested en masse in the Maldives under a succession of queens, shining Money Cowries were moved along trade routes as ship ballast and became the dominant currency of the transatlantic slave trade. Following the cowries brought me and my teenage son to the slave castles of West Africa on the 400th anniversary of the first voyage carrying enslaved people from those shores to America's—seashells revealing as much of human nature as nature.

This book is about seeing what has gone unseen. The life inside the shell; the Maldivian queens and others left out of history books; the connections between the human condition and that of the sea. Just as we've loved seashells for the gorgeous exterior rather than the animals that build them, we've loved the oceans as the beautiful backdrop of life rather than the very source. The narrative also winds through time. It begins with the earliest shelled life known, from nearly a billion years ago, then turns to fossil shells like the spiraling ammonites that left signposts to evolution, extinction, and geologic change.

In the days before the Scientific Revolution, many people just thought the ancient seashells were stones.

Part I

MIRACLE

One

~~~~~~~~~~~~~~~~

FIRST SHELLS

MARINE FOSSILS
Quadrireticulum allisoniae

Along the Alaska–Canada border where the Yukon River hisses with the silts of deep time, an earth scientist hopped out of a helicopter forty years ago onto a forsaken mountain slope, following her hunch about rocks from an ancient sea.

Carol Wagner Allison grew up in the Yukon, where her grandfather settled as a medical missionary in the 1890s, and began her career hunting some of the world's oldest sea life when she was just twenty-one. She was fresh out of Berkeley in 1953 when Shell Oil hired her to work as a micropaleontologist in Bakersfield, California, the petroleum capital of the West. The remains of sea life—the fossils in fossil fuels—could guide oil companies on where to sink their money and wells. So-called bug hunters or micro-men, for they were mostly men, were the hottest scientists in the oil industry in those days before advances in high-resolution seismic imaging gave geologists a 3-D view of the carbonate reservoirs underground that hold much of the world's drillable oil.

In 1960, Allison left fossil fuels for fossil sand dollars. She returned to Berkeley to earn her doctorate in invertebrate paleontology, specializing in the flattened sea urchins that have made their exquisite five-petal shells since before the dinosaurs. But she was keenest on older, tinier life-forms that couldn't be seen with the naked eye. Settled as a professor and museum curator at the University of Alaska in Fairbanks, she spent her summers 400 miles east—"looking for her

bonanza" along the Yukon border, as her husband and fellow paleontologist Dick Allison later put it.

Allison found that bonanza in 1977, chiseling into limestone and
black shale along a remote mountain range known as the Fifteenmile
Group. Back at her lab, magnifying rocks from the haul, she discovered a curious batch of tiny fossils. Single-celled organisms from a
primal sea, the microscopic beings were covered in intricate plates,
like minerals woven into lace. But the fossils hadn't held up well in
the fragile rock. Allison was not able to confirm their age, give or
take a few hundreds of millions of years. Nor was she able to solve
the mystery of the plates—had they been made by the organisms
themselves, as she suspected, or formed after their death?—before she
herself died of cancer at age fifty-nine.

Allison had to abandon her research to fight her illness, and she
never knew just what she had found. It would be decades before
another young bug hunter, not yet born when Allison collected the
plated microfossils, picked up the trail and finally confirmed the late
paleontologist's hunch. The Yukon rocks held the earliest-known
evidence of biomineralization—life's ingenuity to take up minerals
from the surrounding environment and turn them into hard parts.

Allison had discovered Earth's first-known shells.

FOR ALL THEIR color, gloss, and architectural flair, the allure of
seashells may have most to do with the geometric order in their forms.
The intricate patterns follow evolutionary blueprints drafted in those
earlier seas. Seen sideways with their two halves pressed together, the
radial ribs of a cockle shell close like a pair of wings around a great
bird. To stare into the spiral top of a whelk or cone shell is to see the
swirl of the Milky Way; a reminder that Native American people
as widely separated as the Aztecs of Mexico and the Winnebago of
Nebraska equated shells with stars.

Spiral seashells evoke galaxies because of their logarithmic pattern
of growth, best seen in a cross section of the Chambered Nautilus.
Each graceful coil is wider than the next by a constant factor, making

a nautilus shell one of the most recognizable spirals in nature. Life loves logarithmic spirals. They shaped the shells of tiny foraminifera, some of the first marine microfossils studied in microscopes in the seventeenth century. They patterned the ammonites, fossil mollusks long vanished, but close enough to the living nautilus that they emboldened scientists in the same era to think about evolution and geologic change.

Nature's precise aesthetic became ours. The evidence that a shell-inspired Leonardo da Vinci designed the left-handed spiral staircase at France's Château de Blois divides architectural believers and skeptics to this day. Browse the pages of Leonardo's notebooks, full of coiling fossil shells and his own sketched whorls, and count me with the believers.

Seashells were models for the original minaret, the protective portico, the scalloped edge, and countless other iconic forms now moved from sea to skyline: Antoni Gaudí's vaulted rooftops in Catalonia; Frank Lloyd Wright's spiraling Guggenheim Museum in New York; and Jørn Utzon's Sydney Opera House in Australia, the waterfront beauty for which Utzon credits the fierce-looking cockscomb oyster, *Lopha cristagalli*.

Yet appreciating seashells apart from the life that evolved to build them is like appreciating Leonardo for his notebook sketches while overlooking his living, breathing paintings.

Indeed, some mollusks have two retractable eyes, mounted at the tip of curious tentacles, that seem to follow you like the *Mona Lisa*. Others have a hundred electric blue eyes, set in dazzling rows. They are animals with rapacious tongues and rows of teeth to feed big, wolf-hungry stomachs. They are animals that dive and leap. Animals that scurry across the ocean floor, burrow down into sand, climb up rocks, turn corners, and flip somersaults. Animals that leave tracks like paws in mud. Animals that swim—propelled by wings graceful as butterflies or clapping shells, clunky like cartoon clams. Animals that ascend and descend in the water column; the Chambered Nautilus filling its sections with liquid and gas like a master diver who spent half a billion years perfecting buoyancy.

They are animals that breathe and bleed and have a beating heart. Yet our infatuation with them generally strikes only after the heart has stopped.

THIRTY YEARS AFTER Allison collected her curious fossils at the Alaska–Yukon border, in the summer of 2007, another helicopter startled a grizzly bear off the same rocky outcrop and dropped off a pair of earth scientists shouldering a shotgun along with their rock hammers and packs.

Phoebe Cohen and Francis Macdonald, then graduate students at Harvard, set out to retrace Allison's steps to see if they could find more of the plated beings, better preserved. It was an unseasonably cold June, and snow still streaked the mountainside. For weeks, the scientists spent daylight that lasted through midnight hammering along the steep slopes and filling their backpacks with rocks—making for an ungainly run when the bear returned.

They collected the shining black shale and the hardier limestone. The latter, it turned out, had preserved its makers. Back in their Cambridge lab, boring into the rocks with weak acid, Cohen and Macdonald found the microbes, impeccably preserved. They were just 20 microns wide; one-fifth the width of a strand of hair. Supersizing them to three dimensions with a scanning electron microscope, the scientists saw just what they'd been looking for—the latticed plates described by Allison.

They also thought they saw a revolution on the primal path to life. Like microscopic medieval knights in elaborate chain mail, the organisms seemed to be decked out in armor, studded with sharp spines jutting out along the perimeter. Some look like the spikes on that protozoan pincushion the thorny oyster. Cohen and Macdonald suspected they were looking at the origins of biomineralization—the ability of organisms to construct hard physical structures like shells, bones, and teeth from minerals in the surrounding environment.

Previous research had placed biomineralization at about 750 million years ago, but only tenuously. Like trees in a petrified forest, mineral structures that looked like the work of living creatures might instead have turned to stone after they died. Cohen's doctoral advisor thought this was what happened to the Alaskan microfossils. Cohen—who'd loved seashells since childhood, their forms inspiring her countless hours in the lab researching and admiring microscopic shelled sea algae called coccolithophores—suspected he was wrong. But it took her ten years to prove it.

Cohen, now a professor of geosciences at Williams College, has smiling brown eyes, matching hair in a short pixie, and an admiration for Dana Scully. Like her *X-Files* idol, she stays on the case. Over the next decade, Cohen returned to the steep slopes of the Yukon and, with colleagues at Dartmouth and Oxford zooming in with one of the most powerful microscopes in the world, went on to analyze hundreds more microfossils bearing the mineral plates.

The shapes themselves didn't vary among specimens, suggesting life had built them with a biological blueprint. By comparison, the fossils of soft-bodied creatures become randomly squished and distorted by the forces of geology and time. Analyzing the images over months and then years, Cohen and her colleagues were able to show how each plate was woven in precise order from elongated mineral fibers—as if with purpose.

"The hexagonal order, with lots and lots of details, showed that this was not random, but made under the influence of life," Cohen told me. "These are organisms deciding how they are going to look."

The scientists were ultimately able to date them to around 800 million years old. The tiny beings had lived in a radical geologic era called the Neoproterozoic: the bridge between the largely microbial world and the great diversity of animals to come. Cohen appreciated another bridge, as well: "It makes me so happy to know that in a very male-dominated field, it was another woman climbing over those mountains, fighting off the mosquitoes, keeping an eye out for bears, and collecting those magical rocks."

~~~~~~~~~

WHEN THE FIRST shelled life-forms were coming to be, Earth was like Whoville without Horton: a world of tiny life, nothing big. (It's *still* predominantly a tiny world, only the microbes are now dwarfed by the less populous fungi, plants, and animals—vertebrates like us and invertebrates like the mollusks.) Life had roused in the primitive sea, perhaps in hydrothermal vents as chemicals roiled in fissures cracked open on the ocean floor by grinding tectonic plates. Oozing green, purple, and brown slime and wafting a rotten-egg stink of hydrogen sulfide gas, microbes clung together in vast mats, mounds, and reefs that took shape hundreds of millions of years before the first corals.

In this time before land, single-celled organisms picked up a familiar habit: eating each other. Cohen and other geoscientists find that the earliest marks of predators date to around the same time as the earliest evidence for shells. Microscopic amoebae can be scarred with drill holes not unlike those bored into unsuspecting clams. On the summer afternoon my five-year-old collected shells for her necklaces, she didn't ask—and I didn't offer—how they got their perfectly beveled holes. It wasn't quite the day to explain how some mollusks sneak up on others from under the sand; secrete acid to weaken the top of the clam or cockle shell; drill a hole with a spiked tongue covered with hundreds of teeth; then probe the saw-tongue inside to paralyze their brethren, scrape them from the shell, and devour them alive.

Surrounded by a sea of predators, the first shelled beings armored themselves with materials they could find nearby. The plated microbes discovered by Allison and identified by Cohen used calcium phosphates, a family of minerals abundant in the primordial oceans. A couple hundred million years later, mollusks would claim calcium carbonate for their shells and mix it into building material to make their own famously creative defenses: spikes that look remarkably similar to those of the primal shell-makers. Narrow openings to foil unwelcome visitors. Polished veneers that slip through crab pincers.

Tough little doors, known as opercula, that can slam shut like a castle gate.

When you lift a living conch from the shore and turn it over, if you are lucky and catch a glimpse of the soft animal as it pulls back into its shell, the muscular flesh you see contracting is known as the foot. The mantle—the part of the animal that builds the shell you are holding—is just inside, a thin cloak wrapped around the organs. Like the magical cloaks of folklore, you don't see the mantle in most mollusks, but it's there working miracles. As it protects the animal's soft parts, the mantle also pulls minerals from the surrounding seawater and secretes the shell at its edges.

A mollusk mixes the calcium carbonate with a bit of protein, creating a sticky matrix; then builds the shell layer by layer as it grows. While a human brick-mason works from the bottom up, the mollusk starts at the top and adds new layers at the bottom—constantly widening its aperture, or the part of the shell we hold to our ear, to expand its living space. In a brick wall, the top layer is the newest. In shells, the pointy top of a spiral is the oldest part of a mollusk's home. Known as the apex, it's where the animal fit when it was just born.

Mollusks that build a single spiraling shell are the gastropods (from the Greek: *gaster* for "stomach" and *podos*, "foot"). They coil their shell around an invisible axis as they grow. Hinged bivalves—double-shelled mollusks such as clams—expand their homes at the open edge. Each different species has its own model for arranging the calcium carbonate crystals at different angles to create rough or glossy surfaces, colors, and structures.

Sculpting minerals from the seawater into hard parts like those early armored microbes, mollusks and their shells became part of the ocean's great chemical harmony, and in turn the Earth's. Calcifying sea creatures including mollusks and crustaceans, corals, and sea urchins—joining shelled plants like the vast, sun-loving blooms of coccolithophores that Cohen admired under her microscope—upcycle ocean chemicals to build their forms. They make their calcium carbonate from dissolved calcium and carbon—the backbone of life that becomes too much of a good thing when overloaded into the atmosphere as carbon dioxide. That makes every shell not only a work of art, but a chemical vault; carbon bound up in beauty rather than warming the world.

THE ITALIAN WRITER Italo Calvino imagined only one possible purpose for the creation of something so intensely beautiful as a seashell. In Calvino's short story "The Spiral," 500 million years ago a mollusk falls in love. Sensing his competition among fellow blobs, he begins to secrete a shell to draw the attention of his desired. He expresses himself in his art, winding all of his love into the spiral and color of his shell.

But the mollusk fails to anticipate that when eyes evolve, they will bring sight to other creatures—yet not to his beloved. She never sees the temple he has built. Looking back from modern times, he is rueful. His consolation is that he has also made the world: the honeycombs and the reign of Cleopatra, the telescopes and the ice cream trucks.

"In making the shell," he realized, "I had also made the rest."

As calcifying organisms evolved, flourished, and perished, they created a lot of the rest. We walk on a world of shell—the carbonate remains of all the calcified life that has ever lived. Added up in the sea and on the land, those remains also represent among the largest stores of carbon on Earth.

Shelled planktons and corals and mollusks made some of those oil reservoirs hunted by the micro-men. They made the limestone aquifers that hold freshwater underground. The calcifying life-forms gave us mountains and they gave us marble. Tiny creatures transformed titanic contours in limestone cliffs from Lake Michigan to Moldova; in the karst islands of Vietnam and Greece and the Caribbean; and atop the highest mountains on Earth. Mount Everest is peaked with marine life that plied the Tethys Sea between India and Asia more than 400 million years ago. "If by some fiat I had to restrict all this writing to one sentence," wrote John McPhee in his geological tome

Annals of the Former World, "this is the one I would choose: 'The summit of Mount Everest is marine limestone.'"

Marine life is reincarnated in the sculptures of ancient gods and in the antacid tablets we chew for heartburn. Ancient people burned shells to make slaked lime, one of the first manufactured chemicals. Now limestone is churned into cement in one of the largest manufacturing endeavors in the world. Limestone is woven into our lives in smaller ways too; it is a key ingredient in cereal and toothpaste. Humans have been brushing their teeth with seashells since at least the Ancient Greeks, who crushed oyster shells into their toothpaste as a cleaning abrasive for the same reason Crest adds calcium carbonate today.

In the United States, many of us walked our children to the first day of kindergarten on shells. The shell was in the carbonate bedrock deep underground, and it was in the sidewalk under the new pair of shoes. Compressed by the elements and time, tiny fossil seashells formed the limestone that underlies much of the nation—in turn mined, cut, and built into some of our iconic human spaces: the Pentagon, the Lincoln Memorial, the Washington National Cathedral, the Empire State Building. All owe their power to fragile beings.

From a distance, the walls of Rockefeller Center look creamy smooth. Look closely, and you'll see coils and spirals, fans and curlicues embedded in the limestone, quarried in Indiana and formed from denizens of the shallow sea that covered the Midwest 300 million years ago. In the sprawling lobby at 30 Rockefeller Plaza, flat-coiled mollusks called maclurites swirl in white against dark stone formed from Ordovician seas and quarried from Crown Point Formation in Vermont. The coils gleam from the same stone in the shining black floors of the Maine State Capitol building in Augusta, constant if unnoticed reminders of the negligible span of human affairs.

Like squeezing toothpaste from the tube, geological pressure pushed the remains of shelled creatures up and out of the ground over half a billion years, fusing their hard parts in rough outcrops, smooth marble, and fine chalk. Along England's southeastern coastline, a great bloom of plated coccolithophores drifted to the ocean floor 100 million years ago. They lay buried in a shallow sea until Earth's crust shifted and thrust them upward to form the White Cliffs of Dover, a climax of geomorphology and Shakespeare's *King Lear*:

> There is a cliff, whose high and bending head
> Looks fearfully in the confined deep.
> Bring me to the very brim of it,
> And I'll repair the misery thou dost bear
> With something rich about me.
> From that place I shall no leading need.

From a distance at the high and bending overlook, the cliffs look smooth and flawlessly white. Up close, the chalk is filled with the casings and casts of extinct marine life. Visitors find sharks' teeth, sponges and echinoids, prehistoric lobster-like claws, and ancient bivalves. Coiled ammonites pop from the chalk after high tides and cliff falls, as if brand new.

IN 1868, THE intensely sideburned English biologist Thomas Huxley gave a talk to the working-class men of Norwich, England: "On a Piece of Chalk," to this day one of history's most impressive works of science writing. Launching from "the bit of chalk which every carpenter carries about in his breeches-pocket," Huxley sketched out the entire geological history of Britain beginning with marine life in primordial seas.

He described the "innumerable bodies" visible in chalk under a microscope. The work of tiny plants and animals to separate carbonate of lime from the oceans to build their shells. Their hardening at the bottom of the sea over time, like the calcium crust at the bottom of the teakettle. "There is no escaping the conclusion that the chalk itself is the dried mud of an ancient deep sea," Huxley told the working men. "A great chapter of the history of the world is written in the chalk."

Once home to strange swimming reptiles such as the ichthyosaurs, now to the working men and their carpenter's chalk, only shelled life straddled present and past. "The longest line of human ancestry must hide its diminished head before the pedigree of this insignificant shell-fish," Huxley said. "We Englishmen are proud to have an ancestor who was present at the Battle of Hastings. The ancestors of [lamp shells] may have been present at a battle of Ichthyosauria in that part of the sea which, when the chalk was forming, flowed over the site of Hastings."

Oh, for a modern-day Huxley to simply explain the world we are reverse engineering. Long-buried carbon that would normally take millions of years to return to the atmosphere is released in an instant when we burn fossil fuels. Human dominion unwinds nature's elegant precision. Life-infused limestone girds our highways, seawalls, and parking lots as the main ingredient in cement, in turn the main ingredient in concrete. After water, concrete has become the second-most-consumed material on Earth: Three tons are manufactured annually for every woman, man, and child on the planet. That intensity also makes its manufacture one of the largest sources of carbon dioxide emissions in the world.

Chalk became a symbol of the Covid-19 pandemic in 2020, as artists sketched hopeful messages on sidewalks; children turned to street games like hopscotch; and shopkeepers drew circles or lines to keep customers at safe distances in queues. In a trend that began in France and spread around the world, rebel botanists began to chalk the names of wild plants pushing up through sidewalk cracks and along walls and roadsides. The *Sauvages de ma rue* ("wild things of my street") chalking campaign aimed to help people see the little daisies, dandelions, and other flora too often derided as weeds and doused with pesticides.

The chalk itself—and the concrete canvas—could have been a campaign: *this chalk . . . this sidewalk . . . this wall, brought to you by foraminifera . . . coccolithophores . . . corals . . . mollusks.* Keeping carbon safely buried for 500 million years.

~~~~~~~~

ARISTOTLE MINTED THE word *mollusk*—in the UK it's mollusc—from the Latin *mollis*, or soft, to describe fleshy bodies. The concept of *mollis* shows up frequently, too, in Aristotle's treatise *Rhetoric*, on the art of persuasion. It describes softer arguments, or ways to cushion hard truths; a soft core more palatable than rigidity.

Leaving a half-billion-year record of their evolution, extinctions, and surrounding environment, mollusks became the world's great truth tellers. The accumulation of shells lodged into mountaintops and valleys worldwide were obvious signs of the churns of Earth and its life. Six centuries before Christ, the Greek philosopher and poet Xenophanes of Colophon had seen seashells high in a cliff on Malta and speculated that the sea had once covered the land. Aristotle later conjectured that sea and terra firma must sometimes change places. The waxing and waning of the ocean, and flooding and drying of the land, were part of the world's "vital process."

With the same insights and given far less credit, the Zunis, Native American Pueblo peoples from the Zuni River valley, saw the presence of fossil oysters, clams, ammonites, and other marine life as proof that an ocean had once covered the North American desert. The landlocked Zunis had keen knowledge of the seas bordering the continent when the Smithsonian anthropologist Frank Hamilton Cushing lived among them in the late nineteenth century. They described the Ocean of Sunrise; the Ocean of Sunset; the Ocean of the Place of Everlasting Snow; and the Ocean of Hot Water (the Zunis' accurate description for the hurricane-fueling Gulf of Mexico). Cushing brought a Zuni party on a cross-country train journey to meet their "beloved mother," the Atlantic Ocean. As they sat on rocks at Boston Harbor's Deer Island with the tide rising over their feet, "They recognized in the tide the coming of the beloved gods of the ocean to greet them in token of pleasure at their work."

Stone beings had convinced the Zuni of earlier life-forms from a time long past. They told Cushing that "we often see among the rocks the forms of many beings that live no longer, which shows us that all was different in the 'days of the new.'"

Those convictions were borne out in the fossil record, slowly chiseled to reveal sea life that no longer survived and sea bottoms preserved on clifftops. But, steeped in legends and religious beliefs, modern people took centuries to discover—and then accept—the science of evolution and geological change.

Marine fossils such as the coiled ammonites and dart-like belemnites (from the Greek word for dart, *belemnon*) seemed to pop up in the most auspicious places. In China they were sword stones, *Jien-shih*. Embedded in the Himalayas were ammonites called salagrama in Sanskrit; Hindus worshipped them as holy, avatars of the god Vishnu. In the Americas, Blackfoot and other High Plains Indians called ammonites and other marine fossils "buffalo stones," or *iniskim*, for their resemblance to a sleeping buffalo. The sacred stones were thought to bring forth buffalo herds and other good fortune.

In many parts of the world, people noticed that these talismans tended to appear after storms. In the Middle Ages, belemnites and other fossil shells were often called thunderstones, thought to fall from the sky when lightning struck in a thunderstorm. Dutch legend had it that anyone who found one would be protected from lightning forever more.

By the 1400s, educated Europeans tended to accept one of two theories about seashells lodged in hillsides and mountains. The first was that, true to the Bible, the shells likely rose from the sea in an epic flood. The second, known as spontaneous generation, was that the shells might have grown in the rocks themselves—popping up like crocuses in spring.

Leonardo da Vinci's leather-bound notebooks are full of geological truths that would not become clear to scientists for many more decades. In childhood, he had roamed the Italian hillsides formed by a Jurassic sea. "The shells of oysters and other similar creatures which are born in the mud of the sea," he wrote in his curious mirror script alongside black and red chalk drawings, "testify to us of the change in the earth."

The shells did not magically appear in rockface but had helped make it, as marine mud and shellfish "became changed into stone."

As for Noah's Flood lifting seashells to the mountaintops, "here a doubt arises," for the simple fact that floodwaters flow downhill. Leonardo went on to consider every theory for the existence of shells "within the borders of Italy, far away from the sea and at great heights," and challenge each one: Even if floodwaters had reached the mountaintops, he wrote, the heavier shells would not have. To those who claimed "the shells left their former place and followed the rising waters up to their highest level," he described the slow movement of a live cockle, furrowing through the sand. "With such a motion as this it could not have traveled from the Adriatic Sea as far as Monferrato in Lombardy, a distance of two hundred and fifty miles in forty days."

Leonardo never published his shell-inspired observations, likely knowing they wouldn't be accepted in his time. More than a century later, the Danish anatomist Nicolaus Steno became obsessed with the same seashells while serving in Italy's royal *Accademia del Cimento* ("academy of experiment") in Tuscany.

Steno would become the first scientist to publish the theory that the cockles, clams, and other "figured stones" nestled into mountains must be organic remains. The insight came to him in the middle of a dissection, as he sliced into the head of a huge shark caught off Italy's western coast. The giant's teeth were familiar. They looked like the "tongue stones" often discovered in the countryside with shells— believed to be the petrified tongues of snakes or dragons.

In his classic 1669 work *De solido* ("on solids"), Steno wrote that "an examination of the shell itself proves that these shells were parts of animals at one time living in a fluid." He went on to show how the bivalves in the sea and in the mountains had to be one and the same. The animals themselves, he speculated, built their shells with a fluid secreted at the edge, "like perspiration." His description of how mollusks make their shells would prove remarkably accurate:

All the diversity of hues and of spines which arouse the wonderment in many in the case of shells not only from our own land, but also from other lands . . . has no other origin than the edge of the animal enclosed in the shell. This edge, gradually growing and expanding from something exceedingly small, leaves its impress upon each margin of the subdivisions, since these margins either form from the fluid which is exuded from the outer edge of the animal, or are themselves the creature's outer edges, which, like the teeth of the shark, grow up anew, perhaps, in the place of the earlier edge.

Steno had joined the *Accademia del Cimento* to experiment on muscles, particularly how they contract. (His appointment included a steady supply of cadavers from prisoners sent to the gallows—or strangled by hand if he needed their necks intact.) But, excited by the fossil shells, he abandoned muscles and turned to mussels. In *The Seashell on the Mountaintop*, Alan Cutler describes Steno's increasingly long "mussel journeys" into the Tuscan mountains and mines as he worked out the principles that earn him remembrance as the father of modern geology: that younger layers of rock sit upon older layers, telling a story in sequence.

Geology students learn "Steno's principles" to this day. But two hundred years would go by before his theory of fossilization was accepted in a society that thought of the mysterious shell charms as thunderstones, tongue stones, or snakestones—coiled snakes turned to stone.

Legend had it that a severe snake infestation had plagued the whaling-port town of Whitby in North Yorkshire, England, until the founder of the local abbey, St. Hilda, turned the plague of vipers to stone and hurled them off a cliff. Her feat explained the profusion of ammonites found at the precipice.

IN THE SAME years Nicolaus Steno wandered on his Tuscan mussel journeys, America's early naturalists also wondered over shells—even as they encountered hummingbirds, rattlesnakes, and other strange new life on the continent. Their wax-leather boots crunched marine shells along riverbanks and at the top of the Appalachian Mountains. Their iron spades turned up shells in fields, in ridges, and in the mounds that loomed over much of the Eastern Seaboard.

Beneath the New World lay an old. Riverbanks "are composed of Multitudes of Scallops, Cockles, & other Sea-shells mixed with the Earth," wrote America's unheralded first naturalist, John Banister, from the Colony of Virginia. An Oxford-trained minister with a parish at the mouth of the Appomattox River, Banister compiled the first catalogues of plants and shells from the colonies a century before naturalist William Bartram's better-known *Travels*. He was one of the first scientists on either side of the Atlantic to describe mollusk anatomy—a snail's speck of a contracting and dilating heart—and is believed to have given Europeans their first look at a North American marine fossil. The immense, iron-like scallop appeared in the 1685 *Historiae Conchyliorum* published by the London physician Martin Lister.

Banister would not have imagined that the great scallop with deep ribs and mesmerizing swirls at the margin was extinct; it had lived in the Pliocene 4 to 5 million years ago in what is now Virginia's coastal plain. Extinction was still an unacceptable concept—an imperfection of God's creation. But Banister did reject the popular notion that the devil placed animal likenesses in rocks to tempt Christians' faith. "Now either the Earth by some kind of Salt of its own, that naturally shoots into such figures, produced these," he wrote in his catalogue *Mollusca, Fossils, and Stones*, "or the Sea in the former Ages came further up in the Country which I am rather apt to believe."

Marine shells were found not only in America's inert stone and on its living shores, but in a compelling third realm: Seashells wound, practically and spiritually, through the lives of people long settled on the continent. Native mounds rose along much of the Atlantic and the "Bay of Mexico," up the Mississippi River, and far beyond—hundreds more stood undiscovered across San Francisco Bay and other parts of the West. The coastal mounds, often built of consumed shells, were testaments to the density and size of the oysters, clams, scallops, and other mollusks crowding the estuaries when the Europeans arrived.

Like the bird flocks that darkened the celestial sphere; squirrels that skittered in by the hundreds to pick a cornfield clean overnight; and "legion" mosquitoes that kept colonists locked inside on summer evenings, the abundance of mollusks in early America is hard to imagine today. New England colonists described fortress-like oyster banks running for miles; one barricaded the mouth of the Charles River. The credible English traveler John Josselyn described oysters "nine inches long from the joynt to the toe."

Banister's *Mollusca, Fossils, and Stones* was unusual among early scientific catalogues for including Native shell cups and beads, along with the shells "of which ye Indians to ye Northward make their mony called Wampom Peack." The word *wampum* was an anglicized version of what Native people called their traditional shell beads, hand-carved from North Atlantic channel whelk and quahog clams. Tooled into small white-and-purple tubes and woven into patterned belts, wampum obviously held great value. Europeans often interpreted it as money—source of our slang use of "clams" for dollars. But the shells were closer to language than coin.

Like an extension of the animal that made the purple-and-white luster, shell beads were considered living in the manner of living documents, Chief Irving Powless Jr. of the modern Onondaga Nation explained the year before his death in 2017. Exchanged in story belts or strings, wampum was a system of discourse and diplomacy. Anyone holding the shell-beads was said to speak the truth.

The Iroquois welcomed the Europeans to their territory with the "Two Row Wampum Belt" known as *Guswenta*, whose rows of darker shells symbolized two vessels traveling the same river in peace, each adhering to its own laws and customs and not steering into the other. Yet laws and customs collided as the colonists appropriated shell beads for what they often described as "mercenary transactions." The definition stuck with Carl Linnaeus, the Swedish naturalist who developed the two-name classifications for plant and animal life. He named the American quahog used for purple wampum *Venus mercenaria*—later changed to *Mercenaria mercenaria*. Setting course for the future nation's free-market system, colonists made wampum an official substitute for colonial money and even began to manufacture it themselves. Mercenary it was.

Like the Zuni, Native people in the American East encountered by the colonists believed through their traditions that "these parts are supposed some ages past to have lain under the sea." Banister was among the colonists who wanted to learn from the earlier people. He was working on the Native section of his natural history of Virginia when he died in an apparent accident on a collecting expedition. He was kneeling to collect a small something near the Roanoke River when he was shot and killed by one of the hired men in his party.

Banister died at age thirty-eight, just as he was preparing to spend more time with "ye Indians to take a View of their Towns, Forts, Manner of Living, Customes." His unfinished manuscript "Of the Natives" is a glimpse into the lives of a people settled near the tidewaters; historians believe they were the Appomattoc. His descriptions of their worship, language, food, medicine, music, tools, and clothing reveal a ubiquity of shells. The men shaved their faces with sharpened mussel shells. The women gathered long hair into a single lock wrapped in white shell beads. Worshippers left wampum at a sacred place at the falls.

IT WOULD BE nearly two hundred more years before the Scottish geologist Charles Lyell laid out in his book *Principles of Geology* how Earth had been shaped, not by biblical catastrophes but by slow-moving natural forces of change. Lyell's insights about distinct past eras came largely from fossil seashells—some he learned about during repeated trips to America in the mid-1800s. He traveled with his wife, Mary Elizabeth Lyell, a shell collector who had studied with her geologist father and became an expert in the taxonomy of fossil shells.

Lyell, trained as a lawyer, knew how to find the right witnesses and ask the right questions. During tours of the plantations spread across America's Deep South, he learned that fossil seashells regularly turned up in the cotton fields and popped up in the wells. He tried asking plantation owners what lay beneath their fields, but they often couldn't say. They hadn't dug the wells themselves, or labored in any other excavation work. Enslaved people, Lyell found, knew the land and knew its fossils. When a proprietor couldn't answer his questions, Lyell would meet the enslaved land manager, and "he could usually give me a clear account of the layers of sand, clay, and limestone they had passed through, and of fishes' teeth they had found, some of which had occasionally been preserved."

Enslaved people "were very inquisitive to know my opinion as to the manner in which marine shells, sharks' teeth, sea urchins, and corals could have been buried in the Earth so far from the sea and at such a height," Lyell wrote. "The deluge had occurred to them as a cause, but they were not satisfied with it, observing that they procured these remains not merely near the surface, but from the bottom of deep wells, and that others were in flint stones."

The story reflects how the path of discovery is often slow, looping, and multi-sourced, rather than the linear breakthrough of a lone genius. Science is informed by countless contributors—historic and living, named and unnamed. Finders and fanciers of fossils and living seashells made serious contributions, and still do. The British fossil expert Mary Anning is anonymously known to the generations who grew up with the verse said to have been written about her: "She sells seashells by the seashore." Her real legacies, the fossil discoveries she made in Lyell's time, were often credited to the male paleontologists and wealthy collectors who bought them from her shop on the Jurassic Coast of England.

Anning was eleven when her father died, leaving the family impoverished. She and her brother dug in the cliffs of Lyme Regis for ammonites, then still called snakestones, and other curios so their mother could sell them to seaside visitors. Anning made three of history's great paleontological discoveries before she was thirty, including excavating the first complete ichthyosaur and plesiosaur skeletons. She became known to scientists on both sides of the Atlantic, led them through the cliffs, and shared her finds and ideas with them. As one visitor noted, Anning could identify any fossil bone and shell, and "she understands more of the science than anyone else in this kingdom." She corresponded with leading geologists, including Charles Lyell. In 1829, Lyell enlisted her help in measuring erosion along the Dorset coastline as he worked through his ideas about geologic change. But he and most of the scientists who learned from Anning and her fossils did not credit her.

A few years before she died of breast cancer at the age of forty-seven, Anning wrote with modesty to a friend, that "from what little I have seen of the fossil World and Natural History, I think the connection or analogy between the Creatures of the former and present World" is "much greater than is generally supposed."

It was still twenty years before Charles Darwin published his theory of evolution—the great analogy between creatures of the former and present world.

CAROL ALLISON, THE Alaska paleontologist who had discovered the earliest-known shelled life before she died of cancer, would be credited by the young bug hunter who picked up her trail in the Yukon. Phoebe Cohen named her plated microfossils *Quadrireticulum allisoniae.*

Allison's abiding legacy is not only the novelty of finding Earth's first-known shells. Cohen and a new generation of scientists now drill into the world's oldest layers of shelled life—or into more recent fossil shells, themselves—to understand past climate change and how life survived it or perished. In microscopic foraminifera coiled like ammonites; in corals that show scientists the times the world was warm; in giant clams with rings visible as those on a tree, the shells tell stories. They tell of dramatic temperature shifts, different levels of gas, rising and falling seas, the great swelling and retreating of ice. They tell who lived and who died. The legacy of Allison's discovery,

Cohen says, is the story of coevolution and survival: "Organisms and their environment evolving together through time."

Or going extinct, as it were, because of chemical changes in the ocean. Cohen suspects that was the fate of Allison's ancient, armored Whos.

Two

~~~~~~~~~~~~

EVERYTHING FROM SHELLS

THE CHAMBERED NAUTILUS
Nautilus pompilius

Sitting near the window in his classroom at East Dover Elementary School in New Jersey, the fourth-grader traced leaves on thick paper with his stylus, a tool for writing in braille. His teacher had nailed screening to a wooden board and mounted the paper on top so her blind student could punch out a catalogue of trees in his new country. The maple and oak leaves were strangely different from those he knew back home in the Netherlands; rougher and pungent.

Geerat Vermeij could hear the scratches of lead on paper from his classmates drawing in their sketchbooks. Then he heard a different sound; his teacher, Caroline Colberg, busy at the windowsill setting up a new display. He finished his art assignment before the others and wandered over to investigate.

Mrs. Colberg had vacationed on Sanibel Island in southwest Florida and brought back a trove of seashells. It was 1956, and shell collecting was becoming "the nation's fastest-growing hobby," the *Washington Post* declared. Servicemen stationed in Okinawa, Guam, Midway, and other Pacific islands during World War II had returned home with rucksacks full of tropical shells, quiet beauty that perhaps tempered the horrific memory of the shells of war. Collectors were launching shell clubs from New York to California to share prize specimens and travel tips. Sanibel was their mecca.

The island juts sidelong from the Florida peninsula in a wide, upturned curve, in the shape of the Cheshire Cat's smile. As mollusks and their shells tumble in Gulf currents, tides, and storms, the animals and their coveted homes are trapped by 12 miles of beach sloping gently along the cat's chin. Shells pile up in famed drifts, "sometimes as high as three feet," the *New York Times* gushed in a Sanibel travel article the year of Mrs. Colberg's display, "every shell immaculately fresh and iridescent."

From those drifts Mrs. Colberg picked out a long, tapered Lightning Whelk and a deep-ridged Atlantic Giant Cockle, among other seashells, for her fourth-graders back home. It was right around his tenth birthday when Geerat walked to the classroom window, moved his hands along the sill, and picked up the whelk. The shell was surprisingly heavy. He marveled at its heft and form, narrowing from a knobby crown at the top to a slender funnel at the tip, its interior smooth as his family's modest Dutch plates that had shattered in transit between the Netherlands and their new home in America.

Geerat ran his fingers across the different shells and their spired peaks and saw-toothed edges, comparing the ribbed exteriors and the smooth insides. He was struck by the contrast between these shells and the smaller, coarser finds he'd collected on the beaches of the North Sea. Even more than the American tree leaves, the Florida seashells were wildly different from the Dutch. The shells he knew from Scheveningen Beach were chalky and plain. Those from Sanibel Island were polished and ornate. Even those with the same name—cockles—had little in common besides their fan-shaped ribs. The cockles at home were nubby and much thicker than this sculpted giant engulfing his hand.

"I was overwhelmed by the beauty and the elegance," Professor Vermeij tells me six decades later as we sit with our knees close together in his office at the University of California at Davis, just outside Sacramento. The evolutionary biologist's hair and trim beard are white, but he has the open smile and enthusiasm of the ten-year-old child. His pale blue eyes are captivating and seem to look directly into mine, though I know they are prosthetic.

His office is packed with shells, their tiny labels raised in braille. Yet I can think only of his parents, who in the late 1940s and early 1950s made two life-changing decisions for their youngest son. The first was to have his eyes surgically removed at age three. After the boy endured a string of unsuccessful operations, they decided to sacrifice his slight vision to eliminate his intense pain and risk of brain damage from an unusual childhood glaucoma. The second was to leave their native Netherlands, where blind children were then educated in special boarding schools—Geerat's far from their home. The family chose New Jersey for its enlightened policy that the blind should be educated with sighted peers in classrooms like Mrs. Colberg's.

The boy's questions grew as numerous as the shells on the windowsill while his classmates, many with fathers who had fought in the Pacific, began to bring in others from home. Why did the cowries seem to be covered in a thick coat of varnish? Why would a tiny mollusk build a prickly spinning top like the *Tectarius coronatus* a classmate's father had found in the Philippines?

The answers had to do with mollusks' extraordinary survival over 500 million years—thanks in large part to their shells. The soft animals averted predators and endured noxious conditions through mass extinctions that killed almost everything else in the sea. Their shells told the story of their evolution, and that of the organisms and oceans changing around them. Vermeij learned to read them like braille.

I LEARNED NOT to ask shell collectors and scientists their favorite shell. It's like asking a parent to name a favorite child. They almost always have a vivid memory of their *first* shell; the one that got them hooked. I was struck by how often the story involved schoolteachers or grandparents.

At the Smithsonian's National Museum of Natural History, Christopher Meyer studies how cowries evolved and spread around the world; I figured his first shell would have been some lustrous Tiger Cowrie. But the mollusk curator traces his life's work to making shell crafts with his grandmother on family visits from his home in Pittsburgh to hers in Wyckoff, New Jersey. She'd stashed boxes of shells from her trips to Sanibel and its sibling island, Captiva, down in the basement. Meyer loved digging gently through them for attached donax—shell people call it "articulated" when the halves of a bivalve are found fused—the colorful clam shells that look like little butterflies. The two spent hours making collages, dioramas, and shell animals with Elmer's glue.

Surrounded by the world's largest and most valuable shell collection, Meyer's most beloved shell keepsake is a little duck he made with his grandmother. A child's excess of glue cements its Moon Snail head with shell beak to its Turkey Wing body with shell feet.

Harry Lee's first shell was a lustrous Tiger Cowrie. The Florida physician is credited with the largest private shell collection in the world. As a child, during extended stays with his grandmother in South Orange, New Jersey, they'd visit the collection amassed by the retired scissors-maker who lived across the street—the seashells took up two floors of his home. Lee was entranced by a dark-gleaming spotted orb. The collector gave it to him. When I met Lee seventy-five years later, he still had the Tiger Cowrie—and a million other shells.

At East Dover Elementary School, Geerat Vermeij told anyone who would listen that he was going to become a conchologist. The word *conchology* came out of the court at Versailles in the eighteenth century to describe "the study of the shells of the sea." Before the French Revolution, the study was much more about the shells than the animals inside. Most of the scientists naming mollusks—including Sweden's Carl Linnaeus—never saw them swimming or scooting, or even got a glimpse of the dead animal by the time seashells arrived on ships from the tropics. An eccentric French naturalist named Constantine Samuel Rafinesque-Schmaltz introduced *malacology* to describe the study of the living animals in 1814, the year before he lost his large shell collection and all else he owned in a shipwreck on the Connecticut coast while on his way to America.

Now mollusk scientists are called malacologists, and shell experts are conchologists. But the shell-obsessed amateurs often know as much as or more than the scientists do, creating a rare degree of reliance and respect between the two. When I shadowed Harry Lee one spring, I imagined we'd scour beaches, go to shell shows, and meet shadowy figures trying to sell him rare shells. He does all those things. But we spent most of our time at Lee's high-powered microscope at the Florida Museum of Natural History, where he volunteers to sort and identify fossil micromollusks. The tiny white conchs, horns, and tulips, many extinct, are delicate miniatures of the shells he has collected since boyhood. They came from a 3-million-year-old layer of the Pliocene, the era named by Charles Lyell to describe when modern marine mollusks in the fossil record—the shell shapes we recognize—begin to outnumber those that are extinct.

Nearly 50,000 living marine mollusks have been identified by science; that's perhaps only a third of those now inching along the oceans. More than half of the five hundred or so new ones named each year are described by citizen experts like Lee.

In a related tradition that has begun to vanish—in some ways for better and in others for worse—malacologists often recruited the next generation of scientists from young shell collectors. The most devoted invariably made their way to museum curators for help identifying shells. The beloved twentieth-century malacologist R. Tucker Abbott, often seen wearing one pair of reading glasses on his nose and two tangled around his neck, was credited in his 1995 *New York Times* obituary with turning America's seashell hobby into "an organized mania." Abbott wrote about his own generous mentor in a letter to a young collector:

> *You may wonder why I've taken so much time to answer your letter, even though I am 300 years behind in my work and writings. It's because I wrote a letter similar to yours about fifty years ago to a Harvard shell professor, Dr. William J. Clench. He patiently answered me in detail and started me on the right track. It's my turn to pass on his kindness. I hope you'll have the chance to take your turn.*

Geerat Vermeij wrote his "Dear Curator" letter at age fourteen. He'd launched his collection with a few of Mrs. Colberg's Florida

shells, and others he gathered at the New Jersey seashore, plunging his
fingers and toes into the coarse sand at Cliffwood Beach on Raritan
Bay to feel for live mollusks and empty shells. When the fifth-grad-
ers at school visited the American Museum of Natural History in
New York City, they returned with a box of shells for Geerat, each
one labeled with its Latin name and the far-flung place it was found.
He punched out new labels in braille. His parents, though they lived
simply and never had much money, bought him every shell guide in
print, including Abbott's 1954 tome *American Seashells.* They and his
older brother, Arie, took turns reading the books aloud.

By eighth grade, Vermeij was asking questions the guides couldn't
answer. He wrote to the American Museum for help identifying
shells. Malacologists William Old and Henry Coomans invited him
to visit and led him, Arie, and their mother, Aaltje, on a private tour of
the shells on the fifth floor where the public wasn't normally allowed.
Vermeij's nose clued him in as they passed dusty wooden shelves
crammed with stuffed birds and specimens preserved in alcohol.
Coomans, a Dutchman, encouraged the boy to study marine fossils,
and to subscribe to Dutch and American malacology journals such as
The Nautilus, published since 1886. Owing to Jules Verne, the name
evokes nautical engineering genius rather than the evolutionary genius
of an animal that rises and sinks along the coral ledges of the Indo-Pacific by filling its chambers with liquid and gas. Verne based Captain
Nemo's submarine in *Twenty Thousand Leagues Under the Sea* on a real
torpedo-toting "diving boat" named *Nautilus*, built by the American
Robert Fulton in Paris in 1800. More recently, the name is grimly
associated with a Canadian deep-sea mining company whose industry
threatens some of the last surviving populations of its brand animal.

Aaltje patiently read her son every scientific journal article, those
in English and Dutch, in German and French. No one, at least in his
childhood, even hinted that blindness might limit Geerat's pursuit
of such a seemingly visual interest. When he graduated from high
school at the top of his class, Princeton offered him a full scholarship,
a great relief to his parents. Outside the opportunities it had given

Geerat, America had been a disappointment. His father never could seem to advance beyond menial jobs, and neither his parents nor Arie ever settled into the culture. They continued to speak Dutch at home, and to season their beef with nutmeg. Arie returned to the Netherlands as soon as he turned eighteen. Their parents followed ten years later in 1973.

By that time, Vermeij had already crawled on his hands and knees along the shores of Guam, Palau, the Philippines, Ecuador, Brazil, the Caribbean, Singapore, and West Africa during his doctoral program at Yale, pursuing an explanation for the question that struck him in fourth grade: Why the polished and the rough, the curved and the spiked? The thick conchs and the ethereal jingles?

What could account for the spectacular and diverse architecture of shells?

IN OUR ERA of meme over matter, the idea that all living things descended from a common ancestor, evolving traits that helped them survive, is branded with Charles Darwin's name. The letters are displayed, in tall type, on those familiar fish-with-feet bumper stickers and T-shirts.

The sloganeering of evolution belies how Darwin's elegant theory emerged slowly, fitfully, uncertainly—beginning generations before his time—and continues to unfold as scientists blend evolution and genomics to reveal how life rose in ancient seas and persisted amid profound environmental change. Darwin's grandfather, a portly physician named Erasmus Darwin, foresaw by two centuries Vermeij's now-famous theory for the adornments of seashells. "The irregular protuberances," he wrote in his poem "The Botanic Garden," "serve them as a fortification against the attacks of their enemies."

Grandfather Darwin was among a number of eighteenth-century philosophers and scientists who spoke up about the fossil evidence of animals that no longer lived, in a time when questioning God's perfect creation was still a radical risk. His poem "The Temple of Nature" describes a big bang—"Ere Time began, from flaming Chaos"—and the rise of minute life in the sea before "successive generations bloom, new powers acquire, and larger limbs assume."

Erasmus believed all life had descended from one tiny shell, with "a single living filament" wriggling inside. Though he lived in the cathedral city of Lichfield in Staffordshire, with deeply religious neighbors and patients, Erasmus was so excited about his shell origin theory that he wanted to share it with others who might also be willing to question the conventional wisdom.

The Darwin family's coat of arms featured three scallop shells, a popular symbol then and now. (Princess Diana's coat of arms, passed down in the Spencer family since the sixteenth century, also included three scallops. When they turned eighteen, her sons William and Harry adopted shells into their own crests in her honor.) Erasmus Darwin decided to add the motto *E conchis omnia*, or "everything from shells," to the Darwin coat of arms. He put it on his personal bookplate, but the wider community wouldn't see it there. And so, like evolution's modern defenders with their walking-fish bumper stickers, Erasmus Darwin in 1770 had the crest and new motto emblazoned on the side of his carriage.

His pious neighbors were scandalized. Over at Lichfield Cathedral, Canon Thomas Seward became so incensed to see Erasmus "renouncing his Creator" that he wrote a satirical poem:

> *Great wizard he! by magic spells*
> *Can all things raise from cockle shells. . . .*
> *O Doctor, change thy foolish motto,*
> *Or keep it for some lady's grotto*
> *Else they poor patients may well quake*
> *If thou no more canst mend than make.*

Not wanting to insult the church or lose patients, Erasmus Darwin painted over the shell motto on his carriage, though he kept it on his bookplate. His heirs, biological and intellectual, found verity in common descent, if not from a primal shell. But today's paleontologists do theorize that the long-lived group of animals we know as the mollusks evolved from a single, shelled ancestor.

Scientists haven't yet found that mother mollusk. But they know it evolved at least 540 million years ago—after the single-cell microbes, some of which had innovated life's first shells, finally began to wriggle into more complex beings. Between the oozing microbial mats that reigned for most of Earth's history and the great rise of animals ahead came two underappreciated waves of life:

First, the earliest multicelled organisms squirmed to existence in soft bodies, now glimpsed only in the burrows and tracks of Earth's oldest rocks. These squishy pioneers discovered how to capture energy in sunlight. But their innovation also helped destroy them. The photosynthesis they developed had a by-product: oxygen, poisonous to most microbes that had evolved in the primal, low-oxygen sea.

Many of these enigmatic organisms died off in a mass extinction that preceded the Big Five in textbooks—and the devastating sixth now underway. Only those that could adapt to the chemically altered Earth pressed on, many by building shells.

The second wave comprised tiny, weakly mineralized beings that scientists call the "small, shelly fossils." They are nicknamed "small shellies" or "small smellies," owing to the only way to collect them, which is to dissolve blocks of limestone in an acid bath. The little mineral-builders included wormy, tubular, and sponge-like creatures as well as some of the earliest mollusks. Extinct *rostroconchs* looked like a clam, but with halves fused together into a single shell. Snail-like beings called *helcionelloids* carried a shell shaped like a wizard's hat. They lived in the shallows of a shifting sea. They were about to have a crush of company.

More oxygen led to more photosynthesis, hiking proteins like collagen that animals need to make their tissues. Volcanic ash may have upped the calcium carbonate in the oceans, creating a ready store of shell-building supplies. Preserved in 555-million-year-old ash beds on the banks of the Onega River in northwestern Russia, a soft, cowrie-shaped animal called *Kimberella* hauled a non-mineral shell several inches long. Scientists following its feeding tracks and crawling trails know that it moved backward, possibly on a creeping foot.

After that soft launch, hard shells—along with skeletons—appeared around the world in the Cambrian rise of animals. Blobby, slow-moving life-forms began to give way to a rowdy marine bestiary of predators and prey. Segmented trilobites crept along the seafloor like their insect and crab descendants. Bigger sea creatures were evolving to eat them, such as two-foot-long, lobster-like *Hurdia victoria* with spiny claws and a lance-like shell protruding from its head.

In the Canadian Rockies, hundreds of a spiky slug called *Wiwaxia* are preserved in the great Burgess Shale fossil deposit from the Cambrian sea. The 505-million-year-old relics are shingled with scales and jutting with spines. As with *Kimberella*, scientists have not settled whether the creature was an early mollusk or a worm. But they do see that many of its spines had been snapped off—likely by predators—and then repaired.

WHEN HE WAS a boy falling in love with seashells, Vermeij's aesthetic senses were offended when his fingers came across broken spines, cracks, and other flaws. At first he blamed the blemishes on waves bashing against rocks—or on bungling human collectors. For Christmas when Vermeij was fourteen, his brother Arie bought him a *Conus litteratus*, a heavy, brown-on-white lettered cone with a striking flat spire. As soon as Vermeij picked up the shell, his fingers moved along a thin break that his brother couldn't see; the fracture ran from the apex to the end. "I tried not to show my disappointment, but I could never like the specimen as I did all the other shells in my growing collection," he remembers. "The scar was a blemish, an insult that some careless human collector must have inflicted on the specimen."

Only when he began to feel along wild shorelines during his graduate research did Vermeij realize just how common scars, breaks, and holes are—especially in the tropics. A shell's construction is partly influenced by geography, in the way our homes reflect our place in the world; adobe construction in the desert, chimneys in the north, hurricane shutters down in Florida. The differences between the cold-water and tropical shells that Vermeij noticed in fourth grade are explained in part by the availability of building materials. Calcium carbonate is easier to take up in warm water, making mollusks' bricks and mortar more abundant—essentially cheaper—in the tropics, where they can afford a bigger, showier house. The Australian marine biologist Sue-Ann Watson has shown how cold-water mollusks build more practically because it takes so much more energy for them to make a shell.

But the environment can't explain all the spines and spikes, ribs and ridges, grooves and whorls, and infinite other adornments. The shell scars were clues. Vermeij found that warm-water shells are not only heavier and showier. They are also more beaten up and more scarred. Splashing with Vermeij through hot tide pools in Guam in 1970, a colleague handed him a Money Cowrie with its glossy dome broken clean away. When Vermeij expressed surprise that the placid waves could have crushed such a sturdy shell, his colleague said he'd seen crabs in his aquarium break shells just that same way with their claws.

Vermeij began to wonder if the artistry of tropical shells—their tight coiling, knobs and spines, narrow openings, tight-shutting doors— evolved with the weaponry of predators. Examining shells in museum collections and along many more miles of coastline—he now explored Indian Ocean shores in Madagascar and Kenya and the beaches of the northern Red Sea with his wife, Edith Zipser, a fellow biologist he'd met at Yale—he felt not only breaks, but impressive repairs.

Mollusks use their calcium carbonate cement to caulk shell cracks and to patch even major holes. Running his fingers across thousands of calcified reconstructions, Vermeij came to understand that shell repairs were not insulting imperfections. They were the battle scars of survival.

THE CAMBRIAN MARINE BESTIARY lasted 50 million years, into the Ordovician that filled the seas with corals, jawless fishes, and more prodigious shell-builders. The ammonites radiated to thousands of species, some as big as semitruck tires. Chunky brachiopods, with paired shells like a clam, spread in every sea, building up ancient reefs and leaving scientists more fossils to study than perhaps any other animal in Earth's history. The Romans named them lamp shells for their resemblance to the clay oil lamps of the ancient world; the clay versions had themselves been modeled from shells that burned oils and animal fats in prehistoric times.

One day in that early Ordovician sea—if you're keeping track, it's now 478 million years ago—a shaggy slug with a combat-style shell cap drifted down into a muddy grave. The mud embalmed its soft body, leaving a fossil imprint that looks like a big toe with a darkened nail—the shell cap. The squishy soldier also had a saw-like tongue with 125 rows of teeth.

Eons of waves and storms buried the creature with hundreds of real sea monsters in sediments that would someday rise as red hills in eastern Morocco and a famed fossil site called the Fezouata. Of all the bygone beasts just one had that toothy tongue, called a radula. Latin for "little scraper," the anatomical rasp is found only in mollusks. That makes bristly *Calvapilosa kroegeri* and its shell cap as close as scientists come so far to guessing what the mother of all mollusks might have looked like. Molecular clock analysis, which measures genetic changes in species to tell evolutionary time, shows that bivalves and gastropods split into different branches of the mollusk family tree around this time.

A third major branch, the cephalopods with arms or tentacles rather than the one big foot, also evolved with shells. They included nautiloids that hurtled toward their prey in missile-shaped shells up to 15 feet long. Evolution would bend the straight nautiloid shells into elegant curves with internal nacre chambers. They heralded one of the longest surviving animals of all time.

The Chambered Nautilus persists in the trove of ocean life between Papua New Guinea, Indonesia, and the Philippines known as the

Coral Triangle. Some ninety white tentacles wave from its tiger-striped shell, or shrink back under the animal's leathery hood at the sign of a predator. Its cephalopod cousins, the octopus and the squid, are sometimes described as having shed their shells for intelligence and speed while the nautilus kept its chambered beauty as some sort of relic. But the shell is an emblem of survival.

The evolutionary biologist Peter D. Ward has devoted nearly seventy years to the nautilus he fell in love with at age five when Disney released its film version of *20,000 Leagues Under the Sea*. Where others see vestiges of prehistory in the creature's pinhole eyes and festoon of tentacles, Ward sees an advanced animal "superbly adapted for existence in the deep sea." The nautilus relies on sophisticated chemical sensors to trawl for food with its tentacles, and to reproduce. Ward describes its ancestors as having risen above the shell-crushing, bottom-hunting fish evolving alongside them in increasingly dangerous seas. The iridescent nacre lining the shell, also known as mother-of-pearl, is one of the strongest materials on earth. Even the shell's colors and patterns—dark-striped on top and light at the bottom, camouflage known as countershading—are designed to foil predators hunting from above and below.

The Boston physician and poet Oliver Wendell Holmes Sr. considered the nautilus the perfect metaphor of life's toil and triumph. In what is arguably America's best-known seashell poem, "The Chambered Nautilus" he described the beauty born of struggle; the "lustrous coil" of the shell and its "frail tenant" that builds each chamber nobler than the last: "Till thou at length are free," Holmes wrote, "Leaving thine outgrown shell by life's unresting sea!"

SOME 445 MILLION years ago, the marine Eden that took root in the Cambrian began to wither in the first of Earth's great die-offs, the Late Ordovician mass extinction. Up to 85 percent of sea-dwelling species perished, including many of the abundant corals and clam-like brachiopods that had spread in every ocean. Its cause is uncertain, but scientists find increasing evidence that volcanic eruptions may have released enough carbon dioxide to warm the world and starve the seas and marine life of oxygen.

That first mass extinction still did not come close to the worst, known as the Great Dying, more certainly caused by global warming. About 250 million years ago, volcanic eruptions spewed lava across what is now Siberia, sending huge amounts of sulfur dioxide and carbon dioxide into the air. Those greenhouse gases warmed the Earth and detonated a more dangerous bomb: methane gas that continued to vent for thousands of years. Recorded in the shells of brachiopods entombed in limestone beds from China to Chile are signs that the gas poisoned the air and seas and pushed temperatures higher.

Oxygen levels in the oceans plummeted, killing off the ancient sea monsters in deep water where there was no exchange with the scant oxygen above. As the water became impossibly warm, soaring past 104 degrees Fahrenheit by the early Triassic, the coral reefs blackened and died. The last of the trilobites perished. Most of the ammonites died off. And the brachiopods—which had burgeoned to more than 30,000 species around the world—plummeted to a few hundred. They survive today, hardly seen; relics of the past like the ancient oil lamps for which they are named.

The Great Dying killed off most life on land, and almost all the wondrous creatures left from the Cambrian sea. Among the small numbers of marine species to survive were a handful of mollusks. They included a few petite gastropods and a greater number of bivalves, also small. Like a band of survivalists in end times, they lost most of their own kind as they endured worsening conditions. As the mass dying continued, the rotting corpses released more carbon dioxide and poisonous hydrogen sulfide into the sea.

In fossil layers from the Great Dying and its 5-million-year aftermath, the darkened stratum of the lifeless marine Eden stretches around the world. On top of it are layers upon layers of increasingly familiar forms, evidence for a newly ascendant group of marine animals.

From that world of death evolved the polished beauty that, countless tides later, would inspire Leonardo da Vinci, Frank Lloyd Wright, and humbler shell-lovers. Mollusks, particularly the bivalves, held on in small numbers, then began to run riot in the emptied ecological Eden.

AS VERMEIJ STUDIED seashell fossils on shores and in museum drawers, he counted their breaks and blemishes—the tiny repairs raised like braille. The battle scars, he found, increased through geologic time. Repairs became especially abundant in the Mesozoic—the Triassic, Jurassic, and Cretaceous periods better known for the dinosaurs.

He also investigated the predators' side of the story. Vermeij and his wife, Zipser, examined fossil crab claws in museums around the world, finding that they grew bigger and stronger through geologic time. Returning to Guam, they caught crabs and mollusks and put them together in aquariums, recording the gory details of the attacks. While some mollusks were killed, leaving only a pile of shell bits, others had such strong apertures that the crab attack failed. A surprising number of crabs gave up, leaving behind a damaged shell—its soft animal alive inside.

Over a lifetime of interpreting the story of mollusks through their tactile homes, Vermeij would show how predators drove the evolution of shells. The rise of mollusks and their artful fortresses that followed the Great Dying coincided with that of shell-breaking fishes and fierce-clawed crabs. Survivors of attacks—those with the best defenses—were more likely to reproduce and pass on the traits that protected them. Surviving species stayed one coil, flourish, nub, or spike ahead of predators that were in turn developing fiercer jaws and claws.

Vermeij named this surge of sea life and shell the Mesozoic marine revolution. At the same time the dinosaurs were evolving on land, mollusks developed not only heavier spines and spikes and other armor, but their thin openings, trap doors, tricks to vanish beneath the sand, toxicity, camouflage, and other flourishes to thwart predators.

The paleontologist Lydia Tackett lives in North Dakota, where fossil seashells from the era are fused into rocks from the floor of a shallow sea and dinosaur bones are hidden along what was once its coastline. Bivalves, Tackett has found, were some of the earliest in the Mesozoic to figure out their clever protections. Brachiopods had begun to vanish because predators could pluck them right off the sea bottom. Other bivalves began to attach themselves to the substrate or go into hiding: Oysters evolved their protein adhesive, clams their ability to bury themselves deep in the sand—long siphons poking up to draw seawater and oxygen. "Burrowing clams go wild," Tackett told me. "Scallops begin to swim when they couldn't before. Oysters evolve the ability to cement themselves to surfaces."

Later in the Cretaceous, gastropods evolved all manner of new spines, knobs, whorls, corrugations, thicker walls, even barricades. Cowries filled out the teeth that defend their apertures like razor wire; their glossed round shell is thought to give the slip to crab pincers. Cones developed their impossibly narrow apertures.

The beautiful fortresses helped mollusks survive to become the second-largest group of animals in the world and in the sea. They are outnumbered only by the arthropods—including the crabs that still crush them or slink away, foiled by the same spires and spikes that attract their human admirers. Their safety is our beauty. And their building materials still depend on the ocean chemistry that brought them to life, made their shells, or banished them to the fossil record like St. Hilda sweeping the snakestones of Whitby, England, off that long-ago cliff.

THE MOST RECENT mass extinction is best known for killing the dinosaurs, a truth so incredible that it draws even toddlers to existential doom. As the die-off at the end of the Cretaceous wiped out more than half of Earth's life, the world also lost the last of its ammonites, among the most successful and diverse groups of animals that ever lived.

Yet more incredible than the story of the animals that died is the triumph of those that survived. They include a sculpture garden of the shell makers humans love most. The modern nautilus is one. Its deep-water habitat, says Ward, became a refuge as many other mollusks perished. The nautilus tucked eggs at the deep rock ledges where its young were born in their shells—somewhat sheltered from acidifying seas, as they may be today—while ammonites sent their delicate lar-vae floating on currents that became too acidic or warm to survive.

As with the grim geology of the Great Dying, scientists can see the End Cretaceous extinction in a physical layer called the K-Pg bound-ary, a band of rock found around the world. The cause was disputed for a long time because scientists couldn't parse two closely timed possible origins: volcanic eruptions in the Deccan Traps of India, which sent the greatest lava flows ever known across the subconti-nent; or the impact of the massive Chicxulub asteroid in the Yucatán that hurled tsunamis across the seas and enough ash into the skies to blot out the sun. The K-Pg boundary contains iridium, also found in asteroids—long bolstering the asteroid theory. Thanks in part to shells, scientists see increasingly that the End-Cretaceous extinction was more complex—and that climate change was an accessory, as it had been in the other great extinctions.

The geochemist Andrea Dutton was the first scientist to show me the feathery gray rings that run through the cross section of a fossil clam shell. Much like tree rings, the wispy bands mark growth, time, and conditions in a mollusk's life. Dutton, who has in common with Vermeij the happy surprise of a MacArthur genius award, reads the climate past in ancient corals and fossil shells. A shell is a log of both the chemicals and conditions in the surrounding sea from which the mollusk drew its building materials.

Cretaceous bivalve fossils collected from the Antarctic Peninsula gave Dutton and her colleagues a rare, multimillion-year record of life and extinction during the dinosaur-era die-off. Half the humble clams and cockles had gone extinct well before the asteroid, the other half after. Measuring the oxygen isotopes in their shells revealed that two great heat spikes pulsed during the long-running extinction. The first corresponded with an early eruption of the Deccan Traps. A couple million years later, the second spike lined up with the asteroid strike. The shells told the story of a one-two punch. They suggested that both the eruptions and the asteroid—along with global warming that followed—had roles in wiping out the dinosaurs and other life. "The analogy," Dutton says, "is that a weakened ecosystem is more prone to subsequent environmental stress."

It is, she says, the analogy for our times: Coral reefs already stressed by pollution and sedimentation from coastal development become less resilient to other environmental burdens such as warming. Clam beds and oyster reefs deprived of their river flows are more vulnerable to acidifying seas and other damage from climate change. The Chambered Nautilus, as we'll see, may succumb to human pressure for jewelry, décor, and trinkets—even having persisted through the dinosaur extinction, and after its forbears' survival of every last calamity in the half-billion-year evolution of mollusks.

GENERATIONS OF AMERICANS have read or been made to study Oliver Wendell Holmes's Chambered Nautilus poem outside its context as dialogue in his beloved series "The Autocrat of the Breakfast-Table" that ran in in *The Atlantic Monthly*. The poem was first published in 1857, the year Holmes helped found the magazine and suggested the name *Atlantic* to invoke a wide separation between this new, American journal and those of the Old World.

In the series, men and women living in a fictional Boston boardinghouse carry on witty philosophical debates about the issues of the day. The poem follows a meditation on human progress. In the preceding repartee, Holmes wrote of the importance of always moving forward, "sometimes with the wind and sometimes against it—but we must sail, and not drift, nor lie at anchor."

Later, when the poem had become a classic, the famed twentieth-century malacologist Tucker Abbott prominently criticized it as "a zoological catastrophe" for mixing up two cephalopods, the Chambered Nautilus and the Paper Nautilus, the lovely parchment egg case made by the octopus cousin *Argonauta*. Tucker's sharp words have been repeated in a half century's worth of marine-science books, and at least one mollusk encyclopedia. "We can excuse Mr. Holmes," Abbott wrote, "for in his day not many scientists, much less poets, appreciated the difference between a paper nautilus, or Argonaut, and a chambered, or pearly nautilus."

But it was the scientist who misunderstood the poetry. Holmes, a professor of anatomy who read and published in the scientific journals and followed Charles Darwin, preceded the poem with an explanation of the two animals and why he put them together in verse—to evoke life's successive journey. "We need not trouble ourselves about the distinction between this and the Paper Nautilus, the *Argonauta* of the ancients," Holmes wrote. The point was the metaphor: a ship of pearl making its way through "life's unresting sea," as the poem ends.

The humanity—the breakfast-table talk and the poetry, the good and the greed—was for Holmes as important as the science. That was also Vermeij's conclusion after a lifetime of interpreting mollusks through their shells, beginning in Mrs. Colberg's fourth-grade class. For five hundred million years, mollusks had modified their shells to survive all manner of changing ocean conditions and more-powerful predators. Vermeij found them unable to adapt in the case of only one. Traits that helped them fend off nonhuman enemies, such as large sizes and well-armored shells, "attracted humans and therefore became disadvantages," Vermeij says. Meanwhile their adaptations to us—to the likes of pollution and coastal destruction—were limiting their ability to grow and reproduce.

Yet Vermeij still sees the ultimate outcome of life's struggle to survive as leading to creativity and innovation—to the ingenuity to build a logarithmic spiral, each chamber bigger than the last. As Holmes pointed out, it was the story of the Chambered Nautilus and the story of us. We are predators and destroyers but also builders and repairers. Saving or sinking our ship of pearl.

Three

THE VOICE OF THE PAST

TRITON'S TRUMPET
Charonia tritonis

Ten thousand feet above sea level in the Andean highlands of ancient Peru, conch-shell trumpets bellowed through a steep river valley, calling worshippers to the temple complex at Chavín de Huántar. Inside the temple, conch voices echoed in the stone walls and deep underground, penetrating subterranean altars and worshippers' hearts in haunting, low notes that seemed to come from everywhere and nowhere all at once.

In lonely strains reminiscent of whale songs in the sea, the calls of the conchs could carry more than a mile through the valley, auguring to travelers that they were nearing Chavín, a religious complex that thrived between roughly 1,500 and 500 BCE. The architectural wonder predates the Inca and Peru's Machu Picchu by more than two thousand years.

Early Andean people are believed to have made the mountainous trek to Chavín to seek the wisdom of an oracle. Artifacts also reveal offerings and hallucinogenic rites in vaulted belowground chambers. Beneath its flat-topped temples, terraces, and sunken plazas, a labyrinth built of thick limestone slabs winds in darkness for more than a mile. Twenty ritual galleries connect through narrow passageways and stairs, small rooms and hidden alcoves, air vents, and aqueducts. The further archaeologists dig into the humid underworld, the more evidence they find that the space was designed for stomach-churning special effects. Water running through granite troughs above and below the worshippers mimicked a clamorous roar. Vents could be shuttered to create total blindness. Mirrors made of anthracite coal could direct dust-sparkling shafts of light onto idols. Echoing through it all were some of the most commanding sounds of the prehistoric world.

The conch trumpets of Chavín de Huántar could sound thunderous and terrifying; distant or distorted; solemn or even serene. They could roar like a jaguar. The origin of the sound was impossible to discern, particularly if pilgrims had downed psychotropics like the San Pedro cactus prominent in Chavín artwork. Carved stone human and animal heads at Chavín depict the upturned eyes, grimaces, and nasal mucus trails of a hallucinogenic trip. Scientists find clues to sound effects worthy of *The Wizard of Oz*; architectural flourishes to give voice to the oracle. In imagery, the oracle was half-man, half-beast. He smiled with the fangs of a jaguar. And evidence suggests that the oracle of Chavín de Huántar spoke in the voice of a conch.

WE THINK OF ancient life as quiet compared with our sound-blasted society and its amplifiers and pulsating car-stereo speakers loud enough to create their own wind. But early people also amplified their messages and music. They did so with tusk and bone horns, stone lithophones, wooden bullroarers, bird-bone flutes, natural and built echo chambers, alligator-hide drums—and gut-stirring trumpets made of shells.

Silence is a virtue for mollusks, which aren't aggressive or fast enough to take on hearing enemies like fish and turtles: Evolutionary scientists say animals need to offset the benefits of making a racket with the means of fighting or fleeing any predators attracted by the noise. While they may live quiet lives, the world's biggest sea snails leave behind shells that make up for it by sending forth a foghorn bellow.

The interior coil of a large gastropod shell—a triton, conch, or helmet shell—creates the ideal windway for blowing a clear and powerful tone. Low and long, urgent and mesmerizing, there is nothing in the soundscape quite like it. Their gravitas led early people to blow shell horns to sound birth and death, battle and hunt, and not uncommonly, the divine. Shell trumpets seemed to call from another world. They were revered around ours.

Shell horns, their surfaces often mouth-worn, have been found in rock shelters and caves on almost every continent, including long distances from the sea. Stone Age people blew the high-spired Triton's Trumpet—*Charonia tritonis* from the Indo-Pacific, or *Charonia nodifera* from the Mediterranean—one of the oldest instruments known. Growing easily a foot long, the tritons are among the largest gastropods in the sea. Their swollen shells evoke plumage in handsome buff and brown; early shell aficionados often compared them to partridges or pheasants.

Hearing a shell trumpet blown in a quiet academic hall at the turn of the twentieth century, the biblical archaeologist John Arthur Thompson described the sound as so emotive it "makes one a little ashamed." Thompson imagined "it was not perplexing to our forefathers that what was official and symbolic one day should be a foghorn or a cattle call the next. What served to scare off evil spirits would also serve to frighten thieves."

But modern scientists say shell trumpets' frequent discovery in places like Neolithic burial sites, temple complexes such as Chavín, and later palaces, city walls, and fortified towers suggests that they were rarely if ever trivial. Shells carried the gravitas of the podium or pulpit.

Hundreds of Triton's Trumpets, often with the apex worked off for the mouthpiece, have been excavated alongside human bones in late Stone Age caves and settlements. Tritons from the coast of Liguria were passed on throughout Neolithic Italy, across the Alps to Switzerland, and even to ancient hearths around landlocked Hungary's Lake Balaton. Their discovery almost exclusively with human remains points to their religious significance, according to the British archaeologist Robin Skeates. They were crafted not so much for music, but for messages, Skeates writes, "directing and impelling men and gods to action."

In the Americas long before the Inca; in the Buddhist and Hindu cultures of ancient India; in China during the Han and Tang dynasties; in Triton myth of ancient Greece; in war and religious traditions of classical Japan; across Oceania from five thousand years ago to this day, the sea horns sounded potent calls that connected people and the oceans across time, place, and belief.

For the Hopi and Zuni people of the American Southwest, conch-shell horns manifested the voice of an underground-dwelling plumed serpent that could quake the earth or spew volcanoes. In Hinduism the divine conch resonated the sound of *Om*—the essential sound and breath of the universe.

"The murmur of the shell was the voice of the god," wrote the British conchologist J. Wilfrid Jackson a century ago.

So the pilgrims to the underground labyrinth at Chavín de Huántar were led to believe.

BUILT OF HULKING, quarried stones at the confluence of Peru's Mosna and Huachecsa rivers in fifteen phases that continued for centuries, Chavín reached its zenith a thousand years before the Spanish arrived. It was still legendary in 1616 when the friar Antonio Vázquez de Espinosa described the ruins as a "sanctuary, one of the most famous of the gentiles, like Rome or Jerusalem among us."

Chavín still stands today, a UNESCO World Heritage Site, in worn resilience to time and the disastrous floods, landslides, and earthquakes that have left it damaged and diminished. The temple complex is being steadily consumed by its verdant mountain valley, covering up its subterranean secrets and silencing its heart-thrumming voices.

The Peruvian archaeologist Julio César Tello began excavating the site in 1919, uncovering the galleries and artworks that often fit human and wild-animal parts together in monstrous jigsaw. On carvings, sculptures, and pottery, features of jaguars, eagles, hawks, anacondas, and black caimans fuse with human bodies in ancient metaphor: Faces framed by wings. Hair as writhing snakes. Append-ages as tongues and fangs and talons.

Tello came to interpret Chavín as *la cultura matriz*, the mother cul-ture from which the Inca and other pre-Hispanic Andean civilizations grew. Later research revealed it was influenced by yet older centers in the highlands and along the coast, perhaps settled by people chased inland during relentless years of El Niño–related storms seen in ancient climate records. A pulse of warm water and air born in the western Pacific, the great weathermaker El Niño wreaks predictable havoc in the atmosphere. The ocean warming disrupts large-scale air movements in the tropics and hurls dramatic weather around the world, starting with fierce rains and flash floods in this part of South America.

There is little question mountain-tucked Chavín culture was inspired in part by the sea. Chavín artists often added giant conchs and *Spondylus*, or thorny oysters, to their mashups of animals, plants, and people. Species of both mollusks live in the Pacific Ocean, hun-dreds of miles up and over mountains to the west. They recur in the rituals of many pre-Columbian cultures. In 3000 BCE in Ecua-dor, children of the Valdivian culture wore masks made of *Spondylus* shells; two small eyeholes pierced for young faces. The red, orange, and purple orbs were offered to many deities. They were a crunchy favorite of the Andean fertility goddess and Earth mother, Pacham-ama. They were buried with the dead for communication in the after-life, or in fields to bring needed rain.

The spiked *Spondylus* and supersized conchs were also conspicuous in the rituals at Chavín de Huántar. On the cornice stones that once encircled the Old Temple outside, a winged man leads a processional, conch trumpet in his hand. The next man in line clutches a thorny oyster. Another conch, decorated with a gigantic eye, is centered on the 8-foot granite Tello Obelisk found by its namesake at Chavín. The cyclops-conch joins fanged caimans, healing plants, and the trademark mélange of human and wild-animal parts.

Most prominently, a "Smiling God" engraved on a slab found at the patio of the New Temple—fanged but grinning, with hair of snakes and talon toes—grasps a conch shell in one hand and a thorny oyster shell in the other.

The shell-clutching figure depicts the *axis mundi* of Chavín de Huántar. Inside the Old Temple at the lowest level of the complex, deep within the labyrinth in its own cruciform stone chamber, stands a 15-foot granite monolith with a fiendish smiling face, eyes rolled upward and two jutting fangs. The sculpture penetrates both floor and ceiling, the tapered top rising into a second-story gallery. Its central worship spot at Chavín, along with its size, workmanship, and iconography, suggest it was the principal cult image of the original temple.

Nearly a century after Tello named the figure El Lanzón (Spanish for a large spear or lance), another generation of archaeologists uncovered its conch-shell voice. An entire choir of conchs once roared and whispered at Chavín.

IF JIM HENSON had dreamed up an archaeologist to dig around Sesame Street, his name might be Professor John W. Rick. The Stanford University professor of archaeology and anthropology with unfastidious gray hair and beard shares his work with the excitement of a six-year-old who has found a mummy in the Peruvian desert. Which is exactly how old he was when he found a mummy in the Peruvian desert. Rick's father was a renowned wild-tomato expert at UC–Davis who brought his family on summer seed-collecting trips up and down the Andes. Exploring the coast with his mother one afternoon, young Rick found a textile-wrapped finger, which led him to a perfectly preserved mummy and a lifelong search to understand the lives of ancient people.

Rick has spent nearly every summer since in the highlands of Peru. He was there as a barefoot graduate student in the 1970s. He was there again as a young professor in the 1980s, leading his own students on excavations of early hunter-gatherer societies in the central grasslands. Those societies may have been his life's work, if not for a terrifying night in 1987 when he and eight graduate students were confronted by

Shining Path guerrillas. Rick told their leader that he and his students were Canadian, and he expressed sympathy for the rebel cause. The ruse likely saved their lives. Later the same night, the rebels gunned down a local leader just steps away from Rick's door. At sunrise, he rushed his students out of the mountains and never returned.

A few years later, the director of Peru's National Museum of Anthropology and Archaeology, Luis G. Lumbreras, invited Rick to spend a week at Chavín. Leading excavations at the temple complex since the 1960s, Lumbreras had pieced together how Chavín's spiritual leaders used landscape, architecture, and special effects to inspire awe—and belief in the power of an oracle—across a broad swath of the highlands.

Its stone construction along a channelized riverbed and alignment with the surrounding mountain peaks and rising and setting sun conjured a powerful spirit of nature as worshippers approached. Chavín's engineers sent water uphill into the temple's hidden pipes and troughs a thousand years before the Romans siphoned water uphill. Lumbreras discovered how, in rainy season, the directed water would have roared and echoed through the subterranean complex. Its twisting galleries, built across multiple floors with colorful walls and niches for "multimedia" displays including light, smoke, and sound, seemed designed to disorient. During ritual drug use, scientists theorize, Chavín "would have provided a full sensory impact—and probably overload."

Chavín's leaders were determined to establish their otherworldly power. Lumbreras hypothesized that priests cinched their authority, and belief in an oracle, by predicting environmental events. No written Chavín language has been found. But the Chavín culture, like others connected and disparate, associated their conch trumpets, and by extension their oracle, with water, rain, climate; perhaps to call or to banish what we now understand as super El Niño cycles that bring drought-busting rains or disastrous floods to the Andes. Myth and reality fused.

Lumbreras speculated that Chavín priests may have learned to forecast the arrival of severe weather cycles by reading ecological signs. The southward movement of tropical marine life including *Spondylus* in years of a warming Pacific Ocean might have signaled climatic changes yet to arrive in the central Andes. The insights could have allowed priests to foretell storms, severe weather, and disasters like mudslides.

The theory would help explain an ancient highlands obsession with shells. In 1972, Lumbreras began to excavate the smallest gallery at Chavín. It stands off by itself just south of Circular Plaza, and lacks the dramatic entrance leading to the Lanzón. The humble site turned up fragment upon fragment of shell. The pieces were Eastern Pacific Giant Conch, large marine snails that live in the mangrove shallows to the northwest, even now a considerable trek from Chavín.

Lumbreras named the site Galería de las Caracolas—Gallery of the Sea Snails. It would be three decades before Stanford's John Rick dug deep enough to learn it was the perfect name.

Rick began mapping and excavating at Chavín de Huántar in 1995, now hauling his graduate students to the high sierra instead of the grasslands in summertime. About three years into the project, they hit a sizable stone slab on the west side of the Old Temple. It turned out to be a broken piece of the first of the cornice stones that once circled the temple—the stone with a procession leader holding a conch trumpet to his lips. They couldn't make out the rest of the carving where the stone had broken. But something about the break was familiar to Rick. He used his new digital camera, his first, to photograph the carving and its broken corner.

Back in California that fall, he painstakingly compared those images with hundreds of others he'd taken of Chavín artifacts. He finally found the missing piece of the ancient jigsaw. One of his archival photos perfectly matched up with the broken piece—revealing the next figure in the processional line. That figure was holding a thorny oyster.

Rick remembered Lumbreras's speculation about the sacred nature of giant conchs and thorny oysters at Chavín, and the possibility that their abundance at certain times could have helped the temple priests predict El Niño. "It was beginning to make beautiful sense," Rick says.

Unearthing and matching up the cornice pieces paled before what was the discovery of Rick's career—a treasure buried in the Galería de las Caracolas under a few thousand years of sediment, broken black pottery shards, and llama bones. In the summer of 2001, Rick and his students had dug through the layers of silt and shard when they finally reached the original floor of the sea-snail gallery. They began to unearth fragments of large shells. Then, they came upon what appeared to be an intact object. After a half day of gentle excavation, they saw it was an ancient conch. As they worked around it, they reached what should have been the spire. They brushed off the dirt, and the top was missing. But it hadn't been broken.

"It had a beautifully smoothed mouthpiece," Rick says. The horn-blower engraved on the cornice came immediately to his mind. "I said, 'Oh my God, we've got one!' We had an intact trumpet. It was unbelievable."

It took another half a day to lift the worked conch from its tomb, brush out the interior, and clear the mouthpiece of diluvial dirt. As he handled the shell, Rick, who'd played bugle as a boy, felt an irresistible urge.

He lifted the ancient horn to his lips. It took him a number of tries to blow a sound. Finally, the conch bellowed—for the first time in at least three thousand years. The tone was haunting, somewhere between the musical notes C and D. Echoing on the stone walls of the small gallery of sea snails, it sounded as if it had been made for the space.

When I asked Rick nearly two decades later how he felt in the moment he heard the sound, he recalled:

It was riveting. It was a gut-wrenching sound—literally gut-wrenching, an actual physical transformation of the human body because of the low frequencies we are not used to. There was this corporal reaction. It was a profound, loud, primary voice.

The feeling was: We are hearing sound from the past. We are hearing what they heard. I'm just thinking: We've got it! We've got it! We've got the voice of the past! It was fantastic and emotional and transporting to hear that and to realize that. Everything became very real to me at that point. The conch, much more than a bugle, has a primary voice.

It's one of the most direct connections to the past that I have ever felt.

I thought of the archaeologist's six-year-old self, finding the textile-wrapped finger in the Peruvian desert. When the past spoke to him from a long-buried conch, Rick thought of the voice and the moment as ephemeral—a single, sonorous thanks for a lifetime of quiet digging in the Peruvian earth. It was more lasting than that. Just as the finger had led to an entire mummy and an entire career, the first conch would unwind a larger story involving the very nature of truth and authority.

Within days, Rick and his team found a second conch horn, even more spectacular than the first. The flared shell was carved with detailed geometric designs. This shell, too, had "a stunning voice," Rick recalls, "a much stronger and attention-grabbing sound that seemed to justify its intensely elaborate engraving."

By summer's end, the team had unearthed twenty intact shell trumpets, tooled from ivory-white conchs. The original craftsmen were the same Eastern Pacific Giant Conchs that Lumbreras had found in pieces in the gallery floor thirty years before. Gentle grazers that eat microalgae, the giants are known for their ability to travel miles across the seafloor over the course of a few months to mate in the mangroves and sandy shallows between the Gulf of California and Peru.

The horns' artful engravings suggest a second set of craftsmen somewhere along the coast tooled the shells into trumpets and decorated them. The shells were then traded another considerable distance—over the world's highest tropical mountain range to the temple priests at Chavín.

A third set of craftsmen, perhaps the horn-blowers, notched a distinctive *V* into the top outer lip of each conch. The notch might have given them a finger grip, or helped them see ahead in a processional. Each conch is rubbed thin from playing by multiple generations of musicians. Each has a well-worn mouthpiece. One shell still shows the imprint of the textile that swathed it before disintegrating with time into the Peruvian dirt.

Galería de las Caracolas seems to have been a special storage room for a ceremonial collection of horns. Rick suspects they had been hung in bags along the underground wall or from the gallery ceiling. They were left behind when the site was abandoned in the middle of what would be its final construction phase, five hundred years before the birth of Christ.

MYTHOLOGY'S BEST-KNOWN SHELL trumpet, blown by the muscled merman Triton of Greek mythology, son of sea god and goddess Poseidon and Amphitrite, saved the world by calling the water. Triton's Trumpet could raise or calm the sea; convene the river deities around their monarch; or frighten off enemy giants who thought the wails were the roars of wild animals drawing near. In the myth of Deucalion's flood, the call of Triton's spiraled shell brings an end to the universal deluge. Ovid described the scene in *The Metamorphoses*:

> *Triton, sea-hued, his shoulders barnacled*
> *With sea-shells, bade him blow his echoing conch*
> *To bid the rivers, waves and floods retire.*
> *He raised his horn, his hollow spiraled whorl,*
> *The horn that, sounded in mid ocean, fills*
> *The shores of dawn and sunset round the world;*
> *And when it touched the god's wet-bearded lips*

And took his breath and sounded the retreat,
All the wide waters of the land and sea
Heard it, and all, hearing its voice, obeyed.
The sea has shores again, the rivers run
Brimming between their banks, the floods subside.

Progeny known as Tritons, the messengers of the sea, were likewise popular in stories and culture. Visible on ancient coins are four
Triton statues blowing shell horns; they are believed to have been
mounted at the corners of the Lighthouse of Alexandria in Egypt.
One of the Seven Wonders of the Ancient World, perhaps it was also
a beacon of truth—like Chavín, felled by the same natural disasters
it forewarned.

India's epic Mahabharata War was famously launched with the
bellow of shell trumpets. Each warrior had his own conch horn, and
each conch its own name carved into its shell. Each shell was personi-
fied in the poem as if it had a life of its own. Krishna blows his conch,
Panchajanya. Arjuna blows his, called Devadatta. Bhima blows "his
mighty conch, Paundra," and so on, and so on, until "That tumultuous uproar rent the hearts of the sons of Dhritarashtra, filling the
earth and sky with sound."

Englishmen who wrote about the mythic named conchs of India
often compared them to the famous swords of European heroes given
immortal names: King Arthur's Excalibur, Charlemagne's Joyeuse.
But perhaps the conch horn names were a rare early acknowledgment
of the lives of the shell-makers, the mollusks.

Revered in India to this day as the sacred chank or shankh, the
divine conch is the life's work of a carnivorous sea snail, *Turbinella
pyrum*, that lives off the southern coasts of the continent in the
Indian Ocean. *T. pyrum* congregate in the patchy grass and sand
shallows to gorge on worms. Alive undersea, they look brown; their
pear-shaped shell is encased in the velvety dark skin known as periostracum. The layer acts like a cast, framing and protecting a mollusk's shell while it's under construction. When the human chank

craftsmen remove the skin and buff the shell, the empty coil shines milky white.

Like the conchs of Chavín and ancient Greece, the sacred chank swirls with the intrinsic human fear of rising waters; life-changing or life-taking floods. Krishna, an incarnation of the Hindu god Vishnu, acquired his conch trumpet when Vishnu dived into the sea during a great flood to rescue divine writings hidden inside a chank by a demon. Four-armed Vishnu holds the shell Panchajanya in his upper left hand; the shell is a symbol, like Krishna's shell trumpet, of the power of good over evil.

The chank is also tied to authentic speech. By the fifth century BCE, the chank had been adopted as one of the eight auspicious symbols of Buddhism, also central to Hinduism and Jainism. The great Vedic gods are said to have granted Buddha the icons upon his enlightenment. The powerful sky god Indra gave Buddha the white conch horn as a symbolic urging that he "proclaim the truth" of the cosmic law of dharma. The white conch came to stand for Buddha's speech; it remains today in Buddhism a fearless symbol of truth.

THE SPEECH CONDUIT appeared to Miriam Kolar the very first time she descended into the narrow gallery of the smiling, fanged god at Chavín de Huántar. The stone gallery was murky and chilly compared with the warm Andean sun outside. The monolith was then lit with disco lights, strangely out of context in the ancient temple. Slow-changing primary colors beamed onto the supernatural face of the oracle, which rose through the floor from its sunken chamber. As she stepped from a corridor into the sanctum of the god, Kolar saw a connection she could not have gleaned in the prior year she'd spent researching this space from Palo Alto. The toothy mouth of El Lanzón leered before her. It lined up precisely with a ventilation shaft located behind her, extending through the corridor and opening onto the Circular Plaza outside.

Rick had shown how this horizontal shaft might have been used to light the god's face. Kolar, then a doctoral student at Stanford with a rarified interest in psychoacoustics, now saw that the shaft might be a conduit for sound. Could the temple's design have given voice to its oracle? The question was high on the list of hundreds that ran through her mind as she set up experiments to map how sounds were transmitted and transformed at Chavín—and how they might have transformed the people who heard them.

Rick's discovery of the conch-shell horns had sparked great excitement back at Stanford, especially in the Center for Computer Research in Music and Acoustics where Kolar was a fellow. The center was founded by John Chowning, a father of digital music who had helped create "surround sound" in the 1970s. When he heard about the ancient horns, Chowning emailed Rick to suggest a full workup on their acoustics. Now, in the summer of 2008, a team from the center arrived in Peru to do fieldwork at the ancient site, and to record and study the horns, on display at the nearby Chavín National Museum.

In ceremony and in protest, conch horns remain a rousing part of Indigenous culture in South America. They are called *pututus*, from the Quechua language of the Inca. On the very same day that Rick unearthed the first conch trumpet at Chavín and blew into its dusty mouthpiece, a *pututu* heralded Alejandro Toledo, Peru's first Indigenous president, during a special 2001 inauguration at Machu Picchu.

To help the Stanford team record sound ranges for the ancient conchs, Rick invited the Peruvian master musician Tito La Rosa to come play them at Chavín. Kolar and her colleagues set up a makeshift acoustical measurement studio at the museum. The Princeton computer-music scholar Perry Cook, an avid shell musician, was among the professionals who came to blow experimental toots and other sounds with the horns. The team fit small microphones into Cook's mouth; into the smooth conch cavities; and all around the shell horns. They would later use custom signal-processing software to characterize each horn's acoustics. Cook and La Rosa played with different techniques to modulate pitch and tone, trying out ways that *pututus* may have been played at prehistoric Chavín.

When La Rosa arrived at the museum, the feel of the experiments shifted from scholarly to sacred. Cradling the first conch, he appraised it "as if it were the most special, precious, sacred, wonderful, miraculous thing that you could come into contact with," Kolar remembers. When he brought it to his lips, everyone in the studio held their breath, bracing for the sound. But for some time, La Rosa only exhaled and inhaled, exchanging air with the conch as if breathing its ancient spirit. When he finally played the *pututu*, "it was like an ocean wave," Kolar told me. "The sound just kept going, and going, and going."

La Rosa's performance in the cavernous concrete museum gave the Stanford team sustained tones and discrete echoes, as if he'd joined his breath with the wind and sent it over the Andes. Kolar next turned to the temple and its iconic landscapes to test how the shells' many voices and tones would have sounded and carried. She was also keen to learn how the conchs' resonance might have made the pilgrims feel and behave.

I met Kolar ten years after that first summer in Chavín. She was presenting at an academic conference, and during her opening comments, she blew into a replica Chavín *pututu*. The graphical sound-level waves that usually undulate gently on my voice recorder hurled themselves off the screen. The note—loud, low, long—startled those of us sitting in the lecture hall, even though we'd been expecting it. The conch's voice filled the institutional-white room with an urgent, colorful spirit, like a child's shout in the middle of a sermon. The shell's insistence drew us all closer. Kolar had everyone's attention.

Kolar's parents were liturgical musicians, and she grew up fascinated by the social impacts of music. A computer-music class at Dartmouth "enabled me to consider the story of the sounds. I fell in love with this idea that music could be about sounds, rather than notes."

Just weeks after her return to Stanford from the Chavín recording sessions, Kolar packed up her boots and books, subwoofers and speakers, and began the first of many return trips that would relocate her from Palo Alto to the small Peruvian town a half a mile from the temple complex. Each morning, she pulled on her hiking boots, tied her straight and shining dark hair back in a pony tail, and hauled her gear—audio equipment, car batteries, power inverters, and more—uphill to the temple in a wheelbarrow. Over the next three years, she crawled through 3,000-year-old slate-lined canals to search for portals of sound. She blew replica *pututus* in Chavín's stone plazas. She scrambled up the surrounding hills to measure how far their sounds could carry.

She set up her subwoofers and loudspeakers to reproduce mathematically generated test signals in every humid corner of the temple, including in the Lanzón Gallery with its ever-watching god. She set up her small microphones inside and outside, in the plazas, galleries, and "everywhere human beings could have been to make and sense sound."

It was while analyzing acoustics from the temple's ventilation ducts that Kolar saw the chilling response graph that confirmed her hunch about the voice of El Lanzón. The duct she had noticed on her very first day in the complex, positioned at the end of a corridor and in line with the mouth of the god, proved to suppress the transmission of most sounds between the gallery and its exterior opening. But the shaft *amplified* the conchs' key sonic frequencies by 10 to 20 decibels above all others. (Adding 10 decibels is like turning up the volume twice as loud.) Perfectly aligned with El Lanzón, the duct served as a functional "line of speech" between the oracle's mouth and the ceremonial plaza outside. Transformed to wind, to jaguar roars, to monstrous or benevolent breath, conch voices could carry from the oracle's mouth to the pilgrims' ears in the plaza outside. Anyone peeking in at the god from the gallery's interior corridors would not be able to see the conch-horns being played in the chamber around it. Those inside would hear only the unnerving sounds—seeming to emanate from a fanged mouth. Those outside would hear only a disembodied voice.

Perhaps, as Kolar speculates, the priests of Chavín created an even grander symbolic performance in which conchs summoned weather and climate, bringing either much needed rain to the valley or flood-waters from a punishing god. At the very least, in architecture, imagery, and sound, Chavín's fabulists brought an oracle to life—and gave the monstrous god one of the most powerful voices in the natural and supernatural worlds.

FROM CHAVÍN AND earlier pre-Columbian cultures, shell trumpets would spread throughout the Americas from the southern tip of Argentina into Canada, often many hundreds of miles inland. While most Mesoamerican sites turn up Pacific species, Atlantic conchs have been unearthed in some Aztec ruins, evidence of trade along both coasts and with far-traveling explorers.

The flood myths are everywhere. Quetzalcoatl, the plumed serpent deity of Mesoamerican culture, was often depicted as having emerged, full-grown, from a gastropod shell. He was said to have lived in a shell palace, and many temples dedicated to him are decorated in shells; one in the ancient city of Teotihuacan with alternating bivalves and gastropods. In Aztec myth, Quetzalcoatl was a god of wind, often depicted wearing a cut conch to symbolize his powers to move rain clouds. Charged with repopulating Earth with humans after the last race had been turned into fish by a flood, he must travel to the dangerous underworld and retrieve human bones from the last creation. He arrives to meet a skeletal ghoul called Mictlantecuhtli, ruler of the underworld. The devious death lord tells Quetzalcoatl he can have the bones if he fulfills an apparently simple task: circling the underworld four times while sounding a conch trumpet. Quetzalcoatl agrees. But then Mictlantecuhtli gives him a conch with no holes to blow into.

Quetzalcoatl turns to nature. He calls upon worms to drill holes in the shell and summons bees to buzz into the cavity and make it roar, fulfilling Mictlantecuhtli's demand. The death lord is furious. He gives up the bones but booby-traps Quetzalcoatl's way back, causing him to drop the bones and break them. The new human race is born, but the broken bones mean that this time around, people are all different sizes.

In the red deserts to the north, long silenced by layers of sand, hundreds of conch-shell trumpets have been unearthed in Native pueblos of the American Southwest, most of them traded inland from the Pacific. The Hopi and Zuni people blew conch shells during their plumed serpent ceremonies to give voice to the breath of life—also evidence of distant trade. The conch's call was said to bring to life the Zuni people's Great Shell, a spiritual being who could help snuff out enemies.

Horsemen traveling with Spanish explorer Francisco Vázquez de Coronado in 1540 described the warning calls of conch-shell trumpets as they approached what they believed to be the first of the Seven Cities of Cibola—actually the Zuni pueblo town of Hawikuh in what is now New Mexico. The Zuni's "Great Shell Society"—*Tsu 'thulanna*—is often described in the Indians' stories of Coronado's attack on their pueblo. (The Zunis were forced to flee but managed to bean Coronado with a stone.)

Only one year before, as he made his way through the wet wilds of interior Florida in 1539, Spanish conquistador Hernando de Soto was met with the sounds of friendly flutes—until the Native Floridians became wise to the true intentions of his men. In the words of one chief: "They are demons, not sons of the sun and moon, for they go about killing and robbing. They do not bring their own women, but prefer to possess the wives and daughters of others. . . . Warn them not to enter my land."

The angry Native people tried to warn the Spaniards away with urgent conch-horn calls. The farther Hernando de Soto pushed on, the louder the eerie wails became.

The conquistadores did not heed the warnings of the conchs. Spanish colonialists would crack down on trumpet shells under the rules of the Inquisition, which banned instruments associated with ancient religions.

Still, the shell horns blew.

IT IS A shell's deep-coiling cavity that creates its attention-grabbing call; the larger the shell, the lower the tone. That makes the power of the sound directly proportional to the energy and years of work the mollusk plowed into its shell.

Shell trumpets likewise reflect the character of the humans behind their voices. Truth and propaganda, good and evil: All are strengthened by the voice of authority.

William Golding captured the shells' command in the opening chapter of his coming-of-age novel *Lord of the Flies*. A group of schoolboys survives a plane crash on a deserted tropical island with no grown-ups to keep things civilized. In the lagoon, a creamy form catches the eye of a boy cruelly nicknamed Piggy.

"It's a shell!" Piggy exclaims to his new compatriot Ralph. "I seen one like that before. On someone's back wall. A conch he called it. He used to blow it and then his mum would come. It's ever so valuable."

The pink-lipped shell is 18 inches from apex to tip. When Ralph blows it to call all the survivors to the beach, boys stream out of the jungle—looking for the authority behind the sound, for a man with a trumpet and a ship that would save them. Stuck with only themselves, they decide to choose a chief. While he isn't the obvious leader, Ralph stands out for having blown the conch.

> "Him with the shell," Piggy suggests.
> "Ralph! Ralph!"
> "Let him be chief with the trumpet-thing."

Later in the novel, the conch is smashed to smithereens when a rival hurls a rock at Piggy, killing him. The crushed shell represents the final cracking of whatever order the boys had managed to keep on the island. In Golding's metaphor it is nature that offers order, and humans who disrupt it with violence. (In his next and favorite novel, *The Inheritors*, about Neanderthals and *Homo sapiens*, Golding envisions the Neanderthals as "the people," and imaginative; one named Fa thinks of using seashells for drinking cups. *Sapiens*, "the new people," are sophisticated but savage, killing off the original people and stealing their last infant.)

Chavín's conchs, and the manipulation of their sound, likewise leave disturbing lessons of human nature, class, and control. The Andes have long been seen as a birthplace for humanity's transition from relatively egalitarian societies to amassed power and empire. John Rick concludes that Chavín's temple complex—choreographed behind the scenes with clear intent to unnerve, mystify, or even terrify—reveals the rise of public manipulation and the lengths elites would go to build and maintain their authority.

Power requires belief for legitimacy. To establish dominion where none existed before, Chavín's leaders designed, built, and managed their temple as part of a "conscious, calculated political strategy" to sow belief in their power, and the oracle's, to predict natural events or even control them.

The temple itself was built and maintained at considerable expense using cut stone unlike any found in the local landscape. Once participants were inside, Chavín's priests isolated them from the outside world, holding ceremonies in the sunken plazas and internal galleries where they could tightly control everything a worshipper saw and heard.

By building a second-story platform around fanged El Lanzón, the priests could guide worshippers to where they would see the oracle's face lit in a beam of light, and hear its eerie voice, without seeing the conch shells being played. The priests were "exceptionally creative in their manipulation of the human mind through landscape, architecture, images, sound, light, and the use of psychoactive drugs," Rick asserts.

"In the end, Chavín makes me ask big questions about humanity and the way culture has evolved," Rick tells me. "I honestly do not rest easily with some of the answers that seem to loom large."

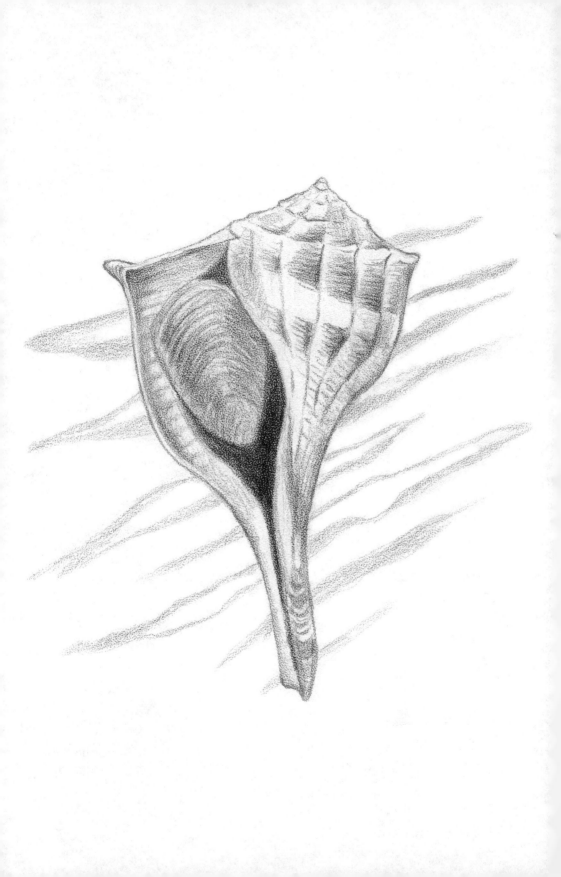

Four

GREAT CITIES OF SHELL

THE LIGHTNING WHELK
Sinistrofulgur sinistrum

North America's first-known great city rose across the Mississippi River from what is now St. Louis, Missouri, a thousand years ago, as striking in its time as the skyline framed by the Gateway Arch today. More than a hundred earthen mounds girded its public buildings and monuments, built around courtyards and a ceremonial center. At the heart of the city, an elevated causeway led to a palatial pyramid, its base larger than the Great Pyramid of Egypt, and a central plaza the size of thirty-five football fields. At the outskirts, neat neighborhoods of pole-and-thatch houses radiated 50 miles in all directions.

Home to 20,000 to 30,000 Indigenous people in its day, the terraced landscape topples the myth that the New World was new; an untouched wilderness roamed by unsettled tribes. This was a political and religious capital—perhaps *the* capital—of the early Mississippian civilization that flourished across the eastern continent from Florida to the Great Lakes until the arrival of the first Europeans.

The eroding pyramid, known as Monks Mound, is the largest Native earthwork in North America, one of eighty remaining mounds in a lonely green vista off Interstate 55/70 in southwestern Illinois, appearing and vanishing in seconds to passing motorists. No one knows what its builders called themselves, or this place, then as now a gateway to the West. Missionaries named it Cahokia, after a later tribe. But the word *tribe* doesn't come close to describing the physical and human expanse, which from 1050 to 1150 rivaled the European capitals of its time.

Thomas Say, a Philadelphia naturalist who would become the father of American conchology, was one of the few explorers who grasped Cahokia's magnitude when he helped map the site as part of the Stephen Long expedition up the Missouri River in 1819. This earth had been moved, Say wrote, by "the labors of *nations of Indians*."

They master-planned a major city on a grid pattern. They traded across the continent. Yet the next city-builders to come along treated the ancient ruins as construction fill. So much of the Cahokia people's history was scooped up, built over, or graded into highway that few Americans know of them. Seventy more hulking mounds in St. Louis and East St. Louis were razed before the Civil War. In the next century, Cahokia's central plaza was developed as a suburban neighborhood. A drive-in movie theater went in just west of Monks Mound, opening to Disney films but over the decades shifting to porn.

Finally, when highway engineers for the new federal interstate system in the 1950s laid their survey markers across what remained, local archaeologists and citizens managed to save the last mounds from the worst of the proposed routes. A portion of federal highway funding went to archaeology. The digs revealed extraordinary details of life in America's first great city, including extensive trade. Cahokia held copper from the Great Lakes, stone tools from the Midwest. And from the Gulf and Atlantic: a trove of marine shells to rival those found in the mounds that had lined America's seas.

In whole shells and fragments, in body ornaments and especially beads, the most abundant shell found in the mounds of Cahokia, by far, has been the Lightning Whelk. Often mistaken for a conch, the cream-colored shell is shaped like a statuesque 1950s actress in a form-fitting evening gown; broad at the top with a long, svelte siphon. It is named for the jagged, vertical streaks that run the entire outer shell, from the apex down the spire and siphon to the tip. Seen from its top, the brown strikes cross the ridged crown in a pattern that is both mesmerizing and vaguely contradictory.

Unlike the vast majority of spiraled shells, the Lightning Whelk winds left instead of right. Hold the shell in front of you with the apex pointed up, and the opening is on your left. Scientists call the species *Sinistrofulgur sinistrum* for the unusual coil. Sinistral, or left-handed, is tied to sinister—at least in the Western mind. The leftward spiral may have had a different meaning for these earlier Americans, as it does for Hindus, who worship the rare chank shell that coils left rather than right. Lightning Whelks also made for an enduring legacy. In temple platforms and walls, in jewelry and sacred cups, the shell has outlasted the people who revered it by hundreds and in some cases thousands of years—symbol of an earlier America contoured and mounded with shell.

EARLY AMERICAN SCIENTISTS knew Lightning Whelks were special to Native people; in his nineteenth-century book *Art in Shell of the Ancient Americans*, William Henry Holmes wrote that they seemed to have been "more extensively used than any other shell," at least in whole form. They are now known to have been the most widely traded large marine shells, by far, during Mississippian times.

Bleached white in ancient sun, darkened to sepia in interment, or blackened in the first American hearths, Lightning Whelks were utilitarian or sacred, a meal or a monument—depending on how far from the sea they ended up. Across the Gulf South, the meat was a staple, the shells honed to common tools and fishing rigs. The farther inland they were traded, the more exotic they seemed and the more elaborate their incarnations. Lightning Whelks were found in 5,000-year-old burials in western Kentucky and throughout Mississippian ruins as far north as Manitoba, Canada. In Canada's densest concentration of ancient mounds, built roughly around the time of Cahokia, First Nations people left gorgets—pendants pierced with two holes to hang at the chest—elaborately carved into faces from the outer whorl. Some surviving gorgets depict a birdman, himself wearing the same whelk pendant.

The most touching clue—make that 20,000 clues—to the status of seashells at Cahokia was unearthed in 1967 from a small ridge-top mound just south of the central plaza. Archaeologists found two bodies, later identified as a man and a woman, interred with 20,000 marine-shell beads. Skilled artisans had honed the beads from larger pieces of shell into flat, coin-sized disks and pierced them with a hole. The deathbed beads had been sewn onto fabric, long turned to dust, in the shape of a huge raptor enclosing the bodies: The bird's head lay at their heads, its wings at their arms, and so on. In the years to come, excavators would unearth the bones of a child and hundreds of sacrificial burials.

The disk beads were not cut from just any shell. They were not sourced from a mollusk living nearby. They were not punched from the largest shell available—or the most abundant. And they were surely not honed from the easiest shell to drill. Most of the disk beads, and tens of thousands of other beads found in Greater Cahokia, were worked from the iron-hard shell of the Lightning Whelk. Staggering numbers of the left-winding artifacts include those found in a shell storehouse and a shell-bead manufacturing center unearthed in the buried city.

Cahokia people also sipped ceremonial drinks from Lightning Whelk cups. Researcher Laura Kozuch at the Illinois State Archaeological Survey found that they even made ceramic shell-cup effigies to look like Lightning Whelks—with left-coiling spirals. Shell beads flourished in the city during Cahokia's rise, while shell-cup effigies are found in sites dated closer to its demise. Increasing conflict may have cut off trade from the coasts, limiting real shells and spurring creation of effigy cups, Kozuch theorizes. Whatever ill fate befell the Cahokia may have led them to new kinds of rituals, she imagines, perhaps to try to end a war or drought, both implicated in Cahokia's collapse by around 1300.

In the common and the ceremonial, in beauty for this life and in requiem for the next, the shell's ubiquity had Kozuch puzzling over the same question, the one that became the oeuvre of her research: "Why Lightning Whelks?"

The answer led her in circles, spiraling counter-clockwise to the left.

~~~~~~~~

MOST MARINE MOLLUSKS begin life in great peril. Hatching from their eggs as free-swimming larvae, the vast majority of them will be eaten before they have a chance to settle and build a shell. Lightning Whelks are among the fortunate few that meet the sea armored. They develop into fully formed little gastropods while still in their egg cases—the cream-colored coils that wash up along Atlantic and Gulf beaches, looking like skin shed from a snake you hope not to meet. The cases are the handiwork of mother Lightning Whelks. During mating season in fall and winter, several males converge on the much-larger female, untucking their penises to fertilize her eggs; scientists have observed as many as nine small males mating with one mother. She lays in late winter or early spring, first anchoring the end of her case deeply in the sea sediments to secure it for the metamorphic months ahead. Over several days she adds connecting capsules, each filled with about a hundred embryos. By the time she finishes the slinky-like case, it can stretch more than a yard and hold thousands of whelks. The protective capsules are stocked with the nourishment embryonic whelks need to grow and build their shell, which takes shape as a translucent bubble. After a few months they begin to chomp out of their coiled nursery, first steps to lives as carnivores that can pry open even the toughest clam.

Young whelks then make their own way to the sand, mudflats, and seagrasses of the Gulf or the Atlantic. They still live where Lightning Whelks evolved left-handedness about 4.5 million years ago during the Pliocene, the era that saw the rise of so many marine mollusks we recognize. Evolutionary scientists are fascinated and frustrated by the question of why the vast majority of gastropods coil to the right; there must be a good reason, otherwise shells would wind both ways. The Scottish biologist D'Arcy Wentworth Thompson, who pioneered the mathematics of shell spirals, egg curvature, and other patterns in nature in his 1917 magnum opus *On Growth and Form*, declared that "no man knows." The late evolutionary biologist Stephen Jay Gould, who loved mollusks even more than dinosaurs, confessed to not having a clue. Scientists still don't. But they do have clues as to why a few marine mollusks wound opposite the crowd. Left-handedness gives Lightning Whelks a slight advantage over stone-crab predators with big, right-handed crushers. But crabs can't be the whole story, says Geerat Vermeij. For one, some left-handed species live in polar regions where no crabs exist. Also, if left-handedness made a big difference in foiling enemies, many or most marine snails would have it. Vermeij's research has found that left-handed marine snails tend to be those born as miniature adults—like the Lightning Whelks, sporting perfect, tiny shells—rather than swimming larvae easily gulped by passing predators. In short, their lower risk of being eaten as embryos removed an evolutionary constraint.

Vermeij calls left-handedness a "tolerance of error"—nature's way of saying it's okay to be different, so long as it's not too risky. Humans have been less tolerant of such difference in their own species. The words themselves—*right* in English, *recht* in German, *droit* in French, and so on—reflect preference for the right and bias against the left. In French, *left* is *gauche* or *maladroit* for awkward or "not right." In Latin, *dexter* referred to manual skill and dexterity while *sinistra*, "left," connoted evil or bad luck—the sinister.

The perceived superiority of right-handedness may connect to ancient warfare; to avoid mortal injury, soldiers held a shield over their heart with the left hand while fighting with the right. Whatever its origins, perception of right as being strong, good, fortunate, skillful arose along with fear of left-handedness. The right hand became the right one for eating in India, Africa, and the Middle East. Around the world, left-handed children were forced to write with their right.

While scientists defined the Lightning Whelk with these sinister-sounding roots, many earlier people looked to sinistral shells with something closer to awe.

~~~~~~~

LAURA KOZUCH, THE now retired curator with the Illinois survey, was born in Illinois to parents struck with wanderlust. In her childhood during the 1970s, they bought and ran several mom-and-pop motels. By high school she could fold hospital corners and operate a telephone switchboard. Just after she graduated, her folks bought the Sea Bird Beach Motel in Florida, on the Gulf at Longboat Key. After shifts at the front desk of a fancier hotel nearby, Kozuch would spend hours floating in the warm sea, walking the beach for shells, and gathering live coquinas for soup.

Her parents' Gulf dreams didn't last. But the sea's hold on their daughter stuck. In an undergraduate archaeology class at the University of Florida, she learned that many Indigenous cultures were also taken with shells. During a field course at Tatham Mound— preserved thanks to its remote location in a swamp—she encountered the same large, left-handed shells she knew from the Gulf. Long-ago people had buried their dead among shells and left Lightning Whelk cups at the summit of the mound, likely following black-drink ceremonies. Southeastern Indians brewed the tea explorers called black drink from yaupon holly, a scrubby evergreen with prodigious red berries still growing wild or trimmed into hedges across the South. They sipped the ceremonial brew from shell dippers or cups, most of them Lightning Whelks—both tea and cup held spiritual importance. But no one seemed to be able to tell Kozuch why Lightning Whelk shells were so special. Horse Conchs, the Florida state shell, are larger. Any number of shells are more colorful.

Kozuch devoted her scholarship to a single shell, focusing in as intensely as her parents had wandered out. She spent her college work-study in zooarchaeology pioneer Elizabeth Wing's lab at the Florida Museum of Natural History, and went on to do her doctoral research with Wing on marine shells from Mississippian sites. Kozuch was reading William Bartram in the Florida sun when she came across his detailed description of the Creek Indians' sacred spiral fire. Circling out from a great central pyre, they used dried cane to build a foot-tall fuse in the shape of a spire, winding it around and around. "Every revolution increases its diameter," Bartram wrote, "and at length extends to the distance of ten or twelve feet from the centre, more or less, according to the length of time the assembly or meeting is to continue."

At nightfall, once the council members had taken their seats, the spire was set ablaze, then "gradually and slowly creeps round the centre pillar, with the course of the sun, feeding on the dry Canes, and affords a cheerful, gentle and sufficient light until the circle is consumed," at which time the council members and guests share "very large conch shells full of black drink."

Bartram's sketch shows the fire spiraling to the left in the path of the sun. It was, for some Southeastern people, a metaphor for life's course: birth at dawn through death at night. Kozuch found the pattern in other Native dances and rituals, including the striking scene of a Natchez war chief's funeral. Eight people carried his body in a series of sinistral spirals while eight others sat at points in the path awaiting strangulation by eight executioners standing behind them.

Working as an archaeologist back in her home state of Illinois, she again found herself surrounded by the sinistral artifacts, this time from the mounds of Greater Cahokia. To try to grasp the scale of beading in the city, she spent hundreds of hours cutting, hammering, and boring into modern shells—Lightning Whelks are literally harder than iron—with age-old techniques and tools such as chert microdrills. The number and precision of Cahokia disk beads and chunkier beads cut from the tough columella—the shell's vertical core at the axis—suggest a skilled workforce devoted full-time to production, Kozuch argues, though other scholars believe bead crafting was a part-time occupation at Cahokia.

Where the massive numbers of marine shells came from was harder to figure. Lightning Whelks live from the Atlantic seaboard around the Gulf states to Texas—which celebrates them as its state shell—and down to the Yucatán. Biochemical tests on the smooth white artifacts proved tricky, and expensive. But it turns out that left-handed whelks from different regions build subtly different shells—most noticeable in the angles of their spires.

It is just the sort of work for which the obsessive collections and field notes of conchologists come in handy. Kozuch and the zooarchaeologist Karen Walker measured hundreds of whelks collected on different coasts and held in natural history museums to confirm that their spire angles can pinpoint their coastal source. Next, she measured hundreds more whelk artifacts from Cahokia, East St. Louis, and Spiro Mounds in Oklahoma. She found that the vast majority hailed from the very same region: the shallow waters of the eastern Gulf of Mexico off the Florida peninsula.

ACROSS A BLUE-GREEN sound from the shell mecca of Sanibel in modern southwest Florida, Native American shell mounds still rise on Pine Island, a throwback of tropical fruit farms and small communities, none of them seeing the need for a traffic light. Past the mango groves on the northwest coast of the island stand the real throwbacks. Overlooking a serene estuary in the shade of gumbo limbo trees, hilly middens glint white with millions of bleached gastropod shells—most of them small Lightning Whelks. To the east, the overgrown remnant of a hand-dug millenary canal, once sluicing 2 miles through the center of Pine Island, still holds water and memory: an ancient thoroughfare of an ancient water people. Beyond the canal, a forested mound stands in solitude and different from the others. Built of sand, Smith Mound is the largest Native burial ground known in southwest Florida.

Protected in the outpost of Pineland, the tranquil mounds represent one of the largest, busiest cities of the Calusa, the Natives who controlled the southern half of Florida for more than 1,500 years before the Europeans found their shores. A tall, handsome people who wore spare hide coverings and grew their dark hair long, the Calusa built a coastal empire in the mosaic of land and water from Charlotte Harbor south to the Ten Thousand Islands of the Everglades.

Alive with scallops and schools of mullet so dense they shaded the water like passing clouds, the bays and estuaries were the Calusa's domain, and the crystal clear distillation of their culture. They were fishers who had no use for agriculture, even as their population grew to Cahokia-scale numbers of more than 20,000. Their most populous cities, to the north at Pineland and to the south at their ceremonial capital on Mound Key, were port-like, built on big islands close to the peninsula and harbored on the Gulf side by natural ramparts of smaller barrier islands.

The Calusa traveled in canoes, dugouts hollowed from pine and cypress trees. The largest were lashed together in doubles with a sheltered platform in the middle, "with decks covered with awnings of hoops and matting," a Spanish chronicler observed in 1566. Their smallest canoes were wee wooden toys, propelled by the hands of their children playing at the shore.

William Marquardt has spent four decades sifting through their sandy grounds to piece together the story of the Calusa, often with the help of their shells. The archaeologist's quiet manner and silent subjects belie his years spent amplifying his electric clarinet in rock bands including the Hollow Bodies and the Neandertones, whose members hung a large wooden bone over the doors of their club venues in the '80s. Marquardt and his colleagues know the Calusa ate more than fifty types of fish. Mollusks alternated as side-dishes or staples when fish were hard to come by. Those living closer to the freshwater-infused bays ate more oysters. Those at Pineland and other estuaries feasted on gastropods— especially the briny meat of the Lightning Whelk.

Marquardt has found thousands of small Lightning Whelks and caches of large ones; and, throughout the Calusa's domain, essential tools made from the left-handed shells. The Calusa hafted a wooden handle through the shell's crown to make a heavy hammer or a sharp ax. They sharpened the outer lip into cutting edges to hollow out logs and carve long canoe paddles. They chiseled the strong columella into adzes at the point of spears, pounders, perforators, and fishing gear. They smoothed the outer whorls into saucers, spoons, ladles, bowls, and cups. Marquardt is increasingly convinced that the

Calusa were also involved in long-distance trade, collecting and moving large Lightning Whelk shells and other goods north by canoe. Perhaps paddling north weighted with huge sinistral whelk shells and smoked fish, returning home laden with exotic minerals including slate and shale, the Calusa reached their zenith not long after the conquistadores arrived in Florida.

In the sixteenth century, Spanish chroniclers described the Calusa living in at least fifty villages. By 1612, a Spanish governor wrote that the Calusa leader controlled "more than sixty villages of his own, not to mention the other very great quantity that pay tribute to him."

Within a century, all the Calusa had been enslaved, killed, or driven from their elevated homes. Within another, virtually all that remained were their shells.

IN THE EARLY years of American science, naturalists like Thomas Say, who would write the first U.S. book of conchology, might as likely find themselves recording Native cultural myths as digging for fossil mollusks or collecting live ones. In the modern jargon, science was interdisciplinary before it splintered into ever-narrower specialties—keeping, say, an engineer from heeding how a dam designed to bring a region electricity could also destroy its food and economic livelihood of fish or shellfish. Among the greatest human-caused extinctions in North America were freshwater mussels in the Southeast, most notably those drowned with the damming of the Tennessee River. Species that survived an ice age, intense Native harvest, and America's early 1900s pearling and shell-button crazes could not survive the electrified world.

A number of those early scientist-humanists were also willing social advocates—another part of Thomas Say's story to come. Frank Hamilton Cushing, the anthropologist who had lived among the Zuni in New Mexico and brought them to meet their "beloved mother" the Atlantic Ocean, also advocated for the Zuni against settlers encroaching on their lands, and against missionary efforts at the pueblo. (While also crossing all sorts of lines, like intruding on their rituals.)

Cushing was self-taught; "a man of genius," said his boss John Wesley Powell, who'd become founding director of the Smithsonian Institution's Bureau of American Ethnology with a vision to organize American anthropology research. Cushing was born to the work. Finding an arrowhead on his family's New York farm before he was ten years old "decided the purpose and calling of my whole life," he wrote. He published his first scientific paper at age seventeen and was recruited to the Smithsonian at nineteen. He'd been celebrated in the press, and at the 1893 Chicago World's Fair, for "going native" with the Zuni, who adopted Cushing and initiated him into their secret society, the Priesthood of the Bow.

He wrote about them with poetic empathy rather than the detached superiority of his peers with college degrees. As the first anthropologist to engage in what's now called participant observation, Cushing also drew the jealous ire of some of those peers. One feud had gotten him pulled off his beloved Zuni project and replaced with a Harvard man. Worse, he'd been accused of fabricating and passing off a Zuni artifact—a jeweled toad.

Cushing was sick over the attacks on his credibility—literally on sick leave from the Smithsonian—in early 1895 when he got word of several tantalizing objects unearthed from the southwest coast of Florida. Digging peat from a mangrove bog in a shell courtyard near his home at Key Marco, a boat captain had found ancient wooden artifacts and "a beautifully shaped and highly polished ladle or cup made from the larger portion of a whelk, or conch shell."

The prospect of finding a lost coastal people restored Cushing's spirits. The more he heard about the objects turning up in the Florida muck, the more he believed they might be "the most important archaeological discovery yet made on any of our coasts." With Powell's blessing, Cushing made his way to southwest Florida by train and then took a steamship down the Peace River. Arriving at Charlotte Harbor in May 1895, he hired the boat captain John Smith and his sloop, the *Florida*, "to explore as many as possible of the islands."

Smith had settled his family on sheltered Pine Island after losing his home twenty years before in a major hurricane. A noted tarpon fishing guide, wiry and tan and often wearing a field hat and suspenders, he knew exactly where to sail Cushing for his first major reconnaissance. Pine Island was an auspicious beginning for a man in search of lost cultures. The Great Freeze that winter had killed back not only the state's manicured citrus crop, but many of the subtropical trees, brush, and vines that tangle Florida's wild islands like J. R. R. Tolkien's Forest of Mirkwood. Beyond dark-green mangroves skirting Pine Island and its islets, Cushing had a stunning view of a past that was normally hidden. Cruising through the clear shallows of the sound, and wading onto keys and salt-marsh shores, he glimpsed what he would later describe as "great cities of shell."

At a little key jutting off Pine Island, Cushing climbed a huge old gumbo limbo tree and barely missed falling onto a platform made entirely of what he described as conch shells. They were Lightning Whelks, extending several feet outward from the base at the tree trunk, "their larger, truncated and spiral ends, laid outward and in courses so regular, that the effect was as of a mural mosaic of volutes," he wrote.

Looking northward from his perch up the Pine Island coast, he saw "a promontory, island-like in appearance, on account of its relative boldness." He asked Smith to head north to check it out—even though their destination at Marco was 90 miles to the south. Wading ashore at low tide, Cushing stepped through settlers' vegetable farms planted among ancient shellworks. He could make out pyramids with leveled-off tops, shell mounds, shell ridges, and shell-walled water courts.

He counted nine hulking structures more than 60 feet high, built around five rectangular courts with short canals leading to the sound. Through the middle, a grand canal ran between the two highest shell formations. Clambering to the top of one, "I was astonished beyond measure with the extent of the works which became visible therefrom," Cushing wrote. "To the southwestward and as far as the eye could reach to the southward, certainly for a mile and a half, stood a succession of these great shell heights with their intermediate water courts, graded-ways, and canals with their shelly surrounding platforms and terraces to the southeastward with yet another series of the gigantic heights."

One enormous, oval mound was particularly impressive to Cushing; built of sand by human hands, "it seemed a more gigantic undertaking than the building of the Capitol of our country," he claimed. Walking its long, spiral path to the top, he found bleached human bones, and a perfect shell-ladle, and he wished he could stay. But Key Marco called.

Cushing had planned for his journey to Marco to take a day. His insistence that Smith stop on every island with an ancient shell structure stretched it to eight. After counting more than seventy-five shell mounds by the time they reached Sanibel Island, he had to quit exploring and hurry south. But that sweeping first view of the "great cities of shell" helped push Cushing's thinking beyond the fantastic artifacts he would find at Marco to conclusions far ahead of his time: The shell mounds were not just refuse, but foundations for temples and homes. Some of the structures had extended out over the water, like our own buildings, on pilings and piers. The people who built them had adapted to the Gulf's bounty, and its hardships.

Cushing returned that December to excavate the Marco site, which he called "Court of the Pile-Dwellers." From the small muck pond covering less than an acre, the team extracted what remains one of the most significant pre-Columbian finds in North American archaeology. Cushing found a thousand wooden artifacts: a few toy canoes; brightly colored face masks; a painted woodpecker so animated it appeared to speak; an artful deer with oversized ears; carved panthers, alligators, and sharp-beaked falcons; and an exquisite feline-human statue 6 inches tall, now kneeling behind glass at the Smithsonian.

Most of the shell artifacts were practical, but there were also shell beads and pendants and ear buttons, and the most intriguing small shellworks: The team found pairs of clamshells, bound closed with palmetto leaves, that when opened revealed a miniature painting inside. The most elaborate one depicted a masked man, drawn simply but for an elaborate headdress and deep palm lines in his front-facing hands.

The team found spoons carved from bivalves and ladles from large conchs. They found all sizes of shell knives and scrapers serrated and smooth. Countless Lightning Whelk hammers, picks, adzes, and gouges were preserved in the muck, many with their wooden handles intact. Shells had also returned to sea duty as maritime tools: Big conchs had become bailers, large clams fishing-net weights. Columellae were fashioned into sinkers. Cushing found "an ingenious anchor" of three triton shells pierced and lashed together with tightly twisted cords of bark and fiber so that the long, spiked ends radiated like the points of a star. They were packed with sand and cement, heavy enough to secure "a good-sized boat."

The abundance of shell tools led Cushing to describe "a Shell Age phase of human development and culture," the seaside equivalent of stone tools. "An art of the sea," he wrote, "for which the sea supplied nearly all the working parts of tools, the land only some of the materials worked upon."

The excavation would prove as tragic as it was monumental. The times predated preservation techniques. Most of the wooden artifacts preserved in the oxygen-free muck fell apart as soon as they hit air, a crisis that brought Powell to Florida unannounced. He urged Cushing to hold the season's work to "this one little court of the pile-dwellers" to limit the damage and plan for a longer-term excavation and preservation project.

To make things worse, an adversary back at the Bureau of American Ethnology accused Cushing of fabricating the miniature painting of the masked man in the clamshell. He dragged in Cushing's old accusers from the Zuni days, as well as some yellow journalists to report the sensational story. Modern archaeologists who have revisited the saga point out that everyone who doubted Cushing was a thousand miles away. Everyone present when the shell was pulled from the muck, including the excavator who found it, swore to the genuineness of the moment and the artifact. Powell and other scientists conducted an inquiry that concluded the specimen was genuine, and they forced out Cushing's accuser.

But then as now, the lie had raced across the nation while the truth was still tying its shoes. The controversy followed Cushing for the short rest of his life and devastated his legacy, at least in the near term. He died after swallowing a fish bone in April 1900, in the midst of his report on the "great cities of shell."

By the time archaeologists returned, "the little court of the pile dwellers" and many of the great mounds of southwest Florida had been built over or hauled off for roads and construction fill, along with the clues they held to people who called themselves the Calusa.

BUILT OVER HUNDREDS of millions of years, mollusks' own shell layers told stories of evolution and geologic change. Built over thousands, Indigenous shell mounds archived the lives and communities of prehistoric people. The new shell construction likewise revealed its builders. By destroying extraordinary evidence of earlier cities, modern industrialists and developers created the impression that it was they who built the American dream, the Florida dream, the California dream—or in St. Louis, the dream of a gateway to the golden West.

Earlier dreams had been built on the same lands, and lived out for thousands of years. With noteworthy exceptions, the new Americans proceeded as if none had come before. Surveying the San Francisco Bay area between 1906 and 1908, the Berkeley archaeologist Nels Nelson found that no fewer than 427 shell mounds had surrounded the bay, but "not a single mound of any size is left in its absolutely pristine condition." Like the Calusa of Florida, Ohlone people in California had turned shell middens, heaped up by Indigenous bay-dwellers for thousands of years, into settlements and burial mounds. The largest shell mound, at what is now Emeryville (between Oakland and Berkeley), was a place of ceremony. In the late nineteenth century, developers built an amusement park, Shell Mound Park, at the base and sheared off the top for a dance floor. Revelers literally danced on graves. In 1924, the mound was destroyed to make way for industrial development; black-and-white photographs show spectators watching a steam shovel tear into the mound and haul it off. Newspaper accounts scoffed at the notion of humans and shells buried together. "As you undoubtedly know, a shell mound was the combination burial ground and garbage dump of California's first settlers," read a column in the *Berkeley Daily Gazette* in 1942.

The perspective reflected a fundamental disconnect that persists to this day, write Michelle LaPena of California's Pit River Tribe and Corrina Gould of the Confederated Villages of Lisjan, one of the Bay Area's distinctive Native groups collectively known as the Ohlone. Where many non-Indians saw a boundary between people and nature, LaPena and Gould write, Native people believed all of nature—plants and animals as well as rocks and shells, which they knew to be forged by life—possessed spiritual power. "The Ohlone consider the remains of *all* who lived before to be sacred," they explain, "whether it is a human or a clam, it is valuable, even sacred.

"To be buried in shells—even today—would be an incredible honor."

In the Gulf South, road builders found shell mounds well-suited material for hardening roads on sandy soils. In 1908, the first federal road agency experimented with a shell base for its earliest highway through New Orleans: 4 inches of clamshell graded onto 7 inches of oyster shell, all mined from local middens.

At Pine Island, some settlers built their homes atop the flat-topped shell mounds or hauled them off by the cartload. They flattened shell ridges and conch walls. They used the shell to build the island's roads, fill its wetlands, and level out its farmland. Shells were mixed into matrix for the bridge to the mainland in the 1920s, and heaped into the Calusa's huge, hand-dug lake to establish avocado and mango fields. Mules hauling the shell-heaped carts had to be fitted in special boots to protect the soft undersides of their hooves from sharp gastropod columellae and spires. Lightning Whelks and Horse Conchs from the mounds sliced right through the rubber tires of cars, trucks, and tractors driving on shell roads or through agricultural fields. It took mechanical rolling by a heavy steel drum to fracture the shell and make the roads and farm fields drivable.

An ad man named Barron Collier, who made his fortune selling advertising on the nation's train, trolley, and subway lines, bought Marco and several other islands of the Calusa, ultimately amassing more than a million acres in the former cities of shell.* The largest landowner in Florida by the 1920s, Collier helped drain the Everglades from the west side, steering the cross-Everglades highway Tamiami Trail through his own holdings. As part of the highway deal, Collier pressured Florida to carve Marco Island and its surroundings from Lee County into a new county bearing his name. Calusa County would have been more apt. Even better would be to give the twentieth-century Florida dream-maker his county, and change the one named for Confederate commander Robert E. Lee to Calusa. But only twenty years after Cushing's death, the Calusa were being forgotten. To complete Tamiami Trail, Florida's road department carried off a massive hunk of Mound Key, the Calusa capital with its raised chief's house where two thousand people could gather comfortably. The bite mark is still clearly visible from the air.

Over the century, demand for shell aggregate for highways, levees, ports, and other hardscapes was so great that excavators next turned to scooping shell out of America's bays. Leasing bay bottom to shell dredgers became a source of income for coastal states. Indian artifacts popping up in oyster-shell driveways in the 1960s made clear that dredgers were scooping up more than fossil shell. They also didn't stick to the bay bottoms. In Grand Lake, Louisiana, nineteenth-century archaeologists had described a colossal effigy mound in the

---

* At one point, Collier offered to sell Marco Island to Florida for $1 million for parkland. In one of conservation history's great missed opportunities, the state turned him down.

shape of an alligator—built entirely of shells. At least 400 feet long, the gator, a conspicuous landmark with "almost perfect symmetry," was built mostly of clams, *Rangia cuneata*, along with some oysters, *Crassostrea virginica*. The alligator was the only effigy shell mound known in the Gulf region. In the 1930s, shell dredgers began to dig into the sacred beast, load it onto barges, and haul it to the mainland for road construction. By 1950, all that remained were some shells on the beach, steadily washing down the banks of Chenier du Fond.

In the 1960s, Collier's heirs sold Marco and its adjoining isles to the Mackle brothers, Florida developers who pioneered large-scale community planning by selling paper lots to finance construction. They advertised tours in northern newspapers and, to give potential buyers panoramic views of the sea beyond the mangroves, built their 20-foot lookout tower atop one of the island's tallest remaining Calusa mounds. Cleaving the landscape with more miles of canals than roads enabled the Mackles to create more than five thousand waterfront lots. A new water people moved in, not knowing the fill for their homesites had come from the Native Floridians' great cities of shell.

Exactly one hundred years after Cushing's first trip to southwest Florida, a billboard appeared on a rare empty lot on Marco, announcing a new condominium. The grass-covered lot lay 350 feet from the Court of the Pile Dwellers. Across the street, trench diggers laying TV cable had unearthed a cache of hundreds of huge Lightning Whelks. Members of the local historical society sprang into action. By May, the archaeologist Dolph Widmer agreed to lead an all-volunteer emergency excavation. Locals readied the site, found hosts for grad students, and trained in excavation techniques. Working six days a week all summer, the group unearthed the remnants of three platform mounds more than a thousand years old and numerous postholes once sunk with pilings. The largest mound, a truncated pyramid, was built on a base of large Lightning Whelks. Multiple layers had been laid, spire to siphon, in a concrete-like matrix that was too tough for a backhoe to penetrate.

Today the site is one of the more modest condominiums in Marco, a three-story with muted-cream block walls. For whatever special place once stood there, Calusa masons had covered the Lightning Whelks with a veneer of nacreous pen shells, giving its exterior a prismatic gleam in the Florida sun.

ON PINE ISLAND in the early 1920s, an earthmoving crew tore into the western side of the great oval burial mound that Cushing had compared with "the Capitol of our country" and carted it away. Half the monument where Cushing had found the bleached bones of Calusa people now filled a wetland.

Another crew showed up in 1926 to pull down the rest of the mound for fill to build a new boat channel and lodge. They were met with a shotgun. The man holding the gun was property owner Captain John Smith, son of the Captain John Smith who, three decades before, had guided Cushing on his momentous first tour of the great cities of shell. Three generations of Smiths helped protect what remained of the Calusa's northern capital at Pineland for the first half of the century. In the 1960s, New York retirees Pat and Don Randell, who had invested in the area with an eye toward development, instead took on the role of caretakers.

The Randells began to find Calusa artifacts in the yard soon after they had assembled 80 acres and settled into their bucolic Florida retirement. They became accidental tour guides, answering questions from curious tourists who came to see the mounds. In the early 1980s, the couple funded the first major archaeological fieldwork in the area since Cushing. They hired Marquardt, then a young archaeologist, to map Josslyn Island, about a mile offshore. The island they'd planned to develop was a seasonal fishing and shell-fishing hub for the Calusa, with shell mounds, water courts, fishing-gear artifacts, and the pits full of Lightning Whelk shells.

Forgoing considerable profit, the Randells sold Josslyn to Florida for preservation. Marquardt's early encounter with the couple sharpened his instinct for paying as much attention to private citizens and landowners—and later the legion of volunteers who came to help him sift through two thousand years' settlement on Pineland—as his own circle of scholars. Over the next four decades, he built a major constituency for Calusa culture and Gulf ecology, tying environmental education with history. He raised millions of dollars in private funding to bring Florida schoolchildren to Pineland. Many had never seen the sea, much less encountered shell mounds. After watching a couple of generations of kids come through, the Randells ultimately donated their property to the Florida Museum of Natural History as an archaeological research and teaching center the public can visit today. Another private donor helped the museum acquire the great mound, named Smith Mound for the man who protected it with a gun, preserving the quiet resting place.

The preservation of Pineland, along with the Calusa's southern capital at Mound Key, has helped scientists reveal a more nuanced story of the Calusa than the fierce narrative implanted by the Spanish. Cushing's work is being carried on with modern techniques—although too late for all the sites buried under highways and condominiums. They now know that massive water courts at Pineland and Mound Key, laid out on similar grids, held surplus live fish; and maybe mollusks and crabs. Construction of the live wells coincided with a sea-level drop that would have sent finfishes to deeper waters, bringing both tough times and engineering innovation. The central canal at Pineland, wide enough to dock the Calusa's double-hulled dugouts, could whisk goods and people from the sound to the mainland in two miles, saving them a 10-mile trip around Greater Pine Island.

Marquardt and Kozuch can't be certain that the Lightning Whelks making their way to Cahokia and beyond came from the Calusa's well-engineered domain. But the evidence is tantalizing. It includes the port-like capitals and cargo canals; the whelk fishing and work sites; Kozuch's statistical analysis of the shell whorls; and modern Lightning Whelk research that finds the animals' greatest density and abundance in Pine Island Sound.

Lightning Whelks remained mostly practical for the Calusa—as food, tools, and commodity exports that became increasingly valuable to the inland people who revered them. For people in Cahokia, Spiro, and yet farther, Kozuch and Marquardt suggest that it was neither color, nor size, nor the mesmerizing lightning strikes that inspired Natives to seek out the shells. It was *S. sinistrum*'s left-handedness. The leftward spiral was an early icon in eastern U.S. spirituality; Lightning Whelk shells "served as both medium and message" of that theme, Kozuch says.

And the left-handed spirals may not have been entirely utilitarian for the Calusa. Their path still winds up Smith Mound, where they could walk to the top to consult their ancestors for wisdom, leaving an offering of tobacco or food. The site is off limits today to all but the white ibis that settle like spirits in the oak trees. But the thousand-year-old footpath is still visible. The trail spirals up the mound to the left.

ONCE IN A great while—once in tens of thousands of shells, depending on the species—a genetic mutation will cause a mollusk to wind its shell opposite the norm. A dextral snail will coil left instead of right, a Lightning Whelk or other sinistral snail will wind right instead of left. Jules Verne captured the excitement sparked by such a wonder in *Twenty Thousand Leagues Under the Sea*. While dredging for rarities with his manservant Conseil off a beach in Papua New Guinea, Verne's narrator, the marine biology professor Pierre Aronnax of the Muséum National d'Histoire Naturelle in Paris, spots an olive shell that opens left:

> "It can't be!" Conseil exclaimed.
> "Yes, my boy, it's a left-handed shell!"
> "A left-handed shell!" Conseil repeated, his heart pounding.
> "Look at its spiral!"
> "Oh, master can trust me on this," Conseil said, taking the valuable shell in trembling hands, "but never have I felt such excitement!"

Shell collectors were "ready to pay their weight in gold" for such a specimen, Verne wrote; true when his undersea adventure was published in 1870 and true today. But no admirer would glimpse Verne's left-coiling olive shell. In the next scene, a Papuan hurls a stone from the shore, making a direct strike on the shell in Conseil's hand and smashing the precious find. Alas, Aronnax cannot turn it over to the Paris museum as he claimed he would.

The shell's fate is identical to that of the conch trumpet smashed to smithereens at the end of *Lord of the Flies.* Two beloved fiction writers commenting, perhaps, on the fragility of nature in human hands.

No reverse-coiling shell was—or is—more valuable than a reversed chank, the divine conch of the Indian Ocean believed to have hidden the Vedic verses. The sacred chank, *Turbinella pyrum*, is also integral to Hindu and Buddhist culture in its common right-coiling form and has been for thousands of years.*

Known in India as *shankh*, the largest of these robust white conchs have a place of honor in Hindu homes, where they take different roles: In Bengal, for instance, a *shankh* is said to sound the divine *Om*. The *Lokkhi-shankh* is meant to bring order. The *jal-shankh* stores water from the sacred Ganges River. The *Lakshmi-shankh* brings the goddess Lakshmi—and her values of fortune and peace—into the household.

Those *T. pyrum* not bound for hearths or temples might be buffed into the pearly bangles worn by married women or ground to powder as a calcium carbonate remedy in Ayurvedic medicine. The traditions link modern Hindus to an ancient past. Like the forgotten city of Cahokia and the Calusa empire of shell, many of the great cities of the ancient Harappan civilization of the Indus Valley in northwest India and Pakistan have been lost to time. Golden sands cover what

---

* People in India describe shell direction opposite to Westerners, holding the shell with the apex pointed down. They call the common coil left-handed and the rare coil right-handed. For consistency I describe chanks here as malacologists do, with the common coil on the right.

was the largest civilization in the ancient world. Archaeologists working to excavate its master-planned cities find massive shell-working sites, many devoted to bangles. They find more remnants of *T. pyrum* than any other species.

In the early 1900s, some five thousand years after Harappan times, James Hornell, a wooden-boat-obsessed marine scientist who oversaw India's chank fishery for the colonial government, saw *T. pyrum* woven into "almost every phase of Hindu life." Based at Madras on the Bay of Bengal, Hornell described numerous villages devoted to the chank industry, some dating to the fourth century BCE.

At the heart of the industry were the chank divers. Ancient Tamil texts immortalized free-divers, who had powers to charm sharks to keep them from attacking. In more recent centuries, the divers came under the poorly paid jurisdiction of the Portuguese, the Dutch, the East India Company, and the British before striking for better conditions in Hornell's time. Long lineages of free-divers, canoe builders and crews, chank cutters, and bangle makers passed their skills through the generations. Their special tools included rock sinkers for the divers, and for the cutters, half-moon saws, big as a lumberjack's bows, to work through the thick shell.

Hornell described the chank cutters in 1912 similarly to the Tamil poets, as "the men seated on the ground with the knees widely spread and depressed outwards almost to the ground to give free play to the great crescentic two-handled saw monotonously droning a single note as it cuts its way laboriously through the hard substance of the shell."

Then as now, chank workers used every bit of *T. pyrum*, saving the meat for their families and grinding the operculum—the animal's trapdoor—for incense sticks. The smallest conchs were smoothed into special cups for feeding milk to babies. By the late 1950s, the divers were bringing up millions of chanks annually. Rarely, one of those millions would coil to the left instead of the right. Hindu myth would come true for the diver, bringing him near-instant fortune. Hindus call the reverse-coiling chank *Dakshinavarti*, said to be the abode of Lakshmi. The Tamil Nadu government paid divers a thousand times more for a left-coiling specimen than for an ordinary chank.

Today, smaller numbers of free-divers still head out into the bays in slender wooden boats, teen boys accompanying fathers and uncles to gather the sacred animals in mesh bags tied at their waists. For each dive, they hold their breath for 90 seconds or as long as 2 minutes, searching for *T. pyrum* in the seagrass shallows. The divers collect the live animals, still covered in their velvety periostracum, and sell them to shell merchants back on the beach. The meat is carved out and marinated for curry or sliced thin for sun-dried chank chips. The merchants next move the shells to the processors, who grind off the skin and buff the exteriors shiny white, piling them by size in heaps under shade trees. Next come the bangle makers, and the wholesalers who will take them on to factories or warehouses, bound for retail shops or the religious market.

A convergence of overharvesting, climate change, and dwindling demand for authentic bangles among young Hindu brides has hit the traditional conch workers, known collectively as *shankharis*, hard. The days of a triumphant diver surfacing with a *Dakshinavarti* seem to have come to an end. The social scientist Aarthi Sridhar, who advocates for local fishers on India's southeastern coast, says older divers can still remember the thrill of finding a left-whirling chank. "They would declare all fishing stopped for the rest of the day," Sridhar told me when I interviewed her by phone. "The excitement would make others put way too much effort into finding one," endangering their lives.

These days, finding a left-coiling chank is almost unheard of. Trawlers scoop up most of the chanks harvested from the bays between India and Sri Lanka. The free-divers have watched their incomes shrink.

Even the common right-spiraling *T. pyrum* have become more difficult for the Native divers to harvest. The catch is no longer counted by individual mollusks, but by the trawled ton. Fisheries scientists have called for a 30 percent reduction in the chank taken each year.

Sridhar says they have done so with no input from the *shankharis*, who have inherited centuries of knowledge about the animal and want the chance to save their own industry. A sacred object believed to bring order, *T. pyrum* is instead at the center of a regulatory tug-of-war, appearing and dropping off India's protected species list with capriciousness.

Left-spiraling *Dakshinavarti* are still in great demand; the shell collector Harry Lee once turned down $15,000 for his impeccable specimen. But retailers offer numbers of them for sale to Hindus at some surprisingly cheap prices for such a rare and sacred object. You can see them for yourself on Amazon, Etsy, and many other online sites, which charge from $20 to $200. Unlike the soft curve of *T. pyrum*, though, most of those advertised have sharp shoulders, and a long, slender funnel at the bottom. They are Lightning Whelks being passed off, their telltale strikes buffed to white.

TAMPA BAY SPARKLED in the early morning sun when I jumped onto the 22-foot research boat *Shuck It* behind Stephen Geiger, a mollusk researcher with the state of Florida. Two young marine scientists, Erica Levine and Wayman Pearson III, were already onboard. They'd packed the boat with snorkel gear, calipers, and bright yellow transect line. It was late November, but the temperature would hit 70 degrees by noon. Only a few clouds patched the postcard-blue sky. A perfect day for counting Lightning Whelks.

Geiger eased the *Shuck It* through the Port of St. Petersburg, then opened up the throttle toward the Sunshine Skyway Bridge and the southern tip of the city at Pinellas Point. Tampa Bay was once edged by jungled shore and flat-topped temple mounds—homes to the Calusa's rivals, the Tocobaga, a smaller coastal tribe that also relied heavily on Lightning Whelks and likely traded their shells. We anchored near a neighborhood I knew only from the land side—the Pink Streets. To draw buyers in the early 1920s, the developer ordered red dye added to the concrete for paving the streets in his new subdivision. He went broke in the Great Depression, but his blushing boulevards lasted; today the Pink Streets are a coveted address.

A zealous power washer buzzed from somewhere beyond the seawall and green lawns that wrap Pinellas Point. Levine and Pearson hopped into the shallows to find the first set of steel mobile-home anchors marking where they count gastropods monthly. Once they attached the yellow transect line, the three scientists paced the bay bottom in a sort of crab walk, feeling for Lightning Whelks and tulip shells with their hands and feet.

Geiger came to Florida in the late 1980s from New York State. As a young scientist, his job involved ensuring that oysters and clams grown in polluted waters in Queens, Brooklyn, and the Hamptons were transferred to what are known as depuration areas—clean waters where the shellfish are placed to filter human waste and other disease-causing microbes out of their bodies before they're trucked to fish markets and restaurants. If no one keeps track, some shellfish might "fall off the truck" before they are cleaned. Such stomach-churning facts are why the vast share of government spending on mollusks goes to the bivalves we eat. The resulting gap has left wild-life regulators with little science on the gastropods that people also harvest, anecdotally in much larger numbers than those reflected in commercial landing data.

The *Dakshinavarti* imposters on Amazon show the prices Lightning Whelk shells can fetch when sold as chanks. Archaeologists also have seen rampant looting of Calusa whelk caches such as those at Josslyn Island. Two men were convicted of stealing state artifacts after being nabbed on Josslyn with burlap sacks full of ancient Lightning Whelk. One of the men, a shell wholesaler, was sentenced to six months in prison for selling the Native American relics as sacred chanks in India.

In his work monitoring Florida shellfish harvests, Geiger began to hear of prodigious hauls of living whelks and Horse Conchs, Florida's huge state shell. That's not illegal; recreational fishers can take up to a hundred pounds a day of Lightning Whelk or any other unreg-ulated mollusk—a lot of snail meat. Commercial fishers can take as much as they want. Aquarists plucked more than 16 million live invertebrates from the state's waters in 2016 alone. Geiger also knew from colleagues that whelks were succumbing to collecting and fishing pressure up the Atlantic Seaboard and across the Gulf in Texas. After a noticeable drop in abundance in Gulf trawl samples, Texas limited the daily catch to no more than two live Lightning Whelks a day. Yet Florida had little science on its historic shell-makers. A decade ago, state marine researchers launched a base study of iconic gastropods. The Tampa Bay count is a follow-up to see if the numbers have changed in the ensuing years. Geiger also wants to learn the sizes at which Lightning Whelks mature. All are insights that could help regulators decide whether to limit the harvest.

Levine finds the first Lightning Whelk, a youngster that fits in her palm. She measures its shell length with the caliper and affixes a tiny, numbered tag, smaller than a fruit sticker, to its outer whorl. She paints the shell edge with a thin streak of purple nail polish—Sally Hansen's *Jam Sesh*—before she returns the whelk. The purple band will show how much their shells have grown if they are recaptured. Slowly, it turns out; a few millimeters a year. Sometimes they get smaller—siphonal tips chipped off, perhaps in a battle with a crab or when stepped on by a wading angler.

The work won't be finished for a while, but Geiger has a few insights. Lightning Whelk densities are not dire, and may not even be much depleted, though he hasn't found the hulking whelks that show up in the Indian mounds and must still live in the eastern Gulf today, judging by the largest shells and egg cases that wash ashore. The older animals may move to deeper waters once their shell is big enough to dissuade enemies. Juvenile whelks seem to hang around these shallow tidal flats where they are born, Geiger says, and mate here as young adults, possibly migrating back from other adventures. The team recently recaptured a Lightning Whelk—tag number 007—within just a few meters of where they first tagged it 634 days before.

~~~~~~~~

LATER IN THE day before I leave St. Petersburg, I decide to drive to the Pink Streets. I want to see a temple mound preserved at Pinellas Point, ceremonial center of a major village of Calusa rivals the Tocobaga. I arrive to find a garden of historic markers and other signs planted at the entrance to the shady mound-park. I'm surprised to see that one of them falsely claims this to be a "Calusa Mound" and attributes a gruesome murder to the shell people. Another sign says this:

> Although some of the signs placed around the mound are histor-
> ically inaccurate, they have been a significant part of the lore sur-
> rounding the mound and its early inhabitants. They have been
> left here to remind us of those who sought to preserve history as
> they perceive it.

There is no explanation of which stories are true and which are false. When I ask Marquardt about the erroneous sign, he responds with the patience of someone who has spent decades trying to correct the Calusa's reputation as a warlike people—lore still taught to fourth-graders. The reputation grew from Calusa defenses that included killing Juan Ponce de León, who was attacking them at the time.

Marquardt also wishes he could change the widely held belief that the Calusa were the consummate conservationists. The myths diminish the Calusa's humanity and their complexity. They, like us, struggled to balance stewardship during times of abundance as well as shortage brought on by natural disasters and the dramatic climate changes that occurred during their 1,500 years on the coast. Like us, they built massive engineering projects, overharvested oysters, and erected larger and larger edifices during prosperous times, making for a greater fall when they faced crushing disasters in the seventeenth century; namely, pandemics and violence by Europeans and other Indians.

Throughout their history, climate changes led to innovations rather than collapse—moving their homes and public buildings from ground level to the tops of their shell mounds, engineering the shell-built seawalls and fish pens. It was fellow humans who drove the Calusa from their watery domain and extinguished them, leaving only their shells—and then, not even those.

Part II

CAPITAL

〜〜〜〜〜〜〜〜〜〜〜〜〜

Five

SHELL MONEY

THE MONEY COWRIE
Monetaria moneta

In the fourteenth century, a queen known as Rehendi Khadijah ruled the islands of the Maldives with epic command. One of the earliest women leaders of an Islamic nation, she derived power from both the sultanate and Islam, even as she declined to cover her head—not to mention other parts. She led the kingdom for a third of the century despite two attempts, both by husbands, to depose her. Neither man survived the effort.

All the more remarkable was the Maldivian queen's role in the dawn of international trade. The chain of atolls, coral reefs, and low-lying islands 600 miles off the tip of India was the center of production for the first global money, making the Maldives something of a Switzerland of medieval times. Queen Khadijah oversaw production and sold the currency to traders who filled up ships bound for Arabia, Persia, Africa, and beyond. The sailing ships arrived on the southwestern monsoons in summertime and hung around the islands until the eastern monsoon winds of winter could carry them home.

The Maldivian tender packed up perfectly and made for excellent ship ballast. It was neither paper nor metal, though it jingled in the pocket and shined up bright as a fresh-minted coin. The first global specie was a species. Hidden beneath rocks and coral ledges in the Indo-Pacific, the world's humblest cowrie in size, color, and pattern had an abashed role in human affairs.

The Money Cowrie—named by Linneaus *Cypraea moneta*, now classified as *Monetaria moneta*—makes a glistening shell in the shape of a small shield, with the cowries' characteristic domed top and flat underside cleaved by a serrated slit. Vaguely toothy, the little ivories range in color from off-white to yellowish. They are enameled, pearly, solid, satisfyingly weighty. Irresistible to pick up and worry, or clack together like dice or coins. The small, durable shells made for ideal currency: Easy to transport and to recognize. Impossible to forge. Perfect for counting—one by one, or by bag or ballast-full. Uniform in shape and size, they yielded a precise value when weighed.

Their adoption as a means of exchange grew from cowries' earlier values as sacred and healing objects, jewelry, and protective amulets. Cowrie beads turn up in Stone Age sites in the Levant. Strings of the distinctive shells, real and gold, are found in Ancient Egyptian graves. They were believed to ward off the Evil Eye, bring fertility, protect in the afterlife. Once they became valuable as currency, they spread quickly across global trade networks. Roman coins were spent far beyond the empire, but the shell money spread farther, the global historian Bin Yang has found, reaching China and Europe before Roman times.

Maldivian cowries were spent like coin in India as early as the fourth century. The little shells filled pockets and purses westward to the Persian Gulf and the Red Sea. The shell money spread to what is now Myanmar and Thailand, and eventually reached mainland Southeast Asia. In the mountains of Yunnan, cowries from the Maldives were the principal currency for nearly a thousand years.

Scholars say the concentration of *M. moneta* living on shallow reefs, along with the auspicious trade winds, helped settle the Maldives much earlier than other outposts in the Indian Ocean. A global culture thrived there in a time when most islands of the Indian and Pacific oceans were still isolated.

By the eleventh century, Maldivian cowries were in demand in West Africa, a staggering journey for the little shells. Arab traders packed them across the Sahara Desert on camels long before the Portuguese and other Europeans began to haul them by sea. In the 1960s, antelope hunters stumbled upon a cache of more than three thousand Money Cowries in a vast, inhospitable sweep of the Sahara at the Mali–Mauritania border. They were linked to a lost caravan; water-borne treasures stashed in waterless dune by someone who never returned for them. Painstakingly measuring these shells and others in museum collections, the British archaeologists Anne Haour and Annalisa Christie have traced the desert cache and other numbers of cowries found in West Africa to the Maldives.

Exchanged globally until around 1850 and in parts of Africa into the twentieth century, cowries circulated longer than any other single coin or paper money in history. Maldivian cowries were as important as gold in their time, and spilled as much blood. They hit their cruel peak in West Africa, where they purchased an estimated third of the enslaved people forced to the Americas. Long before Queen Khadijah's reign and long after, gleaming shell money from her part of the world turned up in striking human spaces—from fourth-century graves north of the Arctic Circle to the slave-house subfloor at Thomas Jefferson's Monticello.

The Maldives controlled shell money for centuries. Yet today, the islands' ancient cowrie history remains buried deep in coral rock and sand. That's owing to its idolatrous beginnings, frowned upon in the Muslim-governed nation where nothing "contrary to the tenets of Islam" is allowed.

Buried with the cowries, too, is the memory of a tenacious queen.

MALDIVES WEATHER IS maniacal in summertime, unhinged by the southwestern monsoon. Most writers who visit struggle to find new ways to describe the turquoise shallows that band the 1,200 islands. Each island is ringed in its own sea-glass swimming pool, a reflection of equatorial sun on white sand. But the blue-greens are only one small detail in the full picture of water in the Maldives. Water weighs down the humid tropical air. Water busts loose from the monsoon clouds. Water rattles the palm fronds. Water pools in the streets. Water presses from the sea into every dock, jetty, seawall, rocky shore, sandy beach, and waterside backyard and boatyard.

Cropped from the renowned blue-green shallows highlighted in travel articles and tourism ads is the nation's encompassing feature: the Indian Ocean in endless slate.

I've brought my seventeen-year-old son, Will, to the islands once known across the world for shell money, now famous as the first nation that could be lost to climate change. The dominant narrative has the islands sinking into the sea that birthed them. They are not sinking, of course. The Indian Ocean is rising over them, a calamity of Earth's warming not aided by the rarified luxury tourism sector to which the republic hitches its fortunes. No one who worries about the climate crisis can travel this distance, or to this place, without wrestling over the carbon emissions. But some stories can't be reported remotely, especially when business or government is working to shape or quash a narrative—the case with both climate change and the matriarchal history of the Maldives.

Two hundred of the 1,200 islands are local communities inhabited by Maldivians. One hundred more are resort islands designated "uninhabited," so that international tourists can don bikinis, down alcoholic drinks, and devour roast pork at the beach barbecues, all illegal in the real Maldives because they are forbidden in the Quran.

Like their guests, the resort islands are constantly fortified, sometimes at the expense of the local islands or their critical coral reefs. The government has permitted dredging of tons of sand to bolster resort islands—and transferring coconut palms and other large trees to them from inhabited islands, essentially moving shade and storm protection from locals who live in modest homes to Russian oligarchs and American pop stars on excursion in rooms that can cost $25,000 a night.

We step out of Malé International Airport to a bank of idling luxury speedboats. The airport connects to an artificial island called Hulhumalé, a fast-growing suburb of apartment towers and hotels ringed by a loop highway and white-sand beach. Hulhumalé, the "city of hope," is built up higher than the natural Maldives from a mountain of sand dredged and piped from the surrounding lagoon. More than 30,000 Maldivians have already moved here; plans call for 200,000 more to relocate to the fake island as seas rise over their own.

Most tourists arriving for a holiday never see a local island—or even meet a Maldivian. They step directly from the airport to the closed-curtain speedboats we walk by on our way to the public ferry. In front of the white Four Seasons Maldives vessel, a white-uniformed Bangladeshi is photographing a blonde woman in a white pantsuit posing atop her Louis Vuitton suitcase. I marvel at the effort she has put into straightness: her back erect on the suitcase, her straightened hair defying the humidity.

The tourists are here to worship the sun. That was the religion of the earliest Maldivians too, according to Naseema Mohamed, a long-time historian with the Maldives' National Centre for Linguistic and Historical Research who retired before new laws prohibited books and papers that "contravene Islamic principles." The islands have been settled for at least 2,500 years. Maldivian legends and ancient scripts tell of an early people, the *Dhivehi*, who came from India and "worshipped objects of nature such as the sun, moon, and stars," Mohamed writes.

The cowries themselves are believed to have drawn a Buddhist high culture that dominated the islands throughout the first millennium. The Buddhists built monuments, temples, and monasteries across the archipelago, along with dome-shaped *stupas* and coral sculptures: elephants, enormous Buddha heads, and representations of a long-ago people, earlobes elongated and hair gathered in a topknot like some of the young Maldivian men wear theirs today in the busy capital city of Malé.

The Buddhist ruins are also full of shining cowries. By many accounts, they are humanity's most revered seashells.

THE NAME *MALDIVES* is thought to derive from the Sanskrit *maladvipa*, "garland of islands," but some scholars also cite *mahiladvipa*, "island of women" for the early queens. The archipelago forms a coral garland looping northward from the equator for 600 miles in the lonely slate sea. The garland's flowers comprise twenty-six atolls, or clusters of islands, each ringed by a major reef. The English word *atoll* was born here, from *atholhu* in the local language of Dhivehi.

Early visitors had another name for the Maldives: "the islands of cow-ries," as the Persian scientist and scholar Abu Raihan Al-Biruni described them around 1030. Then and now, the warm waters and abundant reefs make what the historian Yang calls "a paradise for cowries," particularly the *M. moneta* that live along the shallow rocks and corals.

Once amassed in bags of money, for shell-lovers, cowries are the sea's coveted bag of marbles. High-polished globes of spun gold, a fawn's coat, gaseous rings, creamy maps, copper fishnet, cuts of ame-thyst, oiled mahogany, sundry dots and stripes—colors and markings on just a few of the 250 species known living today—no two look exactly alike. They are rounded on top and flat underneath where the aperture cuts jaggedly across. Depending on the species, the slit at the bottom can be gaping or exceedingly narrow. The teeth might form a harmless, stubby grin like the Money Cowrie, or a fierce-combed trap that looks like it could take a bite of you; the White-toothed Cowrie has such a maw.

Cowries are the most commonly collected seashells, and the ones most commonly found in archaeological sites. They were part of the collection discovered in the rubble at Pompeii, including a Panther Cowrie that had to have come from the Indian Ocean or Red Sea. Nineteenth-century aficionado Edward Donovan captured the pas-sion in his description of the Golden Cowrie, then called the Morn-ing Dawn Cowrie: "We may in truth compare its beauteous fulvous hues fading into white with inexpressive softness, to the warm glow-ing tints and fainter blushes of an opening morning sky in summer." The poetic fervor goes on for a page.

Even octopuses collect cowries in their dens, but not to "decorate" with seashells as their behavior is sometimes described. Cowries are favorite octopus foods. The cephalopod shell piles, like ours, repre-sent meals. Scientists describe the sometimes precise shell cumula-tions as ecosystem engineering; the intelligent octopus no longer has its own shell but can use other mollusks' to build a protected den that also draws prey such as hermit crabs to the front door.

Cowries' marbled mounds rank up there with the baobab trunks, the Matterhorn, and the luxurious coats of big cats among nature's masterworks of art. But the pièce de résistance is to see the live animal wrapped around its shell. Cowries differ from other mollusks not only in their dome, but in spreading their fleshy mantle across it, secreting eternal polish in high gloss. Their mantles have two great flaps shaped like the wings of a manta ray. Settled under a coral rock or ledge, the creature pushes the lobes up each side of its rounded shell until the surface is nearly entirely covered, living much of its life there on the top.

Spreading its glossy matrix over its shell rather than at the opening, the cowrie creates a hump over its spire, each new layer a thicker glaze that hides the last in rich hues that range from creamy to golden to dark chocolate. The animals also evolved striking mantle colors, wildly different from their shells. The soft flesh can be deep purple or pitch black. Some of the animals are camouflaged with the same bright red or orange as the sponges they colonize. Others are colored like a sea cucumber distasteful to fishes. In some species, the mantle flaps are smooth. More often, they are covered with wiggling fingers called papillae that vary in shape and pattern among species. Some wag like separate tentacles. Some grow in tufts.

The mantle patterns of *M. moneta* resemble black-inked fingerprints reaching up their small white humps. Surely no other shell has been as touched by human hands.

AT THE SMITHSONIAN'S National Museum of Natural History, in Washington, D.C., Chris Meyer reminded me of a jeweler showing precious gemstones as he pulled out drawer upon drawer of glossy humps of all colors, patterns, and sizes from the collection of more than 50,000 cowries. The shells are often striped across the mound like a football stitch; this "mantle mark" is where the two flaps didn't quite come together. "They weren't trying to look beautiful," Meyer says. "The patterns reflect the tissue that covered them." Meyer's doctoral research at U.C. Berkeley helped show that the whimsical spots, stripes, rings, and other markings on the mounded shells—or the smooth lack of markings, say, in a Golden Cowrie—correspond to the position and size of the papillae. (He dedicated his dissertation to his grandmother who glued shell animals with him in his childhood.) "The patterns themselves were constructional artifacts," Meyer explains when I visit his lab, which is crowded with small vials of puckered cowrie bodies.

Their ingenious bunkers, prolific reproduction, and great diversity have made cowries fairly resilient to the centuries of human collecting, Meyer says, even in Indonesia where they proliferate amid a vigorous seashell trade. The real crisis for cowries is the loss of their coral reef habitats. Like all animals, cowries need safe places to live and breed. Many lay their eggs on hard surfaces; some even brood like a mother bird to keep their eggs safe. But their homes are imperiled across the tropics as the oceans warm.

Up to now, rising ocean temperature has been a more immediate emergency in the Maldives than rising seas, though both are wreaking greater storm destruction on the islands. Two decades ago, the strongest El Niño ever recorded warmed oceans around the world, bleaching corals a deathly ash-white. The Maldives lost an estimated 95 percent of its shallow-water corals. In 2010, a second turbocharged El Niño warmed the seas again, killing off reefs that were just beginning to recover. Only five years later, the third-ever global bleaching event whitened corals in the Indian, Atlantic, and Pacific oceans. Corals from the Florida Keys to Australia's Great Barrier Reef to the Maldives were drained of their color like the devastating beauty of poached ivory tusk.

Scientists know that El Niños have warmed the shallow waters of the Maldives for millennia. They have found no evidence that the storms ever caused mass coral bleaching until the past few decades, as greenhouse gas emissions increased the global temperature of the oceans.

IN THE NINTH century, the Persian merchant Suleiman described an archipelago he estimated at 1,900 islands, thick with

coconut palms and ruled by a queen who oversaw brisk trade in cowries, ambergris, and coir—coconut husk. "The wealth of the people is
constituted by cowries," he wrote. "Their Queen amasses large quantities of these cowries in the royal depots."

Ancient oral narratives also frequently recalled ruling queens,
reflecting the matriarchy prevalent in the earliest societies on the
islands, writes the Spanish scholar Xavier Romero Frías, who
spent years collecting Maldivian folk tales. Other travelers likewise
described the Maldives' queens, cowrie monopoly, and trade power,
and corroborated Suleiman's description of a most unusual harvest.
The Queen "orders her islanders to cut coco-branches with their
leaves, and to throw them upon the surface of the water," wrote the
Arab historian and geographer al-Mas'udi some time before his death
in 956. "To these the creatures attach themselves, and are then collected and spread upon the sandy beach, where the sun rots them, and
leaves only the empty shells, which are then carried to the Treasury."

Glossy white shells were gathered in the light of the gleaming white
moon. Maldivian women floated the palm branches along shallow
reefs on the full and new moons; scientists say the practice is consis-
tent with cowries' movement at night, and on the lowest tides. The
nocturnal creatures would cover the fronds, and the islanders would
haul them to shore to dry. Some accounts described the women burying the cowries in the sand to speed up their decomposition.

During the tenth century, Middle Eastern seafarers dominated
Indian Ocean trade, stopping in the Maldives to load up on cowries, then trading for rice and other goods in India, Bengal, and the
ancient city of Pegu in what is now Myanmar. Arab traders also
brought a new religion. Ancient stories of how the Buddhist archipelago turned Muslim blend sea monster myth and plausible fact. But
by the end of the twelfth century, the Buddhist temples and stupas
were abandoned and the entire population of the Maldives had converted to Islam.

The Muslim Moroccan scholar Ibn Battuta lived in the islands for
a time in the fourteenth century and wrote the first detailed narra-

tive of life there. He described cowries as sold in a *syah* of 100; a *fâl* of 700; a *cotta* of 12,000; or a *boustou* of 100,000. Traders wheeled and dealed; a dinar of gold exchanged for between four and twelve *boustou*, or sacks, of shells.

Battuta called the Maldives "one of the wonders of the world" for the great numbers of islands gathered in rings, and described the people's devotion to Islam. He was taken aback by the brazenness of the women—though he managed to marry four of them—especially Queen Khadijah, who ascended the throne in 1347 after her brother's assassination.

Battuta described her leadership as another wonder of the Maldives. In Friday observances to Allah, the prayer leader known as the *khatib* always invoked the queen's name, as "an instrument of Thy grace for all Muslims." Yet Battuta complained that neither she nor other Maldivian women would cover their bodies above the navel. Though he tried, "I was quite unable to get them covered entirely," he lamented.

By Battuta's era, treasure fleets from as far away as China were stopping in the Maldives; sailors' logs describe "mountains of cowries" on the islands. When Europeans, led by the Portuguese, began to frequent the Indian Ocean, they saw the Maldives were "a great emporium for all parts." Everyone wanted in on the little shells.

In 1602, a French treasure ship called the *Corbin*, plagued by scurvy and a drunken captain who failed to heed the notorious Maldivian reefs, wrecked on a southern atoll. Of forty initial survivors, only four made it home, including the French navigator Francois Pyrard, who learned the Dhivehi language and recorded the local customs during five years as an unwilling guest. "There is a great trade at the Maldives, and they are much frequented for their commodities," Pyrard wrote. "You see merchants from all quarters . . . Arabs, Persians, men of Bengal, St. Thomas, and Masulipatam, Ceylon, and Sumatra, who bring goods that are in demand there, and take away what the Maldives produce in abundance."

Pyrard described a 400-ton ship from Portugal that came loaded with rice to exchange "solely to load with these shells for the Bengal market."

It was the Portuguese who standardized cowries as currency in the slave trade, filling their ships with shell-ballast later unloaded at burgeoning West African ports as cash. But they failed—three times—to colonize and Christianize the Maldives and control the shell money.

The Maldives slipped colonialism like a glossed cowrie from predator crab claws. By the time the British-Indian archaeologist H. C. P. Bell wrote his monograph of the Maldives in the nineteenth century, the island nation controlled its cowrie trade in partnership with the British, and remained fully Islamic—but with a Maldivian twist.

Maldivians still worshiped nature, "as rife as of old," wrote Bell, "if pursued now-a-days somewhat less obtrusively."

THE PUBLIC FERRY that plies the short, choppy passage between the airport island and the Maldivian capital of Malé is packed with local families. Bright white with blue trim, the boat is curtained like the resort speedboats, except that the ferry drapes are printed with pretty blue roses. They're held open with sashes to let in the sea air.

Malé comes into focus as a dense clutch of tall, brightly colored buildings. The capital island, 3.5 miles square, is home to 215,000 residents, or 40 percent of the national population. We hop off at a waterfront of bustling banks and small cafes, tour operators and airline offices slowed by the monsoon, all just a narrow street and sloshing seawall away from the Indian Ocean. Walking into the city interior, the curvy streets buzz with thousands of motor scooters that stop for no one. Many of the young women wear street clothes with colorful high heels and hijabs, in contrast to the full burkas common on outlying local islands.

The Maldives National Museum is an oasis at the city center

thanks to its air conditioning and shade; the glass and stone museum is one of the few buildings in Malé with large trees, which spread from Sultan Park. A gift from the Chinese, the three-story building is cavernous and spare. Quiet permeates even amid a field trip of primary schoolchildren in bright yellow uniforms.

The kids are listening politely to an extensive history of the Maldives Police Service, one of the museum's marquee displays along with dozens of historic Qurans and a philatelic collection. We browse Maldivian stamps including one featuring the Great Spotted Cowrie, a favorite of collectors. Many of the stamps spotlight famous Westerners: Elvis, JFK, Ronald Reagan, Marilyn Monroe, and Princess Diana.

Lady Di will turn out to be the only woman royal in the entire museum. I figured that Will and I, and the girls and boys on the field trip, would learn at least something of the Maldives matriarchal past—the medieval sultanas well documented from the early years of Islam. But amid hallowed artifacts of many sultans—their beds, carved wooden chairs, tasseled sunshades, lacquered turban containers, and worn Qurans—not one of the ruling sultanas is mentioned.

Ancient Maldivian history is also thin. The displays begin in the eighth century with a few small blocks of limestone carved with raised circular symbols, and some small coral stone caskets. One had held a 6-inch-tall carved elephant and 825 cowrie shells, according to the display. There is no mention of the elephant's whereabouts, no explanation that the circular symbols represented the sun, no hint of the Maldivians' legendary sun-worshipping ancestors.

But we are lucky to see even the small pre-Islamic finds. In 2012, during chaotic protests of the nation's first democratically elected government, as police and the military turned against President Mohamed Nasheed's administration in nearby Republic Square, Islamic fundamentalists ravaged the museum. A half-dozen bearded young men stormed the front doors, ran to the Buddhist displays, and smashed them to bits. They destroyed thirty statues, including a six-faced coral figure and a gigantic Buddha head.

The culprits were (and are) clearly visible on leaked closed-circuit television footage hurling artifacts onto the marble floors—available on YouTube for all to see. Yet they were not brought to justice. The dereliction marked "the start of a climate of impunity for fundamentalism" in the republic, writes the former Maldives news editor J. J. Robinson in his book on Islamic autocracy in this tropical paradise.

The more Maldivians protested the spread of religious extremism, the worse it became. The prominent journalist and critic of radical Islam Ahmed Rilwan was abducted at knifepoint at the public ferry terminal in 2014; he was never seen again. In 2017, a popular blogger named Yameen Rasheed was stabbed to death at age twenty-nine in one of several religious-militant attacks on progressive Maldivian writers with large followings online. Rasheed had frequently lamented the loss of the islands' rich folk legends and ancient history. "The Maldivian child today knows little beyond Pooh Bear and Red Riding Hood. The entire magical world of the sea-faring Djinnis and Maldivian handi stories seems closed to them," he once wrote. "What about the history of our Dhivehi Raajje?! (The country of nature-worshipping Dhivehis.) The land that exported precious cowry shells used as currency in ancient lands as far away as China!"

Unlike the National Museum, the vanishing Maldivian folktales are full of queens and seashells. In the Legend of the Sandara Shell, a queen dives into the sea to recover her husband's gold and jewels, stolen by a hideous demon. A big triggerfish gives her a special shell, the sandara, that she fashions into a necklace. By the end of the story, the shell's good fortune saves both the king and his treasures.

The sandara is actually the round operculum—the trapdoor—of a turban shell. Its polished disc looks like a stunning amber eye.

WITH THEIR SUGGESTIVE shapes and ties to the moon, shells generally, and cowries specifically, were associated with queens and goddesses and femaleness virtually everywhere in the world they lived—and everywhere their shells wound up. Swelled as if with human life on one side, and revealing a soft-folded crevice on the other, their provocative shape often tied them to maternal protection and fertility, and in no small manner to sex. Many mollusks, and cowries in particular, have contoured, involute apertures that "bear a marked resemblance to the folds and curves of external female genitalia," writes Catherine Blackledge in *The Story of V*, her biography of the vagina.

The people of the Vanuatu islands east of Australia believed that the first woman, mother of all mothers, was born from a cowrie. In Ancient Egypt, girdles decorated with cowries were buried with women and girls, and they skirted the pelvic area of female figurines, a talisman for fertility and healthy pregnancy.

In some provinces of Japan, the word for vagina is *kai*—shell. In early Japanese tradition the cowrie was *koyasuigai*, "easy birth shell." Holding one during labor was said to bring an easier birth and good fortune generally.

The Romans called cowries *Concha venerea*, "shell of the act of love." Linnaeus named the genus *Cypraea* for the island of Cyprus, home to Aphrodite, Greek goddess of love.

Virtually every early conchologist writing on cowries had some version of J. Arthur Thompson's "strong reasons for believing that the grip they have taken of mankind has owed a great part of its tenacity to sex-symbolism suggested to the many by the shell's shape." At times the speculation reads as if lady parts were all some colonial scientists could see in cowries. G. Elliot Smith was typical in his extravagant effort to distance men of science like himself from "primitive" ancestors: "The whole of the complex shell cult," he wrote in 1917, "seems to have sprung out of the fanciful resemblance which a particular group of primitive men imagined they could detect between the cowry and the female organs of reproduction."

But cowries draw all kinds of admirers, for all kinds of reasons. Not only "a particular group of primitive men," but a global span of early people believed in cowries as charms for fertility, health, protection, safety, and beauty—as well as for fortune and wealth, which helped lead to their transition to money.

From Africa to India and beyond, cowries were, and often still are, necklaces, bracelets, armbands, aprons, sashes, headdresses, offerings in shrines, talismans in graves. In northern Syria, in one of the earliest settlements on the Euphrates River from 10,000 years ago, archaeologists have unearthed cowrie shells in the graves of adults and children who died of certain diseases. They speculate that the villagers of Tell Halula had faith in cowries to heal.

In the early twentieth century, as male archaeologists honed in on the cowries' sex symbolism and narrow use by women as fertility amulets, Margaret Alice Murray, the first professional woman Egyptologist in Britain, made the case that cowrie meaning and devotion was much broader. Cowries were worn by women, men, and children. They bedecked humans as well as animals. They decorated the bridles of stallions, mares, and geldings alike, Murray wrote, and many other creatures from asses to bulls to camels. The Egyptians adorned their cats. Cowries were regal horse-trappings in China, Persia, Hungary, and Norway. In India, they also bejeweled elephant harnesses.

Cowries were pieces in ancient board games in India. They were lucky charms, built into the walls of homes, cast in amulets, and threaded into hair in Africa. Viewed horizontally, a cowrie's underside also looks like a half-closed eye; the teeth spread up and down in perfect lashes. Writing in the journal *Man* in 1939, Murray suggested the cowrie's ocular shape led to belief that the shell could ward off the evil eye:

> Early occurrences of the cowrie as an ornament or charm are as frequent in men's graves as in those of women. This is markedly the case in ancient Egypt. The usual position to wear it is where it will catch the first glance of the Evil Eye, which is always supposed to constitute the chief danger. If the first glance falls on an

inanimate object, especially if that object is in the likeness of an
eye, the danger is averted.

In 1953, the archaeologist Kathleen Kenyon began to excavate
Ancient Jericho, a Palestinian city in the West Bank and the longest
continuously occupied urban area in the world. From the earliest settlement dating to more than 10,000 years ago, Kenyon discovered
seven plastered skulls crafted to re-create living faces. Teeth still gaped
in the mouths. Shells peered from the eye sockets. Displayed in the
Ashmolean Museum in Oxford, one of the Jericho Skulls looks especially realistic owing to its cowrie eyes. Their slits are positioned in
eternal squint. The figure could be sleeping—or laughing for all time.

But after millennia of human adoration for cowries, religious fundamentalists began to take a harsher view. Some leaders would come
to see the shells as threatening idols. "He who hangs a necklace of
cowries round his neck," admonished one Muslim scholar, "God will
not prosper him."

SMOKY-BLUE MONSOON CLOUDS crush the horizon and
the wind has kicked up to 20 knots on the afternoon Will and I
begin our 55-mile journey north from Malé to the local island that
will be our home for a week. Ancient legend described it as "Kasidu,
one of the most important among the Maldive Islands," a large and
lonely outpost with no land visible from its shore and "so heavily forested that a sailor approaching from the east would not discern any
human habitation."

Now called Kaashidhoo, the rural island lies isolated in its own
atoll in a deep channel once frequented by early traders, who moored
at its natural harbor to load provisions of fresh water and coconuts,
still abundant. Kaashidhoo's land is shaped like a quarter moon, 2
miles long and a half-mile across. A great ring reef, visible at low tide,
traces the edge of the full moon. The island is known for agriculture;
tropical fruit and flower plantations are checkered across its jungle.

At the inner curve of the moon, a town of 2,500 Maldivians spreads from the harbor and lagoon.

Kaashidhoo is also home to a great cowrie mystery. At the outskirts of the modern town lies an ancient Buddhist monastery that held the largest single hoard of *M. moneta* yet found in the Maldives. I've arranged to tour the ruins and several coral reefs—with the hope of seeing history's most famous currency living in its natural home. Like many Maldivian reefs, Kaashidhoo's outlying barrier hasn't recovered from the troika of bleaching episodes; its large table corals are still sickly pale. But color has apparently returned to the shallower reefs along with the denizens that inhabit them, signs of the recovery possible if humankind can stave off warming.

We find our speedboat, the *Endheri Express*, among the many runners bobbing at Malé's crowded Jetty Number 6, and squeeze in with several Maldivian families by an open window. The crew soon unfurls blue canvas window covers and snaps them closed to the rain. A gap remains between the bottom of the covers and the sea. The chasm will become the most important detail on the boat an hour into the trip when one of our seatmates begins to vomit into the white, boiling waves.

Will is increasingly concerned about the weather. Summer monsoon season is typically a brew of storms broken up by the famous Maldivian sun. But the forecast icons have changed from a mix of dark clouds and sunbeams, he stresses, to 100 percent chance of severe storms for "every hour of every day" we've planned on Kaashidhoo. This intelligence annoys me as only that transmitted by teenagers can. But I give him my cheeriest smile as we pull away from the jetty.

The smoky monsoon clouds billow in the distance like paradise caught fire. The *Endheri* races toward the tropical combustion. Thanks to the speedboat we should arrive on Kaashidhoo in two hours rather than the seven it takes on the public ferry. The deep-V hull cuts through the whitecaps. The blue canvas window covers keep us dry from the rain and spray.

We pass countless islands and islets within the Malé atoll. Several forested; some mining islands bare but for dredges and their sand piles; many tourist islands with look-alike tiki huts lining the beach like dominos. The inhabited local islands stand out for their colorful buildings, blue-hulled fishing boats, and seaside mosques.

After an hour we're plowing through open ocean. The sea heaves and our nauseous seatmate follows suit. We are relieved for her when a local island called Gaafaru finally comes into view and she steps up onto the concrete dock with most of the other passengers. They hop onto waiting motorbikes or drop their bags into wheelbarrows, and disappear down the wet sand roads.

The seas are roughest on the final leg to Kaashidhoo. Its famously deep channel was called *Courant de Caridoue* on old French maps, a mysterious term interpreted by some to refer to coconut, but also close to the French cowrie—*cauri*. I hope we get to see the molluscan wealth on the reefs. As we climb swells high as the boat windows and drop into trenches, I question my decision to bring my son to the central Indian Ocean in monsoon season.

We finally pull out of the swells. Kaashidhoo appears like a green sanctuary in the lightening rain. Approaching from the east side, just like the legend said, there is no sign of human habitation—just coconut palm-jungle fronting a lengthy beach. Our young driver rounds the enormous ring reef and expertly swings the boat into moorings at the protected lagoon.

A minaret stands tall at the waterfront, painted burnt yellow like the emerging sun.

THE NORWEGIAN ARCHAEOLOGIST Egil Mikkelsen was tracking tropical cowries discovered in ancient northern European graves when he led the first modern scientific excavations in the Maldives at Kaashidhoo a quarter century ago. *A place called Kuruhinna Tharaagandu*, reported an old tradition connected with a sprawling ruin on the middle of the island. *It is said that this was a house of worship.*

Mikkelsen would find not only a former house of worship, but the largest pre-Islamic religious site in the Maldives—hidden in plain sight in a plantation of coconut, papaya, and banana trees. Beneath the fruit trees, still visible on the surface, lay more than sixty coral stone ruins of all shapes and sizes. They ranged from small, round stupas to a raised platform with sixteen sides and a 60-by-45-foot stone floor—larger than a tennis court—believed to have been the old monastic living area.

The hoard of cowries lay buried by a set of stairs leading to another elevated platform that Mikkelsen called Ruin II. The team uncovered 62,000 *M. moneta* shells under layers of sand. Whether these had been hidden, abandoned, or forgotten by time, cowries "were obviously of great importance in the religious life of the Kaashidhoo Buddhistic period," Mikkelsen concluded. His radiocarbon dating estimated the animals that made them lived between 165 and 345 CE.

We had planned to make the most of our first full day on Kaashidhoo with a visit to *Kuruhinna Tharaagandu* in the morning and a trip to the shallow reefs in the afternoon. But the monsoon rains returned in the night to thrash the island and our small guesthouse. The wind played a heavy mango limb across the metal roof. We were several blocks up from the lagoon but it sounded as if the sea roared at our front door. When we woke it was clear we wouldn't be going out on the water. Dawn was nearly as dark as night. The deluges continued as if the atmosphere had decided to cycle the entire Indian Ocean. The Maldives Meteorological Service had advised against any sea travel, so even if we wanted to return to Malé, neither the speedboat nor the public ferry would be running.

Around midday, the rain eased enough for us to take a soggy walk around town, a grid of small houses and shops radiating from the lagoon along narrow, unpaved lanes, now flooded. Some of the lanes have turned to chalky mud and there is no going around them; we slip off our shoes and wade through.

Children begin to emerge from the houses, which are close together and often share a courtyard shaded by a huge papaya or mango tree. The courtyards are also full of potted plants, and none is without a *jolie*—the Maldives' famously comfy hammock chairs, either built in pew-like rows with a seat for each member of the family, or hung from the trees in individual slings.

Rain-dripping swells of hot-pink bougainvillea and sea-trumpet vines with crinkly orange flowers brighten the unpainted concrete block walls. Several houses and walls are built of coral stone.

As we carefully watch our steps to the waterfront, a familiar hump catches my eye in the mud: a shining Snakehead Cowrie. I'm surprised to find my first Maldivian shell pressed into one of the lanes rather than at the beach. I pry it up and slip it into my messenger bag with a wish that the weather will clear enough for us to see its living relatives out on the reefs.

Another surprise awaits at the yellow minaret by the sea. Overnight, the large public park between the mosque and the lagoon has turned into a massive lake. At the park, long rows of public *jolies* sit submerged. At the waterfront, sandbags have been heaped along the concrete boat dock.

Just up the lane, the Indian Ocean laps at a stunning wooden safari boat three stories high, under construction in a cavernous outbuilding. We talk to the boat builders, who say their customer is a high-end tourism company. Amid so much water on land and sea, I picture Noah and his Ark. In the Quran, Noah is Nuh, a prophet who lived among stone-idol worshippers. The idolaters have forgotten Allah. Nuh warns them to give up their idols and embrace Islam. They refuse, and Allah unleashes a worldwide flood, drowning the infidels in torrential rains. Allah's worshippers are saved.

The monsoon clouds still threaten. But they've held off long enough for us to head to the Buddhist ruins. Our guide pulls up beside a puddle in his solar-powered three-wheeler to drive us to the crumbling stone idols.

Ahmed Saeed is a devout Muslim who, like most Maldivians we meet, doesn't consider ancient stories or Western tourists a threat to his kids. He loves history and riffs on the Maldivian queens. His six-year-old boy, Sayaah, joins us on the three-wheeler, and we bond over having our sons along.

Kaashidhoo is blessedly free of cars. We will see only one in a week, along with one farm truck in the jungle being loaded with golden coconuts. Most islanders walk or bike; no destination is too far. Many zip around on motorcycles, burkas flapping or kids snuggled in front of a parent.

Saeed was born on Kaashidhoo, and his family farmed here for as many generations as they can count back. He moved to the capital at Malé for the opportunity when he was young, as many Maldivians from local islands do. But over time he sensed more missed opportunity—to roam the jungle, breathe fresh air, and later, raise his kids surrounded by nature. He moved back to his home island to raise his family and help bring "guesthouse tourism" to Kaashidhoo.

As the high-end global travel business took root on uninhabited islands beginning in the 1970s, guesthouses on local islands were banned. Ostensibly, the idea was to protect local citizens from dangerous ideas such as "bikinis, bacon, beer, and Bibles," as the journalist Robinson put it. In practice, the ban kept locals from earning a dime from what would become the country's number one industry, now generating more than $2 billion a year. Nasheed, the first democratically elected president, is best known for his underwater cabinet meeting that called the world's attention to Maldivians' vulnerability to sea rise. Nasheed also lifted the guesthouse ban to benefit those living on small and rural islands such as Kaashidhoo. He was soon imprisoned for his "un-Islamic agenda" in the coup d'état of 2012 that also smashed up the National Museum. But there was no pushing guesthouses back into the genie bottle. Saeed seems to join a majority of Maldivians in his optimism that local tourism and Islamic governance can coexist; the problems with each, he tells us, are at the extremes.

Saeed whooshes us through the flooded lanes to the outskirts of town, his wheels throwing off a constant slosh. Ibn Battuta's description of the Maldives from the fourteenth century, far from reality in the crowded capital, still holds true in Kaashidhoo: "The whole country is shaded with trees . . . just as if we were walking in a garden." We pass fields of watermelons and landscape flowers; small papaya and banana plantations; slender areca palms harvested for their popular betel nut. Wild screw pine grows in bunches with its walking-stick roots.

No entrance or sign denotes *Kuruhinna Tharaagandu*, which is tucked behind some houses and a trough full of plastic. Chickens are strutting through bottles and bags to see what they can scratch up. In a large clearing between the houses and an enormous banyan tree, we can make out the former monastery in the soggy sand, as if in a two-dimensional architectural drawing. The moss-cloaked former stupas, sixteen-sided platform, and other forms are much eroded since the photographs of Mikkelsen's excavation between 1996 and 1998. Weeds climb the rubble.

We find remnants of the stairs where the 62,000 cowries had been stashed. The big old banyan tree must have been a hallowed part of the monastery; at some point it was enshrined in a stone border, now overtaken by the huge roots and barely visible.

During his childhood, Saeed says, locals considered the site a quarry: the place to gather coral stones to build houses. After Mikkelsen reported on the historic significance, there was a surge of excitement and some effort to preserve what remained. The community built a short concrete-block wall around the ruins to try to protect them from storms and people trampling through; the site is a shortcut between the jungle and the town. But the small community hasn't the resources to care for the site, and the Islamic national government isn't helping. The little wall clearly isn't protecting *Kuruhinna Tharaagandu*.

Crumbling into the soil, the Buddhist ruins remind me of the constructional artifacts patterning a cowrie shell.

THE MONSOON SWAMPS Kaashidhoo for a few more days. Finally, the famed Maldivian sun overpowers the rain. We putter out of the lagoon in a 16-foot skiff painted bright blue, and anchor at a coral islet known for trapping seashells. The rocky shoreline is covered in cowries tossed up in the storms. We find Arabian Cowries with bubble patterns on their domes; Hundred-eyed Cowries covered with multi-sized circles; Fawn Cowries with spots of delicate white; many more snakeheads; and countless others, big and small, sheen and colors ranging from chalk white to high-gloss brown. There are glistening white *M. moneta* enough to fill a money bag.

Many other tropical seashells are piled on the shore. Strewn among them are dozens of amber eyes. The smooth discs are the operculum of turban snails—the lucky sandara shells of Maldivian folklore. I slip one into the pocket of my board shorts for luck spotting live cowries under the still considerable waves.

The Legend of the Sandara Shell holds true that day and the rest of our time on Kaashidhoo, as we finally get underwater. Unlike the large ring reef that has yet to recover from the triple bleaching, the shallow reefs are full of life and color. Parrot fish nibble burnished copper boulders. Violet branch corals stretch their arms from the white sand. Blue ridge corals stand in wavy folds. I spot more living mollusks clinging to Kaashidhoo's reefs than I've seen anywhere—though I'm probably just getting better at peering in and under small crevices. Giant clams are abundant, at least the small giants known as Crocus, electric-blue and purple grins beaming from the rocks. Maculated Top Shells jut sideways like hot-pink party hats. I spot a stout murex covered in the same yellow sponge as its coral, as if adhering to the color palette of a strict homeowners association.

I'm peering through my dive mask under every ledge in a search for live Money Cowries when I discover two fist-sized Tiger Cowries hunkered near each other. Their mantle is invisible to my eyes; I see only furry papillae undulating from the tops of their dark-dappled shells. Why Linnaeus named them *Cypraea tigris* when they are dramatically spotted like a leopard, neither Meyer nor any other malacologist I've asked can say. Having named 12,000 plant and animal species, maybe Linnaeus was just tired.

Once I finally spot one Money Cowrie I can find them around the smallest rocks. In their habitat they look more like plants than shells—or money. Their mantle patterns remind me of the dark treetops of winter, branching up the sides of their little white shells. Translucent papillae sprig from the mantles like coral polyps.

They seem modest in the face of money's pretention. Yet cowries and the other calcifying life in the ocean have innate value that can hardly be measured in dollars, says the Smithsonian's Meyer. Like most Cypraeidae, Money Cowries are resilient, so far enduring despite rising ocean heat and acidity, and the harvesting that continues for tourist trinkets. Their shallow reef homes on Kaashidhoo are hopeful signs that the world's corals could recover from bleaching if the world's major governments and industries would act to stem emissions and warming.

The human homes on Kaashidhoo are more immediately imperiled. The monsoon-swollen lanes, and the wide lake between the sea and the mosque, have hardly drained by the time Will and I leave the island after a few more sunny days. The persistent flooding was not part of their childhood on Kaashidhoo, Saeed and other residents tell us. The monsoons always came—but the floodwaters always drained with the sun. The islanders are keenly aware of the changes being wrought by warming. Yet the tree lovers among them don't want to relocate to crowded Malé or artificial Hulhumalé. And they can't move inland. While the island is larger than most in the Maldives, the interior is wetter; their farms and jungle border interior wetlands filled with mangroves and pond-apple trees.

Having chosen their farms and car-free lanes over the city, Kaashidhoo's families reveal the inequity of the Anthropocene—the "age of humans." The term makes sense as a proposed new geologic age—when one species began to burn through fossil fuels and produce new materials like plastic so intensely that we irreversibly altered Earth. Just as scientists see the asteroid impact and volcanic eruptions at the end of the Cretaceous, they now see the human signature in a world-wide layer of fossil fuel burning, nuclear debris, and chemical pollution. But the idea of the Anthropocene does not capture the reality that most people in the world do not live in excess or burn tons of fossil fuels.

In their most recent election, Maldives citizens voted out the authoritarian president and replaced him with the challenger who ran on democratic principles and promised to restore human rights. President Ibrahim Mohamed Solih established a Disappearances and Deaths Commission to look into the cases of Rilwan, Rasheed, and twenty-five others. He pledged to redouble climate efforts on the diplomatic and local scales, including stronger environmental protection for inhabited islands; Kaashidhoo's wetlands were among those preserved.

Global cooperation to reduce greenhouse gas emissions remained elusive. Humans could clearly shift values, just as they had come to value the cowrie as currency rather than sacred talisman. Having seen *M. moneta* thriving in the sea, dark branches spreading up the small white shells, I wondered if such a shift was again possible, this time from deifying the money to revering the ocean life. On our clear, moonlit last night on Kaashidhoo, we puttered out in Saeed's boat to see the ocean's bioluminescent plankton, which flashed across the surface like seaborne fireflies. The bright moon recalled Sultana Rehendi Khadijah, collecting the cowries to fill the ships that would upend the world.

Six

SHELL MADNESS

THE PRECIOUS WENTLETRAP
Epitonium scalare

Nearing death at his Amsterdam garden estate in 1644, the Dutch tulip breeder Abraham Casteleyn called for a notary. He had explicit instructions for the handling of his life's prized possessions: His tulip bulbs, lifted from the soil and tucked into drawers that summer after the flowers had bloomed, and his exotic seashell collection—2,389 "rarities of little shells" and "little horns."

The little shells and horns were to be packed in a chest and padlocked, according to his meticulous will unearthed by the historian Anne Goldgar. The chest would have three different locks. Casteleyn's estate would have three executors. Each man would be given one key to one lock. Only together could they open the tropical treasure chest, or show it to a potential buyer. The shell collection was not to be broken up.

Casteleyn's cause of death is unknown. But his shell affliction had a name: conchylomania, a shell-collecting madness that spread across Europe from the sixteenth to eighteenth centuries. Watching the mania seize a relative in the Paris countryside, the philosopher Jean-Jacques Rousseau wrote that "his lively imagination saw nothing but shells in the natural world, and he at last sincerely believed that the universe consisted of nothing but shells, and remains of shells, and that the whole earth was nothing but so much shell sand."

Conchylomania took hold in the Netherlands around the same time as tulip fever (Dutch: *tulpenmanie*). It afflicted many of the same people, who at times paid more for a tropical seashell than for a master painting. The British shell authority S. Peter Dance found that at the same auction in the 1790s, a single Matchless Cone sold for 273 guilders—while Jan Vermeer's *Woman in Blue Reading a Letter* went for 43 guilders.

The undersea artist, *Conus cedonulli*, is still alive and well and making its coveted shell, patterned in burnt-orange and cream maps. The patterns appear raised amid latitude lines stitched across the entire cone, though the shell is actually smooth. Unlike Vermeer, who produced only thirty-six paintings, it is not matchless.

Another Dutch master, Rembrandt himself, caught the fever. His 1650 etching *The Shell* reveres the flush spire and light-on-dark dapples of a Marbled Cone, itself emerging from shadow to light. The world's most admired *Conus marmoreus* is believed to have been part of the artist's own extensive shell collection, noted in an inventory of Rembrandt's house when he lost the brick manor in bankruptcy.

Home to the first stock exchange and first public corporation— the Dutch East India Company, whose ships carried home the tropical shells—it is perhaps no surprise that the Dutch Republic beginning in the seventeenth century also helped spark nature's commodification and modern free-market excess. Shells and other exotica collected like miniature conquests at home tracked the march of colonialism on faraway shores. The rarer, the headier. The more amassed, the more impressive. The more collected, the more desired.

The era also began to bring science to daylight from the era of the snakestones. A network of scholars and explorers sailed and wrote to one another across sea and land in what is known as the Republic of Letters, correspondents sharing scientific ideas and specimens like seashells as they tried to make sense of the natural world.

WHEN MOUNT VESUVIUS erupted in 79 CE, burying Pompeii and many of its citizens in tons of volcanic ash and pulverized rock, it also froze the Roman city in its 2,000-year-old chariot tracks. Preserved in time, the bodies and buildings and art and artifacts offer a glimpse of daily life in the ancient world—including keepsake shells. Archaeologists have unearthed scores of seashells at Pompeii, most of them native tuns, cowries, murex, Triton's Trumpets, and abalones from the surrounding Mediterranean. But at least five other species had to have come from the Red Sea or Indian Ocean: Textile Cones, Gnawed Cowries, giant clams, the black-lip pearl oyster *Pinctada margaritifera*, and some fifty Panther Cowries. Dance, a long-time shell curator at London's Natural History Museum, speculates that they "may have been prized items in the collection of one of the world's earliest conchologists."

The Roman philosopher Cicero extolled two Roman consuls, Laelius and Scipio, for taking mindful breaks from their worries to go shelling together on the beaches of Gaeta and Laurentium and unwind from the demands of the Republic. He stressed *conchas legere*—shell collecting—as a worthy release for the serious mind. Panther Cowries, *Cypraea pantherina*, much bigger than any found in the Mediterranean, also turn up in Roman and Saxon graves, continuing 100,000 years' veneration, seashells worn, seashells carved, seashells buried with a loved one.

Wherever the first shell collections originated, the fever peaked in the Netherlands. Conchylomania arrived, like the plague, on trading vessels. Dutch merchant ships began returning from the East and West Indies in the late 1500s and early 1600s loaded with spices, silks, and other spoils—and soon, the Chinese porcelain that launched a national industry and image in white and Delft blue. Also turning up on the docks were peculiar naturalia: spiky seed pods, scaly fruits, nightmarishly long snakeskins, dried crocodiles, ballooned puffer fish, iridescent butterflies, and especially, strange new kinds of shells. Collectors like Casteleyn went crazy for them.

Tropical seashells had bulk and color and spindles and knobs and whorls and gloss that outshone anything found on the beaches of the North Sea. Men debarking from East India and other exotic shores showed off lavish conchs, swelled volutes, murex with spikes like those on medieval weapons. They brought back giant clamshells resembling bear paws and sea-snail shells that looked like spinning tops, for which the species would soon be named. At first, the sailors and soldiers carried the seashells home as souvenirs. It quickly became clear that the shells could be sold for a small fortune—or given as gifts to curry transactional favor.

Amsterdam grew a thriving curiosity trade, with specialty shops and peddlers hawking tropical shells and other exotic naturalia from the ships. Cornelis de Man's painting *The Curiosity Seller* depicts a silk-robed shell merchant showing off a basket of richly colored conchs, tritons, cowries, and other cabinet-worthy shells. A customer in a white-satin gown looks as if she yearns to own the pearly Chambered Nautilus cradled in her hand. A boy in wool holds a Triton's Trumpet to his ear to hear a warm, exotic sea.

Then as now, while the public pawed through bowls of pretty shells in the seaside shops, more serious trade went on among private collectors and at public auctions like those frequented by Rembrandt. During one auction in the spring of 1637, he paid 11 guilders for one conch shell—*kockeilje horen*—and only one guilder more for an engraved print by the late Italian master Raphael.

By the early seventeenth century, Dutch seashell assemblages rivaled collections of both fine paintings and prized tulips. Rich men and royals who commissioned their own portraits often had their prized shells painted, too. In 1603, the Haarlem master Hendrick Goltzius painted a well-fed textile merchant named Jan Govertsen in a simple black tunic with sumptuous tropical shells at his lap and a pearlescent conical gastropod, called a turbo, in his hand.

Four centuries later, the beauties from the Indian Ocean still steal the scene.

ACROSS EUROPE, a cabinet of curiosities, natural and artificial, became the must-have symbol of status and learnedness. Some wealthy collectors had entire *Wunderkammern*—"wonder rooms"—and prepared catalogues for visiting dignitaries. Scientists would soon begin to name and categorize mollusks and other species, ordering the astonishing ark of animals being found on the continents.

But in this time before taxonomy, shell collectors saw—or at least projected—their hobby as an exalted step closer to God. "Shells and flowers constituted the most glorious specimens from Creation," writes the art historian Leopoldine Prosperetti. Christians could justify collecting as no less than duty to follow the spirals to their own enlightenment.

Scarcity was also at play, at least in the imagination of collectors. Unlike tulips, which could be cultivated, and stamps, which could be printed, only nature's whim—or God's—controlled the rarity or abundance of a seashell. Extinction was imagined but still not articulated—much less the idea that humanity could snuff out a species.

Collectors wanted exceptional shells they'd never seen before: the more exotic, exquisite, and exclusive, the better. The Precious Wentletrap, *Epitonium scalare*, was all that. At least, it seemed to be. The first mention of the small, snow-white spiral—they average only 2 inches long—shows up in 1663. A traveler visiting a doctor's curiosities in Amsterdam noted a "white shell resembling a spirally twisted trumpet and open from top to bottom."

The Dutch named the coiled beauty *wentletrap*, derived from the word for a spiral staircase. The shell is both more compressed and more voluptuous than its namesake; its ivory whorls are plump at the bottom and appear to stack to a pointed peak, like soft-serve ice cream. Opposite a soft-serve of course, the molluscan maker starts at the narrow tip and winds its single tube in ever-larger whorls as it grows. The coils appear to touch, but actually hang suspended, girded by vertical ribs. Each rib represents the former mouth of the aperture—like a festooned growth chart. The overall effect is a powdered Marie Antoinette hair pouf that a royal hairdresser has spent hours twisting into a tower and fitting with elaborate ornamentation.

The shell's perceived rarity increased demand—and prices. The small spectacle was worth more than its weight in gold. Only a few specimens were known, including one each in the collections of Russia's Empress Catherine and Sweden's Queen Louisa Ulrika.

Louisa was patron to the Swedish naturalist Linnaeus, then working through the queen's natural history collection at Drottningholm Palace to build the two-name Latin system that classifies plant and animal life. Linnaeus named 683 mollusks, sometimes awkwardly for having observed only the shells and not the animals that built them. He stuck the swimming scallops, for example, with the stationary oysters. But from Louisa's wentletrap, he gave the species an apropos name: *Turbo scalaris*, the latter word Latin for scaling a flight of stairs.

Among the credible accounts of its market value, the German king and Holy Roman Emperor Franz I was said to have paid 4,000 guilders for a Precious Wentletrap in about 1750, roughly $114,000 U.S. dollars today. The heady prices spun off tales of fakery and forgery. In his *Strange Sea Shells and Their Stories* of 1936, A. Hyatt Verrill reported as fact the common claim that Chinese forgers made imitation wentletraps from rice-flour paste, a substance later discovered by hapless buyers when they tried to wash the shells. A London shell shop owner was sure he had a rice-paper fake in the mid-twentieth century, and came to value it more than his real wentletraps. When he died, he left two surprises. The fake was proved a real Precious Wentletrap. And the shell shop's owner was proved a fake; he was a Nazi fugitive.

Another craze surrounded "golden limpets" in Paris in 1776. The supposedly rare shells were in fact common limpets, *Patinigera deaurata*, "roasted in hot ashes or gently pan-fried to turn their russet color into shining gold," writes Leo Ruickbie in his encyclopedia of mythical beasts.

Real or fake, seashells were rarely a wise investment. When the French ambassador's collection went up for auction in Amsterdam in 1757, his Precious Wentletrap sold for 1,611 livres. Had the buyer purchased gold, he would have about 540 grams—worth more than $27,000 today. But wentletraps didn't hold their value once collectors found them hunkered below the sand in the southwestern Pacific Ocean.

Auction records show the price plummeting, to 50 guilders for a "fine and bright" specimen sold in Amsterdam in 1792; to 8 pounds for a "very perfect specimen" sold in London in 1822. Today, you can pick one up on eBay for $10.

If the queens and kings had only known *Epitonium scalare*'s true distinction, hidden under tropical waves and sand. The artist behind one of the world's most beautiful shells is a parasite, settling beneath a sea anemone it can live off for years. When the Precious Wentletrap gets hungry, it drags its lovely self out of the sand, draws out an extraordinary trunk four times the length of its shell, chooses an appetizing anemone tentacle, tightens its appendage around the anemone's, and shoots in paralyzing venom so the tentacle can't move. Next, it pulls out its toothy radula, saws off a bit of tentacle, and eats the soft flesh.

The wentletrap feasts on its host's tentacles this way for four or five hours before burrowing back down in the sand. Between two and eight weeks later when its hunger returns, the animal will work its way back up and choose another tentacle or two to saw and devour.

Mollusk and shell plump beautifully, coiling wider and wider as they grow ivory white whorls, coveted by a truly strange creature up on the land.

THE LONDON PHYSICIAN Martin Lister was a seventeenth-century spider man: an early arachnologist who figured out how some spiders disperse by ballooning—weaving tiny parachutes of fine-spun silk to catch the wind. He was also a well-known doctor to the rich and famous, including Queen Anne, and a polymath: Inventor of the soil survey and the histogram. Flower aficionado who planted 3,000 bulbs. Collector, chemist and dye-maker, prolific writer, vice president of the Royal Society, French ambassador. (Of that assignment, Lister admitted to taking far more pleasure in "the meanest" nature at Languedoc in the south of France than "the finest alley at Versailles," and his preference "to learn the names and physiognomy of a hundred plants than five or six princes.")

Lister is also credited with publishing the first scientifically rigorous encyclopedia of seashells, *Historiae Conchyliorum*. Others had published seashell plates and books such as the Italian Jesuit scholar Filippo Buonanni's 1681 atlas of known mollusks, *Ricreatione dell'occhio e della mente*, a "recreation for the eyes and mind."

Buonanni divided shells into univalves and bivalves, and in a few cases sketched the mollusks within—which worked well for those he had seen, such as garden snails, not so much for the tropical creatures he could only imagine. Like almost all the earliest shell guides, Buonanni's spirals were also reversed in the printing process, yielding mirror images that would appear inaccurate for someone trying to identify a shell. (Meaning anyone who saw them would think shells generally spiral left like a Lightning Whelk, when the sinistral spiral was exceedingly rare.)

Lister's *Conchyliorum* included more than a thousand copperplate shell engravings, each carefully reversed before printing to ensure its accuracy. The guide was a massive endeavor with four volumes published over six years between 1685 and 1692. How could the accomplished physician and arachnologist do so much? Like his beloved spiders—moreover like so many scientists of his time—he had extra arms: women partners in science, in his case two brilliant daughters named Anna and Susanna.

Conchyliorum was also monumental for its scientific illustrators. The gorgeous seashells—detailed, shadowed line drawings lush like Rembrandt's cone shell—were created by Anna and Susanna beginning when they were only thirteen and fourteen years old. The family's biographer, the science historian Anna Marie Roos, describe how, in a time when girls were excluded from formal schooling, their shell drawings and engravings were meticulous, accurate, and correctly reversed on plate.

Even Rembrandt's famous still life of a Marbled Cone is incorrect; an aesthetic choice, art historians conclude, since his signature is properly rendered. The mollusk man Stephen Jay Gould puzzled over why his forebears would have printed shells in mirror image when they surely knew how to correct them on plate. Perhaps, like Rembrandt, they were moved by shadow and aesthetics in a time when science and art were still fluid.

Two teenage girls reversed erroneous convention by reversing the plates. Anna and Susanna also mastered etching and engraving and were among the first women to use microscopes to produce images—for the journal of the Royal Society as well as for *Conchyliorum*. A Royal Society history published in 1812 lauded the Lister sisters and proclaimed *Conchyliorum* "formed a new era in the science . . . still retains its value, and is still indispensable to every student of conchology."

Conchyliorum is also noteworthy for giving Europe its first look at North American shells. In Virginia, the early-American naturalist John Banister sent Lister detailed descriptions, drawings, and specimens of American mollusks, along with fossils and flies and fungi. Banister's last letter to Lister was dated May 12, 1692. He waxed about snail-shell apertures. The same day, he left for a collecting expedition with William Byrd I, plantation owner and founder of Richmond who was a patron of Banister's science. Four days later, while kneeling at the Roanoke River, Banister was accidentally shot and killed by one of Byrd's hired men. Banister never completed the natural history of Virginia in which he had also described the Appomattoc people. His legacy survives in the many natural icons named for him in the United States; in Linnaeus's early American species; and in Lister's *Conchyliorum*. The first scientific encyclopedia of shells featured seventeen specimens sent by Banister or copied from his drawings, labeled "Vir" for Virginia.

Lister was also the first naturalist to describe and illustrate North American fossil shells. Banister's huge scallop with mesmerizing swirls at the margin was large enough to get its own page in *Conchyliorum*. Lister described it as "the biggest Scallop I have ever seen," blue-clay in color, and so old and weathered it had lost its beauty. The shell looked like the modern pecten—the scallops—and he included it among them with the marine bivalves, speculating it might be a grand Lion's Paw. The colossal bivalve was actually extinct. Thomas

Say, who would write the first American shell book—also with his indispensable woman illustrator—later named the fossil scallop *Chesapecten jeffersonius* in honor of Thomas Jefferson.

With every fossil found, the idea of extinction was soaking up more ink in the Republic of Letters. But for most, the concept was still anathema—incompatible with God's perfection and nature's order. Lister categorized fossils, including a wagon-wheel-sized ammonite, in a separate section of *Conchyliorum*, under the heading "shell-stones." Unlike those being unearthed in Italy and America, England's fossils were so unlike living mollusks that Lister was not convinced they had ever been alive.

He'd met the famous dissector Nicolaus Steno in France while a medical student at Montpellier, and found him "very much of ye Galant & honest man as ye French say, as well as ye scholar." But when he read Steno's theory that shell-stones nestled into mountains were organic remains, he wrote a rebuttal to the journal of the Royal Society. He doubted the possibility because he had yet to see a shell-stone "cast in any animal mold whose species or race is yet to be found in being at this day." Lister said he might be able to accept Steno's theory if he had evidence of fossil shells resembling living mollusks: "My argument will fall," he wrote, "and I shall be happily convinced of an Errour."

The true age of the big, blue-gray scallop from America, 4 to 5 million years old, surely would have convinced him.

fol·ly

Noun

1. lack of good sense; foolishness.
2. a costly ornamental building with no practical purpose, especially a tower or mock-Gothic ruin built in a large garden or park.

Origin

Middle English from Old French *folie*, "madness," in modern French also "delight, favorite dwelling."

While collectors like Lister pursued shells for science, others were in it for fashion or folly—none more so than kings and queens amassing the royal seashell collections of Europe. At Dresden Castle in Germany, Augustus II the Strong oversaw a shell cabinet that grew into a room that grew into a grand hallway. Visitors would see a large monogram of the king embellished with shells, beneath a crown made entirely of shells, surrounded by a shell mural, set off by bouquets of shell flowers and plates of shell fruit. Ornamental shell scrolls, grotesque shell masks, and shell animal figures covered the walls. A shell-encrusted model of Mount Helicon was built into the center of the room, complete with the Hippocrene fountain in mother-of-pearl. The king's prized cones, cowries, and other species were displayed in nine glass cabinets, five bearing shell sculptures of dolphin heads and other sea ornaments.

The rage for shells among the rich and fashionable spilled outside to garden grottoes. Outdoor refuges called *locus amoenus* ("a pleasant spot") had been popular with the Ancient Greeks, who built cave altars to nymphs, mythic female embodiments of springs, rivers, and lakes. Attica, the historical region surrounding Athens, is pocked with dozens of limestone caves sacred to the Ancients. Altars still stand in a few of them, and terra-cotta figurines still turn up, some with bellies swollen in pregnancy. Expectant mothers would leave an offering to the nymphs before giving birth, as a protective measure, or afterward, in thanks for a safe delivery.

Natural shrines arose at ponds and springs, lakes and riversides. The Romans transformed rustic altars into hydraulic edifices, adding details like the lion's-head spouts that funneled water in the Empire. New generations piled on colonnades, arcades, statues, paintings, and mosaics until finally, the grottoes themselves were artificial, complete with caves for al fresco summer dining, fake stalactites and stalagmites piercing the scene.

Real shells grounded the increasingly artificial world to nature and for some, to God. In the mid-1500s, the Italian architect Galeazzo Alessi designed an elaborate grotto for the estate of Genoese Admiral Andrea Doria (namesake of the doomed Italian ocean liner) with mosaic walls made entirely of shells, telling stories of the sea. Centuries later when train tracks were cut through Genoa, the vivid grotto, with its depictions of Perseus killing a sea monster and the sea nymph Galateia steering a dolphin-powered shell, was severed from the rest of the estate and left to deteriorate. Modern visitors to Villa del Principe who know to ask about it can hike uphill beyond the tracks to see the crumbling molluscan walls.

As shell madness swept Europe, architects walled garden grottoes with shells inside and out, recreating limestone wellsprings in the cities. Grottoes "swallowed vast quantities of shells and many wealthy grotto-builders spent small fortunes importing exotic shells from around the world," writes Hazelle Jackson in her book devoted to the shell houses and grottoes of Europe. Artists pieced thousands upon thousands of shells, tiny as periwinkles and big as tritons, into life-sized murals and dramatic mosaics, often harkening the ocean.

History's most elaborate shell grotto exists today only in 300-year-old etchings and poetry. It stood for just two decades at the Palace of Versailles. From the beginning of his reign in 1661 to his death in 1715, King Louis XIV employed France's greatest architects and artists to constantly remodel Versailles and project a glorious image of himself as the "Sun King" to the world. In 1665, work began to redesign a water-tower pavilion on the palace's north side into an allegorical fountain showing the descent of the sun god, Apollo, into the sea at the end of his daily trek across the heavens. The Grotto of Tethys on the opposite end of the palace gardens would provide the parallel, where the sun god rises from the sea in his horse-drawn chariot.

Along the front façade, three enormous arches were hung with gilded grills slanted into rays of light from a central, human-faced sun. Inside, thousands upon thousands of shells, bits of coral, and mother-of-pearl covered an expansive single chamber, conjuring up an undersea cave. Countless mirrors expanded the sense of flowing water. In a central group of marble statues, *Apollo Attended by the Nymphs*, six women bathed a resting Apollo at day's end, including one kneeling to scrub his feet. In two side groups, muscular Tritons washed down Apollo's horses with water flowing from scallop-shell basins and conch jugs.

The water features, including a hydraulic organ, pulsed from a hidden reservoir in the attic. Some flowed steadily and some would surprise visitors, who might get soaked by a streaming conch. Along the walls, strange and wondrous ocean figures swam and played in shell mosaics and carved marble fountains. Tethys housed some of history's most famous shell art, now lost, including six molluscan masks shaped into fantastically evil faces with sinister popping snail-shell eyes, evil gastropod beards, and pointy auger-shell horns.

In a statement of piousness that seems more of an unimaginable waste, Louis XIV demolished the Grotto of Tethys in 1684 to make way for the north wing of the expanding palace, including his gothic Royal Chapel. Only the marble sculptures were preserved and transferred to the gardens. The nymphs and Tritons still do their washing there in an artificial cave.

A century later in the Rambouillet forest to the southwest of Versailles, Marie Antoinette enjoyed and ultimately possessed the two-room shell cottage—*la chaumière aux coquillage*—that had been built for her close friend and confidant the Princess of Lamballe by the Duke of Penthièvre. The thatched roof and rustic outside belied a stunning circular shell interior with mother-of-pearl domes and walls encrusted with thousands of seashells collected from Brittany and Normandy, interspersed with rare species the duke inherited from his grandfather, the Sun King himself.

The duke lost Rambouillet, and the princess her shell cottage, in 1783 when King Louis XVI demanded that his cousin sell it to him. Not content with the forest estate and its shell lair, Marie Antoinette that same year commissioned her own rustic hamlet, le Hameau de la Reine—a mock peasant village near Versailles. Along with artificially distressed farmhouses, vegetable gardens, and hand-chosen peasants to work them, Hameau featured a dairy where the queen could role-play as a simple milkmaid; an engineered bluff called

Montagne de l'Escargot—"mountain of snails"; and a rugged grotto where she would *retraite* and contemplate nature as championed by Jean-Jacques Rousseau.

Scholars do not know whether Marie Antoinette directly read Rousseau, whose writings influenced the French Revolution that ended her life. They say she was hiding in her grotto when the mob from Paris stormed the gates of Versailles.

Little shell caves and huts and larger shell houses had by then become a craze in English gardens as well. In Hertfordshire, England, in the 1750s and '60s, the Quaker poet John Scott excavated a series of chambers 70 feet into his chalk-hillside garden and encrusted them with all manner of conchs, scallops, and other shells. London society flocked to see the spectacle, and tourists still do.

The English poet Alexander Pope, who had a riverside garden on the Thames in Twickenham, tunneled an artificial cave under his villa and obsessed for two decades in studding the walls with shells, crystals, fossils, minerals, flints, stalagmites, and other geologic decor, all leading to a shell temple. Pope spent the last years of his life in the subterranean lair, which still exists, shuttered to the public and the light. "Were it to have nymphs as well," Pope wrote in a letter to a friend, "it would be complete in everything."

THE LONDON INTELLECTUAL Judith Drake would have rolled her eyes. In 1696, the year the Dutch master Adriaen Coorte painted his worshipful still life *Five Shells Mounted on a Slab of Stone*, Drake published one of the great works of early modern feminism, "An Essay in Defence of the Female Sex." Implausibly, it was also a piercing critique of shell madness.

A writer and medical practitioner to women and children— an unauthorized activity for which she was summoned to the Royal College of Physicians—Drake was married to the physician and political writer James Drake. He was a Tory pamphleteer whose work defined the origins of "fiery" for their public burnings. Dr. Drake died when the couple's two children were young, amid almost constant prosecution and government burning of his books.

Drake's anonymous pamphlet, formerly attributed to Mary Astell, is written from "A Lady" to fellow women "or any Man of Sense." By turns witty and fierce, the 25,000-word essay champions the intellectual equality and value of learning for women and challenges the male-centered cult of the Ancients. Girls and boys are taught to speak, read, and write by the same people, at the same time, she observes. Why are they later separated, one sent to learn needlework, the other Greek? She "can find no such disparity in Nature" to justify it.

Perhaps women were denied equal access to education by men, she suggested, due to women's potential to "become their superiours." She builds satirical male archetypes to show that men are equally or more prone to the follies for which women are constantly accused. They include the Pedant, Country squire, News-monger, Bully, City Critick, Beau and—the "most Egregious" of all—the Virtuoso.

Sparing no expense, the Virtuoso buys and collects seashells and other naturalia for the sake of amassing an enviable collection rather than for science and the greater good. He "sold an Estate in land to purchase one in scallop, conch, mussel, cockle shells, periwinkles, sea shrubs, weeds, mosses, sponges, corals, corallines, sea fans, pebbles, marchasites, and flint stones."

His travels "are not designed as Visits to the Inhabitants of any Place, but to the Pits, Shores and Hills; from whence he fetches not the Treasure, but the Trumpery. He is ravished at finding an uncommon shell, or an odd shaped Stone. . . .

"He traffics to all places, and has his Correspondents in every part of the World; yet his Merchandisers serve not to promote our Luxury nor increase our Trade, and neither enrich the Nation, nor himself."

He would "give more for the Shell of a Star-fish or Sea Urchin entire, than for a whole Dutch Herring fleet."

It must have been maddening to confront the exclusion of girls and women from education in those dual times of excess and exhilarating scientific ascendance. Drake herself could not train in medicine. Her defense before the Royal College of Physicians Board of Censors for practicing without a license was that "she only treated women and children, and then without asking for a fee." Martin Lister being an exception, training daughters was rare even in families. The ethos for girls' education was closer to that of Linnaeus: When the great naturalist's wife placed one of their daughters in school, Linnaeus pulled her out, putting a halt to what he considered "nonsensical" education for girls. He also refused Queen Louisa's offer to bring one of his daughters into her court, thinking the environment would corrupt the girl's morals.

Drake supported the cause of science. Her essay compliments the Royal Society and stresses biology, chemistry, mathematics, and "intelligent inquiry into Nature" as heartily "as any Virtuoso of 'em all." She observes a "vast difference" between Virtuosos driven by their greed to amass trifles, and those pursuing biological questions— the answers to which she thinks will prove her case that women and men have the same capacity to learn.

A "tyranny of custom" handed down from the Greeks and Romans maintained inequality unjustified by modern science and medicine. Physicians had informed her that "there is no Difference in the Organization of those Parts which have any Relation to, or Influence over, the Minds."

For Drake, women in science would make a better world. One in which the mollusk is valued more for cure than curio, and its place in the life of the Earth, rather than solely in the collections of men. She was prescient in her criticism of the time and money lavished "to ransack all parts both of Earth and Sea" to procure and name every shell and insect, with little inquiry into how they live and how they might be important for humankind. What improvements or useful arts, "what noble Remedies, what serviceable instruments, have the Mushroom and Cockle Shell Hunters oblg'd the World with?" the healer asked.

Eight thousand miles away on a tropical island in the Indies, one of history's greatest malacologists was working against the odds to finish his lifelong compilation of the noble Remedies: ancient molluscan cures for calming children, for salve, for colic, for stomachaches, for better dreams and better sleep, for the "Mother's Sickness" this gentle man called morning sickness.

But in his time, the only things that seemed to matter to the rest of the world were his shells.

THE FIRST PUBLISHED print of a Precious Wentletrap appears in a 1705 Dutch seashell guide called *The Ambonese Curiosity Cabinet*. The book endures today as a classic of malacology. Its author, the shell-lover and brilliant field naturalist known as Rumphius, spent most of his life on Ambon in the Spice Islands, studying the native flora and fauna with an eye to nature's intrinsic rather than commercial value.

Rumphius's lovingly detailed biography of shells and their makers would have been published before Lister's, if not for an unimaginable string of personal and professional tragedies. Self-exiled from the seventeenth century's rising cities and commercialism, he may not have cared.

Georgius Everhardus Rumphius was born in 1627 in Germany's Hanau on the River Main, a town more famous for producing the brothers Grimm. His story unfolds like one of their bleak fairy tales. "Burning with an insatiable desire to know foreign lands," as he put it in an autobiographical poem, Rumphius enlisted at age eighteen with the Dutch West India Co., understanding that he would sail to Venice. He'd been deceived by a common type of swindler then known as a "soul merchant," who lured adventuresome young men onto ships and essentially indentured servitude. Rumphius's ship headed to Brazil, where the Dutch were fighting the Portuguese. It was captured and the duped passengers hauled to Portugal, where he spent three years as a soldier.

Undeterred and back in Hanau in 1652, Rumphius, then twenty-five, signed on with the Dutch East India Co., sailing away this time—as promised—to Java. He first served as a military engineer at the company's major trade center at Batavia (the capital of Java,

so named by the Dutch after conquering it in 1619; now Jakarta).
More comfortable with local than military affairs, he sought a trans-
fer to the civilian side of the company and was granted a position at
Ambon 1,700 miles to the east.

Thirty-two miles long and 10 miles wide, the mountainous island
was crossed in rivers and trussed in rain forest, treetops aflame in
orchids—"the Aristocrats of wild plants," Rumphius called them.
Shelled animals covered the hued corals visible in the shallows, plod-
ded along the beaches, and congregated in the mangroves, which also
captivated Rumphius. He described more than a dozen mangrove
species using the term *Mangium*, after what the locals called mangi
trees in their language of Malay. Rumphius wrote to his superiors
that he had

> begun a work that describes in Latin all such plants, field prod-
> ucts, animals, etc., which I have seen or which have been reported
> to me during this time of my residence in the Indies, as well as
> others in the future. I have sorted out their proper names, both
> from ancient Greek, Arabic, and Latin authors, as well as from
> recent ones, compared them with and differentiated them from
> one another, and drawn fitting pictures of them from life; I sup-
> plied all of them with their characteristics and powers, derived
> from the aforementioned ancient authors but particularly from
> carefully wrought personal experience.

That kind of intelligence appealed to the Dutch East India men.
Clove, nutmeg, and mace trees were helping build the company into
a colonial and commercial leviathan. Though he worked for the company for half a century—until the day he died—Rumphius never
ascribed to its dominion over native species, culture, or people. He
spent his time listening to all of them.

Rumphius fell in love with Ambonese nature and a local woman,
Susanna, "my first Companion and Helpmate," who helped him
find, identify, and describe the island's seashells, flowers, plants, and

trees. A half-century before Linnaeus, Rumphius named hundreds of species with descriptive gift and loyalty to Malay language. A "dazzling array of lyrical metonymies," his biographer and translator, the late Monty Beekman, said of his subject's nomenclature. Rumphius's shells were dolphins, bearded men, Bishop's miter, Babylonian towers, Cinderella, ghosts, gray monks, old wives, music shells, Pope shells—covered with rows of red globes "as if they were the precious stones on a Papal crown."

He also wrote far more intimately than anyone up to then on the animals inside the shells. His attention to molluscan ecology made him malacology's true pioneer, in Peter Dance's estimation. Hermit crabs, Rumphius wrote, "live in the house of strangers." Cone snails, he warned, have a "little bone," a radular tooth that injects deadly venom into its prey—an observation not duplicated for nearly two centuries. He was the first to describe a living nautilus, and to note *Argonauta*'s habit of latching onto pieces of passing debris to employ as cover.

Rumphius worked daily on these volumes and his administrative duties while also contributing generously to the Republic of Letters, sending prize seashells and new species descriptions to his many correspondents across Europe. One scientific society made him a member based on his letters alone, declaring him *Plinius indicus*, "Pliny of the Indies," after the great Roman naturalist Pliny the Elder. (Who died in the volcano that destroyed Pompeii while trying to rescue a friend and his family. Some speculate the exotic shell collection found in the rubble belonged to him.)

Like Drake's, Rumphius's writings reveal a growing disdain for the dilettantes back home amassing exotic seashells and other natural wonders as showpieces. He sounded an early warning of mass overconsumption with the allegory of a diamond mine: "Today's covetousness and pomp wants (diamonds) in such profusion, that every Tradesman is capable of wearing one, so that there is no time for the stone to become old, while before it might have had 1 or 2 thousand years to rest in its mine."

As he adopted Indonesian culture, Rumphius also came to accept its

ethos that natural objects held special power only if found or given—
not when sold. In a section on Native belief and naturalia he stressed
that "they will only be lucky for the one who found them, or the person who received one as a gift, but not for someone who buys them."

The superstition did not prove true for Rumphius. In the spring
of 1670 at age forty-three, he went blind from what is now known as
glaucoma. He blamed his affliction on his time in the tropical sun.
He wrote that the "terrible misfortune . . . suddenly took away from
me the entire world and all its creatures . . . compelling me to sit in
sad darkness."

Still he pressed on with his work. I imagine his understanding of
seashells deepening, like Geerat Vermeij's, with the touch of every rib
and knob, the feel of every perfect gloss and injured repair. His senses
led him to shell properties that modern scientists would not publish
on for centuries, such as their acoustics. Being a keen listener helped
him understand that a large, spiraled shell held to the ear simply
amplified its surroundings:

> They got the name Kinkhooren (whelks) because they cause
> a kinken, or rustling or soughing sound, when you hold their
> mouth against your ear, and the common people (buying and
> selling shells in Europe) persuade each other, that this is a sure
> sign of it being genuine, because you hear therein the soughing
> of the open sea: But there is something wrong here, because one
> does not become aware of this soughing on every day or at all
> hours, but only during the day, when the air is moved by wind,
> rain, or the voices of people; whilst during the silent nights, one
> will not notice any rustling, even though one has the genuine.

Four years later, Rumphius, Susanna, and their daughter were out
for a walk in Ambon's Chinese Street on a moonlit evening in February 1674 to enjoy the Chinese New Year festivities. The women
had popped into a house to visit while Rumphius stayed outside. The
ground began to shake. People, and then the stone houses on Chinese

Street, were thrown to the ground in one of the worst earthquakes in Ambon's earthquake-prone history. The quake killed 2,322 souls on Ambon, including Susanna and the couple's daughter when they were crushed by a wall.

Rumphius's careful biographer Beekman, who also spent part of his life in Indonesia and shared his subject's love for tropical nature, found that, at least in his writing, Rumphius expressed more frequent anguish over his next misfortune. In 1682, under some sort of coercion that likely included threat of losing the position that made life in Ambon possible, Rumphius was forced to sell his own beloved seashell collection—almost all the shells he'd found in twenty-eight years on the island—to Cosimo III de Medici, grand duke of Tuscany, for his curiosity cabinet. By then, Rumphius was well known in Europe as Pliny of the Indies. He'd sent seashells as gifts and written letters about the specimens he was working to describe; his collection had become legendary. "This was not a deliberate sale beneficial to both parties," Beekman wrote. "What vexed Rumphius was the compulsion, backed by some kind of high-pressure method, to relinquish part of his life to a stranger for money."

Misfortune rained on, like the monsoon. When Rumphius became blind, he'd had to start his *magnum opus* on flora, called the *Herbarium*, all over because he'd written it in Latin and there was no one to translate. After three decades' work and with the help of his son, Paulus, taking dictation in Dutch, the *Herbarium* was nearly finished. Fate was not. In January 1687, a fire tore through Ambon, destroying all of Rumphius's drawings.

He oversaw, without sight, their reconstruction by Paulus and other draftsmen. After laboring five more years, they finished the text and illustrations for half the *Herbarium* and sent volumes one through six to Amsterdam on a ship called the *Waterland*. It was sunk by the French. Everything on board was lost. In a rare stroke of fortune, Ambon's new governor, also a naturalist, had ordered a copy made for himself before he sent the volumes onward from Batavia, and the work was salvaged.

Rumphius finally completed what would become one of the world's masterpieces of botanical literature and shipped it successfully to Amsterdam in 1696. But his bad luck did not abate; rather it traveled to the boardroom of the Dutch East India Company. The directors, known as the Heeren XVII, or the Gentlemen Seventeen, quashed its publication, fearing the *Herbarium* held classified secrets of Ambon that would aid their competitors. Spices, after all, represented a core part of the business.

Incredibly, Rumphius pressed on, determined to finish his seashell masterpiece. *The Ambonese Curiosity Cabinet*, with its exquisite drawings and pages of poetic description for each species—his "lyrical metonymies" on the lives of mollusks—was published in 1705, three years after Rumphius died. The botanical manuscript remained locked in the archives of the Dutch East India Company for nearly four decades more before it was finally published in 1741.

When Cosimo III died in 1723, he owned one of the only three known Precious Wentletraps in Europe. Rumphius, the man who probably collected it, died on Ambon without his shells.

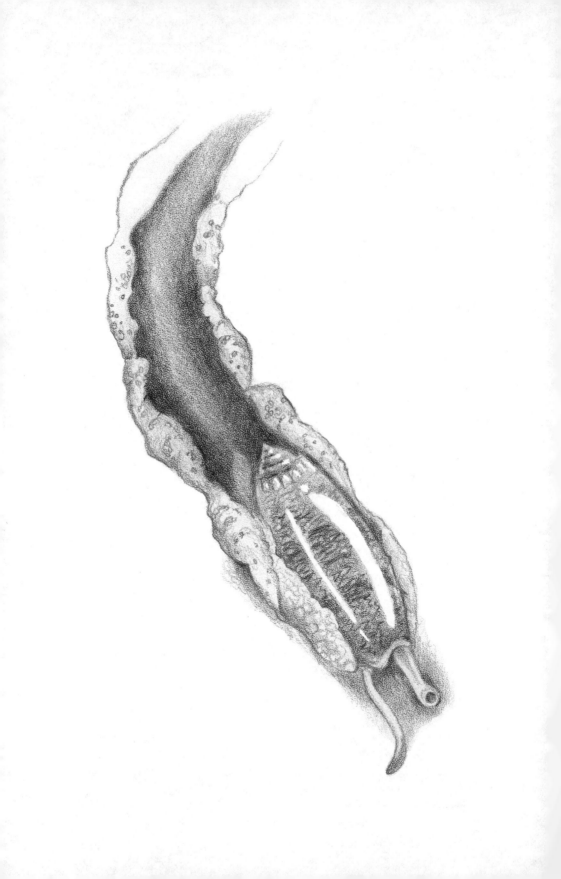

Seven

~~~~~~~~~~~~~

AMERICAN SHELLS

THE LETTERED OLIVE
Oliva sayana

In January 1826, as the United States launched a yearlong celebration of the fiftieth anniversary of the Declaration of Independence, the American naturalist Thomas Say captained a keelboat called the *Philanthropist* down the Ohio River, alternately breaking ice and rowing with an odd menagerie of travelers in search of the country's elusive ideals.

Forty scientists and educators had left the nation's intellectual capital in Philadelphia to board the 85-foot vessel at the Allegheny River in Pittsburgh. They were underway for New Harmony, Indiana, to help build a model society around social and economic reform. Every aspect of the utopia was to be infused with science, from the school to community decision making. The *Philanthropist* listed with the weight of its passengers' books, specimens, and laboratory equipment. The press dubbed it "The Boatload of Knowledge."

Say had not intended to serve as captain. Nor had he planned to stay on at the utopia. Working tirelessly on his own idealistic mission to help legitimize American science, Say had merely agreed to a detour to New Harmony on his way to a collecting expedition.

Say had helped found the Academy of Natural Sciences in Philadelphia in 1812, amid a fervor of scientific patriotism to collect, name, and describe the new flora and fauna being discovered by the day. The first American research institution devoted solely to natural history, its modest founding collection included a handful of shells. INSECTS, as he once wrote in all caps to a correspondent in Europe, were Say's great passion. Mollusks came second. Nonetheless, he unearthed shells everywhere he went. Say would publish the nation's first research paper on conchology and first book on U.S. mollusks and their shells: *American Conchology.*

Six weeks after pushing off from Pittsburgh, Say arrived in New Harmony exhausted by the journey slowed by ice, but exhilarated by a young art teacher who had buoyed his spirits and rowed by his side. Lucy Way Sistare was a gifted artist who'd studied illustration with some of the great naturalists and illustrators of the day, including John James Audubon. She'd joined the boat with her mentor Marie Fretageot, a French educator relocating her school to New Harmony.

Sistare was twenty-five, witty and warm, a good sport. The icy river and the schedule required vigorous rowing, and Say and Sistare were two of the few on the boat who enjoyed it—ostensibly because it kept them warm in the freezing January, perhaps too for the companionship.

By summer, Say had indefinitely postponed his collecting trip. He poured himself into creating New Harmony's school, where children would learn natural history "in its most comprehensive sense." He also planned a library and printing press with colored plates, where the community would publish scientific books and journals.

A year after they had boarded the *Philanthropist,* Thomas and Lucy sneaked off to the courthouse in Mt. Vernon, Indiana, to marry in a civil ceremony. That is how America's first book of conchology was printed in a failed utopia, and stunningly illustrated by a young woman's hand.

THE ERA OF curiosity cabinets and elaborate royal shell displays didn't really come to an end in Europe. Rather, art and natural history collections transitioned to museums and the public trust. Revolutions created museums and colonialism endowed them. Shells in cultural artifact and species lots—groups collected from the same species, time, and location—soared as naturalists, explorers, and colonists pushed farther across the seas. Cosimo III's and the Medici family collections, surely including Rumphius's beloved shells, wound up in the Uffizi Gallery of Florence. The Octagonal Room with its shell-decorated dome displayed the first collection of curios for the public.

After Louis XVI and Marie Antoinette went to the guillotine in 1793, the royal botanical garden in Paris, the Jardin des Plantes, was reorganized as the Muséum National d'Histoire Naturelle, to be run by twelve professors in twelve scientific fields. The naturalist Jean-Baptiste Lamarck was appointed to oversee the "inferior animals" such as insects, worms, and mollusks. He would name this group without backbones the invertebrates; now understood as the vast majority—97 percent—of all animals.

Lamarck, a botanist, is less well remembered as a great shell collector who worked for years on a book he called the "Elements of Conchology." When he became museum professor at age fifty, Lamarck still believed in the accepted wisdom of a constant world; God created each species to live in its permanent perfection. Classifying the shelled animals changed his mind. No other animal group left as much evidence of evolutionary change. He shifted his conchology work into his famous *Philosophie Zoologique*, in which he sketched the first evolutionary tree. Where Linnaeus had identified two classes of invertebrates—insects and worms—Lamarck distinguished ten. They included the *mollusques*, with their own special branch.

Across the Atlantic, early American naturalists like Banister had long shipped key specimens to Europe. In Philadelphia, Say's great-grandfather, the botanist John Bartram, packed up what became known as Bartram's Boxes with American seeds and cuttings, and sometimes "curious stones figured with sea shells." The Quaker collector explored the hazy peaks of the Alleghenies and drew a detailed map in iron gall ink, rivers branching like his cursive handwriting that marked where he found "sea shels in stone" and "limestone & sea shels in it." Benjamin Franklin scribbled a note on the back: "Mr. Bartram's Map very curious," he wrote to a friend as he described the seashells layering inland strata and fused into mountaintops. Franklin imagined earlier eras of life, if not a half-billion years of mollusk evolution: " 'Tis certainly the Wreck of a World we live on!" he wrote.

American plants and bugs and seashells had so far been named by scientists such as Linnaeus and Lamarck, who had never seen them in their natural habitat. Now, American scientists were trying to build their own authority to describe the species rooted in and crawling across the soils and waters of the young nation.

The colonists began to set up academic societies and museums to burnish national identity and their versions of history and scientific prowess. South Carolina colonists established America's first museum in Charleston in 1773 with "many specimens of natural history," open only to its members. The colony was by then known for the most sensational fossils yet found in the New World. Enslaved Africans had unearthed mammoth teeth while digging in a swamp near Charleston in 1725, and immediately concurred that they were the grinders of some type of elephant. It would be more than eighty years before Georges Cuvier, Lamarck's rival at the Paris museum, confirmed that "les nègres" had correctly identified a fossil elephant species before any European naturalist connected extinct mammoths to living elephants.

Franklin and other founders envisioned Philadelphia as an "Athens of America," based on scientific inquiry and social betterment. The city became home to America's first hospital; the nation's largest lending library also holding some of Bartram's curious seashell fossils; Franklin's American Philosophical Society and what became the University of Pennsylvania; and Bartram's 5-acre botanical garden on the Schuylkill River. The artist and showman Charles Willson Peale moved to the city in the momentous year of 1776 and painted portraits of the men of the Revolution. Peale founded the country's first successful public museum of natural history at his home on Third and Lombard Streets: a fantastical hybrid of the pompous curiosity displays of Europe and the natural history collections they were becoming.

THE FATHER OF American conchology was born in Philadelphia in 1787 to a well-known Quaker family. Thomas Say was the son of a doctor and the great-grandson of John Bartram on his mother's side. None of these privileges could protect young Say from the tragedy of yellow fever that ravaged the city in his childhood.

When the epidemic broke out in 1793, federal officials abandoned Philadelphia. Many wealthy families fled as the city descended into chaos. Say's family remained so his father could try to save lives with purges and bloodletting. When he was six years old, Say lost both his mother and his older sister in one of the worst epidemics in American history. More than five thousand died in the city in just four months; about one in ten Philadelphians.

Say grew close to his mother's uncle, William Bartram, who took him exploring and inspired his love for beetles, butterflies, and other insects they collected together. Say also collected with the Peale children, who lived nearby. But he had far more interest in live animals than in those stuffed and shelved at his neighbors' museum. As a child he would show up at the dinner table with his pockets full of little creatures, writes his biographer, Patricia Stroud. A wee snake sleeping in his warm pocket "would be aroused by the clatter of knives and forks, and raising its head and poking its neck forward, would send the family in terror from the board."

By the time he was fifteen, Say had given up both school and Quaker meetings, though he remained devout like the Bartram family in social consciousness and "disdain of riches." His father tried to set up Thomas and his brother Benjamin in an apothecary business, but neither son had the knack. "The thrift of trade, and the art of buying and selling, were either disdained by him or neglected," wrote Say's friend and fellow naturalist George Ord. In contrast, while working on bugs and shells and establishing the Academy of Natural Sciences, Say was so attentive he often forgot to eat. His gaunt lankiness was part of his handsomeness, closer to beautiful with tousled dark hair, hazel eyes, an angular face. Ord lamented that Say put so much time into launching the institution and helping other naturalists that he couldn't begin his own work until midnight, and often kept at it until dawn. Ord claimed that this "injudicious application

to study," along with Say's ascetic habits like eating only a bare minimum of food, ruined his physical health and led to his early death.

Say, always uncomfortable with his family's social status, was
the only wealthy "friend of science" among the academy's founders, who included immigrants, chemists, and a distiller. They
envisioned a research enterprise more democratic than Franklin's
American Philosophical Society, which welcomed only wealthy and
socially prominent members. No one else had a path to science.
Even connected men like Say had no academic track for studying
animals; the university taught natural history such as botany as it
related to medicine. The founders also wanted to distinguish themselves from Peale's showy commercialism. Their first big debate was
about whether they should ban all mention of religion. In a com-
promise, they agreed to admit all "lovers or cultivators of science,"
while warning, "it is not for sectarian or political purposes that the
academy is instituted, therefore no person admitted as a member
ought to introduce the weapons of dogmatic reproof, persecution,
or proselytism."

As the first full-time curator, Say was so devoted to the collections
and the duty he felt to publish research that he slept in one of the
academy's rooms, tossing a blanket over a horse skeleton to make a
tent. He was quoted saying he wished for a hole in his side to deposit
his food and save time. Along with throwing off yokes of social class
and religion, the founders took special pride in making the academy
a center for describing American species, rather than shipping them
across the Atlantic. Say wrote a friend of this patriotic obligation,
"directed to the honour & support of *American* science . . . not so
much however in a *personal* as a *national* sense."

AS THE FOUNDERS of the academy had hoped, their ideals for
a democratic American science institute drew donations of money,
books, journal subscriptions, and specimens from admirers at home
and abroad. Eminent European scientists including Lamarck became
members. None would influence the academy's future—or Say's—

like the geologist William Maclure, a tall Scotsman with red hair and cheeks who was on fire with ideas about social reform.

Like many gentlemen scientists then, Maclure straddled disparate worlds—constantly traveling between Europe and America for his mercantile trade, finding rocks to climb and classify along the way. He made a fortune before he turned thirty-five, which enabled him to relocate to Philadelphia to crusade for his causes of science, democracy, and liberating the working classes. "Since the plots and conspiracies of the great and privileged orders, against the peace, comforts, and happiness of the industrious productive classes have succeeded in Europe," Maclure wrote, "I am mortified beyond measure . . . that I belong to the species and am forced for consolation to extend my views across the Atlantic, that I may be an eyewitness to the prosperity of the United States, and enjoy the gratifying sensation of beholding man in the most dignified attitude which he has yet attained."

He saw science and education for all children as keys to human betterment. He found allies in the academy's scientists and threw himself into their work. He became the academy's most generous donor. He established its journal; filled its library with 1,500 new volumes of books; and funded expeditions, including the institution's first, into the coastal wilds of Georgia and Florida.

Say joined the trek that followed in the bootsteps of his great-uncle William Bartram, who'd described the region in his *Travels*. Maclure secured permission from Spain to enter Florida, not yet an American territory, with Say, Ord, and Titian Ramsay Peale, then seventeen and already a talented artist like his father. Say thought the watery realm would be "the promised land" for collecting. They explored Georgia's barrier islands to search for mollusks and crabs. In the evenings, Peale later recalled, they enjoyed a nighttime sea "filled with multitudes of beautiful Medusae and other Molluscous animals, their phosphoric light giving brilliancy to the water."

On Cumberland Island they visited a four-story "shell mansion" built of tabby, a handy seaside concrete made of oyster shell, lime, and sand. Aesthetic-minded Peale found it "incongruous, almost ludicrous" to sit on elegant imported furniture in a vast dining hall, a sumptuous meal before them, and see "oyster shells sticking out of the walls in every direction."

In Florida they sailed through dense dolphin pods and dug into great shell mounds near St. Augustine, finding Native tools and a conch they believed extinct. The concept that species could die out was gaining acceptance thanks to Georges Cuvier at the Paris museum and his work on the bones of woolly mammoths and American mastodons—also displayed by Peale at the Philadelphia Museum. Cuvier ridiculed Lamarck's evidence for evolution; Lamarck refused to accept Cuvier's for extinction. The two theories would later fit together like mastodon bones.

Say was crushed when the party had to leave Florida early because of the violence escalating in the first Seminole War—"this most cruel & inhuman war that our government is unrighteously & unconstitutionally waging." But after he returned to Philadelphia, he was tapped for the most important expedition of the day: Major Stephen Long's 1819 trek up the Missouri River to explore the American West for settlement. As zoologist on the expedition, Say collected thousands of insects and described numerous American animals including the coyote and gray wolf. It was on this trip that Say and Titian Peale mapped Cahokia and contemplated the once-great city's magnitude. Say continued trying to interpret shells in three realms: living, extinct, and wound with the lives of Indigenous people.

During these years, Say published the first scientific paper on American mollusks, which appeared in the *British Encyclopedia*. He made the case that, contrary to Linnaeus, the study of mollusks should be organized around the living creatures—rather than their shells.

"The animals that inhabit them should guide us in our researches," Say wrote. "They alone are the fabricators of the shell, and the shell is only their habitation, to which they give the form, the bulk, hardness, colors, and all the peculiarities of elegance we admire."

~~~~~~

SAY'S WRITINGS DON'T make clear where he first saw the Lettered Olives stunningly illustrated in his *American Conchology*, the nation's first book of shells. But the glossy-smooth cylinders, moving with linear purpose through the wet sand like slow-motion bullets, are common sights on the beaches of Georgia, Florida, and the Carolinas. Lamarck christened them Lettered Olives, for their brown hieroglyphics rendered on cream-colored cylinders.

The Lowcountry coast was already known for seashells, which also left their mark on the region in another way: tracing a spiritual connection between enslaved people and shells. In many African traditions, seashells had symbolized immortality and transition from this life to the next. Since the flood-prone coastline was undesirable property then, families could bury enslaved loved ones near the sea, "so that their souls might easily return to Africa," according to the Gullah Geechee tradition with African roots in the region.

The graves were often covered in seashells, a practice in many other African-American burials around the country. "The shells stand for the sea," the late Gullah Geechee folk singer Bessie Jones recalled in the Sea Islands of Georgia. "The sea brought us, the sea shall take us back. So the shells upon our graves stand for water, the means of glory and the land of demise."

In South Carolina, Say's friend and correspondent Edmund Ravenel, a physician and a plantation and slave owner, was also a religious man who saw seashells as evidence of the hand of God in nature. Distinguishing the American Lettered Olive from Lamarck's *Oliva litterata*, Ravenel named the American species *sayana* after his friend and fellow conchologist.

Ravenel had homes in Charleston and on Sullivan's Island, a three-mile beachhead at the mouth of Charleston Harbor that is home to Fort Moultrie. Walking the beaches in 1828, he almost certainly befriended a soldier and fellow nature lover named Edgar Perry, who was stationed at the base. The dark-haired soldier was eighteen but had enlisted with a fake age and name after being disowned by his father back in Boston. His real name was Edgar Allan Poe.

Lowcountry citizens have made much of their famous resident's yearlong stint on Sullivan's Island, naming streets Poe Avenue and Raven Drive and attributing Poe's tragic poem "Annabel Lee" to an imagined gothic romance between Poe and Ravenel's daughter. Legend goes that Poe fell in love with a fourteen-year-old Annabel, and the disapproving Ravenel secreted her at a home in Charleston, where she died of yellow fever. But Ravenel didn't have a daughter named Annabel or Anna, nor even a teenager, when Poe was at Fort Moultrie. The first of Ravenel's children was a year old then.

Poe's true enchantment on the Carolina beaches, if he had one, was more likely seashells. His short story *The Gold Bug*, set on Sullivan's Island, follows William Legrand, a naturalist often "in quest of shells," who is bitten by a gold-colored bug. Poe biographer Arthur Hobson Quinn found it "probable" that Legrand was inspired by Ravenel. The story revolves around a shell with a secret message that must be deciphered, a common belief surrounding the mysterious hieroglyphics of the glossy Lettered Olive shell.

Just before Poe turned 30, broke, often drunk, and facing poor sales of his novel *The Narrative of Arthur Gordon Pym*, he agreed to condense an existing book, the *Manual of Conchology*, by his friend the English writer Thomas Wyatt, into a less expensive version for students and the general public. *The Conchologist's First Book*, a scientific guide to seashells, would be his only book that sold well during his lifetime.

As Wyatt recounted the story, his *Manual* proved too costly for popular use, and his publisher, Harper's, refused to bring it out in cheaper form. He was determined to publish for a broader audience, but he had to make it different enough that he wouldn't be sued for copyright infringement. He hired Poe for $50 to change it just enough, and Wyatt sold Poe's version at his public talks. The deal backfired on Poe later, when he was accused of plagiarizing two other European seashell guides. His biographers don't doubt Wyatt's version: Quinn, wrote that *The Conchologist's First Book* "was not entirely a piece of hack work."

"It is grimly ironic, however," Quinn wrote, "that it is probably the only volume by Poe that went into a second edition in the United States during Poe's lifetime."

Stephen Jay Gould came down on the side of plagiarism, blaming Wyatt as much as Poe. But he lauded Poe for conceiving the book's "ruling feature"—"giving an anatomical account of each animal together with a description of the shell which it inhabits," as Poe wrote. *The Conchologist's First Book* was one of the earliest to contrast malacology and conchology for the general reader, although these pursuits didn't branch into distinct fields and use for another hundred years.

The prose is not Annabel Lee gothic. But Poe's description of a living Lettered Olive could read creepily if you use your imagination:

> Oval, involute, mouth somewhat thin at its edges, prolonged to the two angles of the branchial aperture in a tentacular band, and, anteriorly, by a long branchial tube; foot very large, oval, subarticulated, with a transverse cleft anteriorly; head small, with a labial proboscis.

IN 1824, WILLIAM Maclure, the red-haired Scottish geologist, met a fellow wealthy reformer, Robert Owen, who was working a grand social experiment at his company's mill town of New Lanark, Scotland. Owen had experienced the human misery rising with the smoke of factories as a child, apprenticed to drapers in England. He later moved to Manchester, formed a partnership with a mechanic to make cotton-spinning machinery, and imported the first Sea Island Cotton from the United States; it is quite possible the bales were picked by enslaved people on Ravenel's plantation. By nineteen, Owen employed five hundred workers. He built nursery and primary schools, libraries, and other educational opportunities for their children at New Lanark—such a novelty that it became a tourist destination. Maclure was among the admirers who visited from across the world. Owen told him he wanted to build an entire community as a model for reform, ideally in America.

Owen's flawed dream brings to mind that of Silicon Valley's tech titans today: He convinced himself that the same technology filling his pockets could lead to a perfect society—if only deployed with humanity and scientific rationality. He found his proving ground in New Harmony, a neat Lutheran settlement on the Indiana side of the Wabash River. He found his educational partner in Maclure, who became swept up in Owen's vision that America could avoid the profit-driven social ills infecting Europe by building "communities of cooperation" where people would grow their own food and manufacture their own goods in state-of-the-art factories. Boys *and* girls would have the best education in the world, infancy through university, for free—at least white boys and girls. Black families were excluded despite the antislavery work of many New Harmony members.

Thanks to Maclure and the French educator Marie Fretageot, the school was always the strongest part of New Harmony. Maclure had for several years invested in the school reform ideas of the Swiss educator Johann Pestalozzi, building from Rousseau's adage to "let him not be taught science, let him discover it," and delivering object lessons on the likes of seashells, seeds, and minerals. Maclure financed Pestalozzian teachers to found schools in the United States, including relocating Fretageot from Paris to Philadelphia to open her school for girls. One of her pupils was Lucy Way Sistare, a skilled artist and apprentice teacher.

Maclure, who had been named president of the Academy of Natural Sciences amid his considerable donations, now funded Fretageot's move to New Harmony—along with several academy scientists. When the forty educators and scientists and their families boarded the *Philanthropist* and launched downriver to New Harmony, Ord and many others in Philadelphia never forgave the poaching.

Say packed none of his books or collections. He never planned to leave Philadelphia or the academy, but joined Maclure on the journey as a curious detour en route to their next planned trek, to Mexico. Say could hardly wait for the expedition. But meeting Lucy Sistare changed his plans. He would never live in Philadelphia again.

OWEN EXCELLED AT expressing his dream of a New Moral World, but he had little knack for the details. Marie Fretageot, a force, had the management skills. It was Fretageot who convinced Maclure, her close friend until her death in 1833, to recruit scientists to New Harmony and invest in its school. It was she who kept the school thriving after the larger experiment crashed. It's easy to imagine a different outcome for New Harmony had she been given the opportunity to manage the entire community. Fretageot also recognized the genius in Lucy Way Sistare.

Sistare was born in Connecticut in 1801, one of ten children of Joseph and Nancy Way Sistare. She grew up in New York City until her parents sent her and two sisters to Fretageot's experimental school for girls in Philadelphia, popular among the Quaker families that could accept the lack of religious studies. Sistare thrived with the Pestalozzian curriculum, which included hands-on natural history at Peale's museum and scientific lectures from members of the academy. The French naturalist and illustrator Charles Alexandre Lesueur taught drawing three times a week; Fretageot also arranged for Sistare to study with Audubon. By the time they boarded the so-called Boatload of Knowledge, Sistare served as Fretageot's assistant. Fretageot assigned her to oversee art instruction at New Harmony.

As Sistare applied herself to that work, Thomas Say was falling in love with her—and also the mission of New Harmony's school. He had imagined a detour. Instead, he could not help but invest his energy in "the singular spectacle which this place presented," as he would later write. "We were involved in the vortex of experiment to realize the dreams of perfection in human association, which had been so confidently and imposingly promulgated."

Too confidently, it turned out. New Harmony's ideals began to crack almost immediately, as Owen's grandstanding proved utterly out of proportion to his management abilities. Nearly a thousand people showed up the first year, many of them down-on-their-luck drifters who wanted to move in—but had no farming, manufacturing, or other skills. The sawmill and watermill lay idle. The cashless trade system didn't work. Owen didn't help matters when he used the fiftieth anniversary of the Declaration of Independence on July 4, 1826, to give an incendiary speech, "A Declaration of Mental Independence." He called for abolishment of "the trinity of the most monstrous evils" of private property, organized religion, and marriage. Owen lost the support of admirers. Lucy Sistare's mother was among many who wrote letters of alarm.

As the community collapsed, however, Maclure, Fretageot, Say, and others worked to save its most promising dream. Fretageot had created a first-rate school that enrolled four hundred children its first year. Amid great rancor, Maclure ultimately convinced Owen to transfer ownership to community members in three separate sectors—education, farming, and manufacturing—and disentangle the school. Say and Sistare were among the educators and scientists who stayed on behalf of the children of New Harmony, a beacon of nature and science education in an outpost on the Wabash River.

Fretageot, in an argument that echoed Judith Drake, often complained that the scientists' research cut them off from the community. They were "shut up in their cabinet" with "Fish, Shells, Birds, Drawings, perfectly useless to the happiness of mankind," as she wrote to Maclure. "Calculate the expense they carry with them and tell me what benefit will arise from their work to the present and even the future generations."

But she had more appreciation for Say, who labored in the community and taught science to the children. He was somewhat stuck; while he had called for his collections and research papers from Philadelphia, they had been delayed, causing him to postpone his own research. Maclure kept delaying the trip to Mexico, too.

Say and Sistare married in January 1827, a year after they'd arrived in New Harmony. They skipped the formalities of a wedding and the newlywed portraits popular in the early nineteenth century. No known portrait exists of them as a couple. They sat for a pair of modest silhouettes cut by Titian Peale. The humble profiles contrast dramatically with the oil portraits of the Dutch scientific virtuosos across the Atlantic, who almost seemed to try to outdress their sea-shell props.

Say's books and specimens finally arrived at New Harmony, along with a printing press purchased by Maclure, and the couple set out to write and illustrate *American Conchology*. The title page reflects their philosophy of nature education with a line by the eighteenth-century British poet Edward Young: *Read Nature; Nature is a friend to truth.*

IN THE 1890s, a Des Moines biology teacher named Julia Ellen Rogers wrote a gentle treatise on nature study as part of a statewide effort to make its "revelations of truth and beauty" a fundamental part of education in the public schools across Iowa. Madame Freta-geot, the Says, and their New Harmony School came seventy years too soon for the American nature-study movement that swept the country during the rise of the Progressive Era at the turn of the century. Its advocates stressed direct contact with nature as the best foundation for understanding science and natural history. Moreover, if children dug in the soil, studied the seasons, and closely observed the lives of animals—mollusks and their shells, frogs and tadpoles, bird parents raising chicks—they would develop good character and an ethos to care for the world around them.

"To feel an intimacy growing up between yourself and the world of plant and animal life all about you is to feel also an intellectual warmth and joy that is unlike anything felt before," Rogers wrote as she introduced the lesson plans on soils, plants, insects, and school gardens. "A feeling that binds you to nature by cords that strengthen every day."

"Let the books alone for a while," she advised her fellow teachers. "Come out into the fields and woods."

And for Rogers, to the sea. Within a year of the Iowa report, she joined an orbit of nature study and nature writing circling the horticulturalist-philosopher Liberty Hyde Bailey at Cornell University in New York. Within five years, she would travel to the three American seas and publish the first popular book of American seashells of the twentieth century. *The Shell Book* of 1908 endured for half that century.

Rogers visited the pearl-button factories of Iowa, grizzled oystermen on Long Island, clam diggers at Cape Cod, conchologists and collectors in California, scientists at Woods Hole Biological Lab, the bulging shell drawers of America's natural history museums, and the mollusk-reigning islands of Florida's Gulf coast to write what became the best-selling shell guide of her time. *The Shell Book* brought the world of seashells to Americans during the national zeal for nature as a hobby that preceded World War I. The book also helped spark a love for mollusks and their shells in a key generation of marine scientists. If Thomas Say is the father of American conchology, Julia Ellen Rogers is its mother. That's how one of her many shell-collector fans described her in a remembrance after Rogers died at her home in California in 1958 at the age of ninety-two.

I DISCOVERED ROGERS and *The Shell Book* in the cavernous library of an estate overlooking Lake Michigan in Traverse City, where I had the gift of a few quiet writing days. The library at Pine Hollow is stacked with nature writing both well known and obscure. I first read the century-old book in a place that was far from the sea, but hearkened back to Rogers' midwestern farm roots.

In the coming years, I came to understand the extent to which Rogers and *The Shell Book* influenced twentieth-century American malacology and indeed marine science. Among many scientists inspired by the book, the marine biologist Ed Ricketts twice laments in his journal that he didn't have *The Shell Book* along during his Gulf of California expedition with John Steinbeck, immortalized in *The Log from the Sea of Cortez*. The best-known shell man of the twentieth century, the scientist-popularizer Tucker Abbott, told how he found *The Shell Book* in the Montreal public library as a teenager, and relied on it to build his massive first and only shell collection. Abbott's collection grew so large that he and a friend opened a science museum in his parents' basement—"The Boys' Museum of Westmount." Another summer, as Abbott told the story, the boys pedaled their bicycles 2,000 miles on a collecting trip. He came home determined to become a malacologist, but he'd flunked freshman year of high school. "I spent too much time outdoors looking for shells or weird insects," he once told a reporter. Julia Ellen Rogers would not have been surprised that he made it into Harvard.

Over those years, I also tried to piece together Rogers the person. I slowly tracked down members of the famous family she was born into; the nature-study progressives she joined during her years in New York; and the photographer who spent weeks with her on the Gulf but never took one picture of her. I wondered if that was because he didn't find her pretty. I was so excited one night to open an email from her elderly grandniece to finally get a look at Rogers beyond posed portraits. I saw the urgency of the cause for which fellow students remembered her in the Iowa State annual, calling her "Pedagogue and Dress Reformer"—the movement then trying to emancipate women from heavy Victorian corsets and skirts. The blurry family shots show her weighed down in substantial dresses, both on a sunny day at the beach and at the countryside with her bicycle.

I mostly saw her verve: in the five bangles widespread up her forearm in the turn-of-the-century beach picture; in the towering bas relief goddess that loomed over the fireplace in her two-story Craftsman house in Long Beach, where she lived for fifty-three years, a half mile from the Pacific Ocean.

ROGERS WAS A proud granddaughter of the radical abolitionist Nathaniel Peabody Rogers of Concord, New Hampshire, who gave up his law practice to edit the anti-slavery newspaper *Herald of Freedom*. Her grandfather was also known for his poetic nature writing about the White Mountains, which he published in the persona of a well-known natural landmark, "The Old Man of the Mountain" at Franconia Notch, New Hampshire.

The next generation, Julia Ellen Rogers's parents, were socially committed teachers who revered nature over the church that had rejected the elder Rogers for his anti-racism work. Julia was born in Illinois in 1866. Her father, Daniel Farrand Rogers, bought a wild tract of prairie near Iowa's Raccoon River, built a house, and surrounded it with larches, "the glory of the countryside," to raise his seven children amid trees. The family moved from Illinois to the Iowa home they called The Larches when Julia was three. She began teaching in country schools as a teenager and went on to formal teacher training, then a bachelor's from Iowa State, before landing as a high school principal in Minnesota. Educational beadledom didn't suit her. Rogers returned to the classroom to teach high school biology in Des Moines. But as she worked on the nature-studies plans for Iowa's schools, she was already pursuing her nature-writing ambitions. She began her second career in double byline with her father in an adventuresome magazine story about their monthlong camping and climbing trip in the Bighorn mountains of Wyoming. She rode horseback in her petticoats without luxury of a sidesaddle.

Rogers seems to have gotten to know Liberty Hyde Bailey through her younger sister, Mary, who went to Cornell for college and remained to work with Bailey while he cultivated nature studies as a discipline and a movement. Bailey, a son of farmers, was dismayed about the loss of agrarian and environmental values as Americans flooded into the polluted cities in the depression of the 1890s. He believed human character was shaped by how society used natural resources. He optimistically envisioned America building out in three phases: collecting; mining; and then, regenerative production—the latter an enlightened stage where we will arrive, "wasting little, harming not."

"At first man sweeps the earth to see what he may gather—game, wood, fruits, fish, fur, feathers, shells on the shore," resulting in a character that's strong and self-sufficient but also dogmatic and superstitious, Bailey wrote in his book *The Holy Earth*. In the next stage, humans drill beneath the surface for the likes of gold and coal. "In both these stages the elements of waste and disregard have been heavy." But finally, "we begin to enter the productive stage, whereby we secure supplies by controlling the conditions under which they grow, wasting little, harming not," raising crops and animals and planting fish in lakes and streams "with precision." Bailey had a vision that university extension could help build this enlightened character of "earth righteousness."

The outreach began in the New York public schools. Bailey hired the scientific illustrator Anna Botsford Comstock as Cornell's first woman faculty member to lead the nature study program in which she was volunteering. Over the next three decades, Comstock's efforts would touch tens of thousands across the nation: children in Junior Naturalist Clubs; teachers in summer nature-study courses; and many more Americans in home-based programs taught through the mail.

Even as they eschewed textbooks, nature writing was a key part of the movement. Comstock, Bailey, and others wrote charming nature-study leaflets to draw teachers and parents across the nation to brooks and bird-watching. Comstock collected the classic essays in the *Handbook of Nature Study*, which became the best-selling book in the history of Cornell University Press. It is now in its twenty-fourth edition.

Julia Ellen Rogers arrived at Cornell in 1899 to work with Bailey, who encouraged her writing and suggested her first book, *Among Green Trees*. She wrote it in her two years at Cornell while teaching with Comstock in the summer nature-study program and completing her thesis on how to teach nature studies through winter—expressing remarkable sympathy for cockroaches.

Among Green Trees expounds on the cultivation and lives of trees, from how to read their knotholes to a detailed description of how foliage droops to sleep—a phenomenon scientists reported on for the first time more than a century later. The book opens with an ode to trees by her famous grandfather, Nathaniel Peabody Rogers. He called them the architecture of God.

IT WAS HER fellow nature writer Anthony Dimock who suggested Rogers do for seashells what she had done for trees. Say's *American Conchology* and the classic scientific books that followed were locked up in glass library cases and out-of-date. Newer treatises were too expensive and technical for the weekend shell lover. Dimock promised to show Rogers islands populated by more mollusks than men, in the wilds of southwest Florida where scallops massed by the thousands and glorious shells paraded across the sand flats on the backs of their living makers.

After Cornell, Rogers had joined the staff of a new magazine edited by Bailey, *Country Life in America*—"for the Home-Maker, the Vacation-seeker, the Gardener, the Farmer, the Nature-teacher, the Naturalist." The country life movement, which President Theodore Roosevelt would codify with a commission headed by Bailey, commingled reformers set out to improve conditions in rural America and city romantics enthralled with growing chrysanthemums. Published by Doubleday and cross-marketed with the company's nature books and clubs, *Country Life* catered to the flower growers. Rogers appeared in the early volumes with better-known and older essayists including John Burroughs and Dimock; she eventually became editor of the nature desk.

Anthony Dimock was the child of a preacher who had scrimped and saved and—to his son's everlasting horror—accepted parishioners' donations to send the boy to Phillips Academy in Andover, Massachusetts. Determined to overcome the humiliation of hand-me-down clothes and pennies from village boys in his father's churches, Dimock eschewed college for Wall Street. He was a trader before his twenty-first birthday, and "dominated the gold market of the country at the age of twenty-three," according to the *New York Times*. He made spectacular fortunes as a young man, and he lost them spectacularly.

By the time Dimock met Rogers, he had given up finance for outdoor writing and his gun for a camera. He had been humbled: by a fourth bankruptcy and fraud charges on Wall Street, and by the moment he came face-to-face with a great bull elk near Jackson Hole, Wyoming, "so near that it was wicked to shoot it with the rifle which I had ready instead of the camera which was slung to my shoulder." Dimock became an evangelist for outdoorsmen to trade in their rifles, expounding in an 1890 essay on the greater skill required for pictures. "To stalk an elk successfully with the rifle is not difficult," he wrote. "To get within camera range requires the patience of an Indian." When a fire destroyed his cameras and collection of 10,000 negatives and 1,200 glass lantern slides, Dimock made another abrupt change. He left the photography to his talented son, Julian Dimock, to stick to writing.

In 1905, father and son invited Rogers to accompany them on their new houseboat, the *Irene*, for a "long and leisurely summer cruise" along Florida's southwestern barrier islands to see nature's best shell collection while they trolled the Gulf shallows for tarpon. Rogers was thirty-nine and had already signed a contract for *The Shell Book* when they set off from Marco Island, the former Calusa domain. At work on their own book about the silver kings, the Dimocks spent their days staging America's first fish porn, having learned that the smaller the boat, the larger the vaulting tarpon would appear on the page. Dimock played the shining acrobats from a wooden canoe while Julian photographed their leaps above his father's head from a small skiff called the *Green Pea*, built especially for him and his seventeen-pound camera.

Rogers hooked her share of tarpon—175 of them during her travels with the family between May and July. But she owed her real debt to the Dimocks for the times they left her alone, dropping her off on unsettled islands to study the life at the tideline. "On those tide-washed shores, I found at home the bright-hued creatures I had met before only in books," she wrote, "the most varied and beautiful assemblage of shells to be found on any beach belonging to the United States."

Many decades later as an elderly woman—even having spent the rest of her years surrounded by seashells and their aficionados and traveling to beaches around the globe—Rogers would recall that the most brilliant colors and abundance of shells she had seen in her life lay strewn on Sanibel Island that summer in the wake of a storm.

Doubleday published *The Shell Book* in 1908, part of its New Nature Library collection catering to the nature-hobby craze. It remained the leading popular guide to shells through its revised edition in 1951, when Smithsonian malacologists helped Rogers update the species names for a reissue tied to her eighty-fifth birthday.

Alongside her scientific descriptions of more than a thousand mollusks, Rogers rendered the beings in all their spectacle. The bivalve-sucking moon snail is a "businesslike mollusk butcher." The conch is a "master of the situation." She had a special appreciation for the scallops, "gaily painted shells, full of life and grace of motion, sometimes trailing behind them plumes of seaweed."

The book features more than one hundred plates of detailed shell portraits, most of them photographed from the collection of the American Museum of Natural History. It also includes some of Julian Dimock's black-and-white photos. Julian's shot of masses of coon oysters growing on the aerial roots of red mangroves in southern Florida reflects his father's view that the mangroves and the oysters held Florida together at its coast. The damage being wrought by sea rise and storms along mangrove-stripped beaches is proving him prophetic.

There is no photo of the author, herself. Nor does her name or image appear in the Dimocks' *Book of the Tarpon*, which includes ninety-two of Julian's photographs. Almost all of them frame a muscled silver king airborne at the end of his father's line. In other books and magazine articles, Dimock refers to Rogers only as "the tree lady."

I PUZZLED OVER how Julia Ellen Rogers could have vanished outside the shell world—especially having been in Liberty Hyde Bailey and Anna Comstock's influential sphere of nature study at Cornell. Her name never appears in Comstock's autobiography, published by Cornell University Press in 1953. In the wearying pandemic summer of 2020, amid headlines filled with public distrust of science, the answer finally revealed itself. A Cornell researcher and nature-study historian named Karen Penders St. Clair uncovered an egregious tale of how a chain of editors and colleagues suppressed and rewrote Comstock's story, and that of the early nature-study movement.

The first woman faculty member at Cornell had spent fifteen years on a 760-page manuscript of her life and work, intertwined with that of her husband and fellow professor who founded the university's entomology department. The 267-page book it became, St. Clair found, relentlessly abridged nature study; Comstock's insights about how children learn science and environmental care; anything controversial; any language considered too emotional—and much more, including erasing Julia Ellen Rogers and others who helped build the discipline. The published version highlighted Comstock's husband and the entomology department, beginning with the solitary photo of him that opens the book.

St. Clair told me that the redactions not only erased and minimized Julia Ellen Rogers, her sister, and others, but left them out of further scholarship. She has added up scores of books that cite the 1953 edition of *Comstocks of Cornell* as scholars thought they had read everything Anna Comstock had to say. St. Clair has painstakingly restored her voice for an updated edition. But the omissions that St. Clair discovered from a book referenced repeatedly over sixty years skewed the history of nature study and its place in the story of American science education.

Nature study became part of public education in every state in the nation before its undoing during World War I. The progressive movement was giving way to what one historian called "the corporate values of mass consumption and commercialized leisure that became the popular ethos of the 1920s." But gendered attacks were no small part of nature study's fall. Rogers's mollusk descriptions that charmed generations of teachers and students, families and kids, and amateur shell collectors were just the sort of writing that set off disputes between scientists and schoolteachers in the early twentieth century over how science and natural history should be taught in schools. The child psychology pioneer G. Stanley Hall claimed that "many modern nature books suffer from what might be called effeminization." He and others insisted that boys needed books written by men. Ricketts, Abbott, and many other twentieth-century scientists could have told him otherwise.

In 1912, Rogers settled in what was then the village of Long Beach, California, where she helped bring nature and agricultural studies to the public schools, sponsored the Long Beach Shell Club, and served on the Long Beach Board of Education for a decade. As she continued to write books and magazine articles, she taught in Comstock's summer programs for teachers and carried her green gospel across the country in programs celebrating shells, birds, and trees. In a nature study program for California elementary school teachers in 1921, she "urged the teachers to have an aquarium in their rooms, and to acquaint the children with the beauties to be found on the coast," suggesting a lending library for shells.

If there is any question that nature study made a difference in the short time it was systematic in American schools, the answer jumps out in the environmental leaders whose shoes it muddied in childhood. Rachel Carson's mother, Maria, used Comstock's *Handbook of Nature Study* as she led young Rachel on woods hikes and bird-watching while her older siblings were in school. Aldo Leopold was exposed at home and in the schools of Burlington,

Iowa. He raved about "bird study" in his composition book at age eleven, listing thirty-nine species he had identified. Leopold's Iowa teachers brought nature study into their classrooms in the late 1890s and early 1900s—very likely with Julia Ellen Rogers's lesson plans.

BEFORE THOMAS AND Lucy Say had been married a year, he finally made his trip to Mexico with Maclure. The expedition was a boon for collecting many more specimens for *American Conchology*, if not for Say's health, which had always been frail. He described his affliction as "his old enemy, the bilious," a sick stomach that had plagued him since his early years.

Lucy Say worked on shell illustrations late into the evenings while her husband was away, her habits reminiscent of his own late nights at the academy. New Harmony's engraver prepared the plates from her drawings, and when they were printed she took over the tedious task of coloring them by hand. Her front and back views of the Lettered Olive on Plate 3—the perfect ridges and shading on the spire, plump-curving aperture, detailed hieroglyphics, and familiar polished shine—reveal her precise talent.

But for the shell work, Lucy Say was not happy staying behind in New Harmony—or remaining in the post-utopian chaos even after Say returned. Fretageot's letters, revealing friction, complain of Say's attempts to avoid her household chores. Say's own letters express her anxiety about her husband's health and his profession; it was becoming clear that Thomas Say was at a disadvantage living far from the latest books, museum collections, and his fellow academy members. It seems out of loyalty to Maclure that the couple stayed on.

They finished the first and second volumes of *American Conchology* in 1828, "which will, I hope, be published sometime previous to the day of judgement," Thomas Say wrote to a friend. They continued their work at New Harmony for six more years, finishing a total of six volumes. Lucy Say drew and colored 66 of the 68 shell plates. The other two are by the naturalist Lesueur, with whom she had studied as a girl; she said she persuaded him to draw them because he had been her first teacher, "and always my friend," and wanted him associated with her work.

When their engraver fell ill, Lucy Say tried etching herself. One image is inscribed, "first attempt, August 1834." She goes on to write, "I could have learned etching very soon—but an engraver was procured."

Thomas Say died two months later at the age of forty-seven, after years of neglecting his health in pursuit of his work. Lucy Say was devastated, first by her husband's death and then by the loss of her illustration career.

She left New Harmony to join her mother and sisters in New York, where she set out to complete the seventh volume of *American Conchology*—most of which Thomas Say had finished—by taking up engraving. She wrote to friends in New Harmony:

> I am looked upon as being very singular, particularly since I have commenced Engraving—a gentleman remarked "Well! At what do you think the ladies will stop?" I replied, I hoped at nothing, short of breaking up the Monopoly so long held by the Gentlemen—that we were tired of cramping our genius over the needle and distaff.

Say's biographer Patricia Stroud determined that three of the plates in the volume must have been engraved by Lucy Say. Hers is the only name on those pages—and listed in the right-hand corner, the spot traditionally reserved for the engraver. But despite the acclaim of *American Conchology* and her skill as an illustrator, Say did not realize her dream to continue professional work. As a widow in New York, she couldn't even attend a scientific lecture given by an old academy friend, she lamented in a letter to Maclure, "as no one of my acquaintance whose protection I could avail myself of, attended, I remained at home, lamenting the dependence of females in large communities."

She pursued conchology for the half century she lived alone, keeping a small curiosity cabinet and trading shell specimens with a successor to her husband's research. In 1841, members of the Academy of Natural Sciences elected her the first woman member.

DESPITE HAVING LOST their dedicated first curator to a fleeting utopia, the academy became a beacon of American natural sciences. The brick museum on Benjamin Franklin Parkway holds the nation's oldest mollusk collection, though, like those in many natural history museums, the shells are hidden away in drawers and not part of the public displays.

When I visited, I learned that the collections manager, Paul Callomon, is an admirer of Anna Comstock and the old nature study progressives. He grew up in England, the son of a paleontologist who specialized in ammonites, and got his first shells during the family's annual summer drive to Switzerland in their pale-blue Ford Consul DeLuxe. The shells were not from one of the Jurassic-era quarries where his father stopped to explore, but from the Shell service station where he stopped to gas up the Consul. Shell was giving kids real seashells in a promotion; they changed Callomon's life.

Even as he and his colleagues work to digitize the museum's ten million specimens, Callomon is always thinking about how to get shells into the hands of the children. "There is ultimately no substitute," he told me, "for the power of the real."

Many of those specimens were carried back to Philadelphia in the trunks and alcohol jars of some of the great conchologists who came up in the academy. Conservator of Conchology George W. Tryon Jr. worked in the music publishing business—he edited and published librettos of fifty-two operas—but mollusks were his passion. He described and named more than five thousand species in his forty-nine years. He'd planned to describe all mollusks recent and fossil, their anatomy and development, their geography, and even "their relations to man and other animals" in a *Manual of Conchology*. When he died in 1888, he had published the first nine volumes. A fellow scientist wrote a poem to him with the lines,

> *Even Neptune mourns the loss of one who knew*
> *His sea-born children all by sight and name*

Henry Pilsbry had been working at a New York publishing house when he wrote his "Dear Curator" letter to Tryon at age twenty-four. Tryon invited him for Thanksgiving in Philadelphia, saw Pilsbry's brilliance in taxonomy, and convinced him to leave New York immediately to become his assistant. Tryon died two months later, leaving Pilsbry in charge of conchology.

The small, cheery man with an extra-large mustache would describe and name more mollusks than any other scientist or citizen—5,680—and power through what became forty-five volumes of Tryon's *Manual of Conchology*. Pilsbry helped globalize American marine science in expeditions across the world's seas. He was so moved upon reaching the Galápagos and seeing one of Darwin's famous tortoises that, "I fell upon his shell and embraced him." The tortoise hissed and drew its head and legs inside its shell.

Pilsbry worked at the academy until he died at ninety-four. He suffered a heart attack at the desk that had been crowded with his microscope, monographs, and mollusks for seventy years. He had burnished the American reputation for natural science, just as Thomas Say and the other academy founders dreamed.

There was no better proof of that than the "brief and bewildering" telegram of August 1945 sent around the time of Japan's surrender to the Allies from Emperor Hirohito to U.S. Army General Douglas MacArthur. Hirohito had seen the ultimate devastation wrought by American science just weeks before, in the annihilations at Hiroshima and Nagasaki. Yet for as long as he lived, Hirohito, a serious marine invertebrate biologist with a renowned shell collection, believed that natural science was a means for uniting humanity. His telegraph asked "if Dr. Henry Augustus Pilsbry was still alive."

After a flurry of cables, the Army confirmed that Pilsbry was indeed, and still working on shells at the Philadelphia Academy.

Eight

~~~~~~~~~~~~~~~~~~~~~~~~

SHELL OIL

THE MUREX
Hexaplex trunculus

J ust after they married in the fall of 1833, Abigail and Marcus
Samuel opened a small curio shop in Sailors Town north of the
River Thames in London's East End, where the narrow, uneven
streets smelled of tar and salt and surged with working families and
maritime trades. Every sort of seaman and shellback—sailors who
have crossed the equator—came and went from all the oceans to
what was then the largest city in the world.

Marcus and Abbie both grew up in Jewish merchant families that
had emigrated from Holland and Bavaria nearly a century before,
traders in antiques, curios, and bric-a-brac. Then not permitted to
own land or open shops in the city proper, the Samuels and other
Jewish traders—along with all the noxious manufacturers Londoners wanted to keep downwind—had to find their niche in the over-
crowded East End.

The couple found theirs in seashells. Curiosity cabinets, shell
rooms, and shell grottoes had brought the shell cult to its ostenta-
tious hilt among the upper classes and nobility. Now, the middle
classes burned with shell fever. Tropical seashells decorated parlors
and studded parlor crafts. Victorian women, "idled" by the industrial revolution but excluded from its realm, spent hours on crafts
such as shell boxes—grottoes-in-miniature encrusted with tiny limpets or pearly turbos, topped with a nosegay of cowries, cones, or
other exotic shells. The Samuels sold "small Shells for Ladies' Work,"
along with nautiluses and other specimens for still-life drawing. They

also advertised helmet shells for cutting cameos, a craze in Victorian England propelled by Her Majesty, who commissioned a shell brooch for her coronation and distributed shell cameos carved with her silhouette for her wedding in 1840.

Marcus and Abbie did well enough their first few years to move from a backstreet to the waterside. They rented a narrow house near the Tower of London, lived upstairs, and ran their shop from the ground floor across from the new St. Katharine Docks. Marcus could run out to meet sailors coming ashore with all manner of exotica for sale, including tropical shells.

The neighborhood, the decade, and the trade were out of Charles Dickens's *The Old Curiosity Shop*, full of strange goods "that might have been designed in dreams." Marcus Samuel was said to have had the dream that changed his fortunes during a beach holiday at Margate. Families flocked to the seaside resort on steamers from the Thames—men joining for weekends on the Saturday "husbands' boats." The Samuels had nine children who survived past infanthood—six girls and three boys. Lore had it that the kids were making shell boxes at Margate when it occurred to Marcus Samuel that the pieces could be sold as seaside souvenirs.

The Samuels began making the trinket boxes for beach resorts and shops that stocked seashells for sale to tourists. They proved to be the family treasure chests. The little boxes sold so well that the Samuels soon added shell sewing boxes, shell needle cases, shell portholes, shell frames, and other varnished mementos, all first conceived by women as household arts and crafts. They made crown pincushions in honor of Queen Victoria, bejeweled with small shells and stuffed with deep red velvet.

So were born the shell-craft souvenirs sold to this day in beach shops in England and around the world. Samuel also is credited with the ubiquitous labels still in use: "A Gift from Brighton." The family's fortunes rose—"from near-poverty to relative wealth," a British writer who married into the family later put it—in direct relation to the popularity of the Victorian shell box. Samuel eventually had forty women on the payroll manufacturing the boxes in the East End. He and Abbie kept the shop there, but in 1857 they bought a house in Finsbury Square and moved the family to more civilized Central London.

Their middle son and Samuel's namesake, Marcus Samuel Jr., often returned to the docks with his father, who expanded the family's trade with partners across the Far East. The British East India Company had dominated the region since its Charter of 1601. The monopoly fell just after Marcus and Abbie married. Merchants could finally establish trade partnerships, including in China and Japan. Marcus Samuel Sr., known for his good word and making friends near and far, imported the likes of china bowls, olive oil, goatskins, ostrich feathers, sandalwood—and always, an exotic assortment of tropical shells.

Thanks to his interest in the fanciful species of Japan, Samuel had deep connections with business partners in that country during its transition from shogun seclusion to market economy. He is believed to have sent Japan its first mechanical looms. When his namesake took over in the next generation, those time-honored relationships would help Marcus Samuel Jr. upend nineteenth-century global trade as markedly as the first East India Co. ships centuries before.

Samuel Sr. died twenty years before his son's audacious multinational deal. Samuel Jr. honored him when he named the new family business: the Shell Transport and Trading Company, now known as Royal Dutch Shell.

MARCUS SAMUEL SR. listed himself as a "curiosity dealer" in the London business directory of 1841, but by the Census of 1851 had settled into his specialty, "shell dealer and importer." When he died twenty years later at the age of seventy-three, Samuel's assets amounted to £40,000, or nearly £5 million in today's value, though little of that was in cash. A quarter of it was in seashells. His three sons—Joseph, Marcus Jr., and Sam—were beneficiaries of not only the shells and other business assets, but their father's network of trusted partners in the Far East; and some incredible timing.

The Suez Canal had opened the year before Samuel Sr.'s death, connecting the Indian Ocean and Mediterranean through the Red Sea. Increased availability of coal meant that steam-powered ships could compete economically with sail. European vessels no longer had to plow down the west coast of Africa and around the Cape of Good Hope—or spend months awaiting the monsoon winds in outposts like the Maldives—to travel to and from the Far East.

The oldest son, Joseph, took over the business according to their father's will, which also expressed hope that he would bring on his brothers once they turned twenty-five, or earlier if he saw fit. "And that on his doing so all my said sons will be united, loving, and considerate and keep the good name of Marcus Samuel from reproach," the will stressed.

Marcus Jr. and Sam were younger than Joseph by sixteen and eighteen years, and better educated. Their parents had earned enough to send them to Jewish boarding schools, though the family had not broken the ranks of the very few Jews finally attending elite universities like the sons of other British traders. The boys were still teenagers when they joined Joseph in the business, which he ran in his father's mold. Even as popularity for the Victorian boxes waned, "shells still commanded most of his interest," wrote Robert Henriques, a grandson-in-law to Marcus Samuel Jr. and his biographer.

Marcus agitated for new trade ventures, the younger Sam always his ally. But the two had no luck convincing Joseph. The brothers remained "united, loving, and considerate," even as Marcus pined to chase new deals abroad. Abbie made her venturous middle son work for Joseph in the London office for two years before she let him travel to the Far East.

In summer 1873 at the age of nineteen, Samuel Jr. finally boarded a steamship and made his first journey through the Suez Canal. He called on his father's old business partners in Ceylon (now Sri Lanka, between the Maldives and India); Singapore; and Siam (now Thailand). It was an uncommonly hot monsoon season with disastrous extremes in the rains—record flooding near the Ganges River, crop-killing drought in Bengal. He arrived in India that fall to a rising famine. His first international deal was both a humanitarian and a financial coup.

India's colonial government had put up £3 million in emergency food relief from the crown, valued at roughly £300 million today. Using "his father's name and his brother's credit, with or without permission from home," Samuel Jr. found surplus rice in Siam and arranged through a series of regional deals to ship it directly to the famine area. It's hard to imagine today, but direct trade within Asia was unheard of then; almost all merchants were importing bulk goods to England. Marcus bought low, sold high, and came out a hero. He also emerged with a fresh vision for global trade logistics, like the children on the beach at Margate had opened their father's eyes to new markets for shells.

THE FIRST GLOBAL capitalists were the Phoenicians, trading across the ancient world in their broad-bottomed cargo ships a thousand years before Christ. Around the Mediterranean basin and down the Red Sea, they built a network of colonies and trade centers full of papyrus and pomegranates, silver and spices, wood and wool, and endless other goods. They acquired common items like pottery and carried them to faraway people who thought them rare, now trading for ivory or the next local specialty, and making enormous gains. New markets lured them past the limestone Rock of Gibraltar—until then believed to mark the end of the world. The seafaring merchants traded up to ancient Britain and down the west coast of Africa. Jewelry from Greece, linen from Egypt, and carpets from Mesopotamia were traded for Celtic tin and African gold. Like those who came before and after, the Phoenicians also enslaved and traded people by the tens of thousands. They built some of the earliest networks to enslave and trade African people over global borders, bondage continued by their Roman conquerors and shamefully into modern memory.

Like the multinational corporations that followed three millennia later, the Phoenicians did more than trade, transport, and plunder. They were also manufacturers, heating quartz sand into glass, honing the cedars of Lebanon into timber. They were best known for "prestige goods," accessible only to the elite. No good was more prestigious than the one behind their name. The Greek *Phoinikes*, or Phoenicians, derived from the word for purple—*phoinix*; a rich hue closer to blood than the violet in a rainbow. The "purple people" were named for their striking purple dyes, produced in foul-smelling factories at Tyre and other port cities around the Mediterranean. The hue was so sumptuous—and the manufacturing process so arduous—that Tyrian purple was restricted by law to royalty. The era gave us the names "imperial purple" and "royal purple," and the phrase "born to the purple" for an aristocratic pedigree.

The ancestry fits the regal animals whose small, soft bodies were crushed for the dye. The prized purple was wrung from sea snails known as murex. They are among the largest living family of marine gastropods, the Muricidae, whose more than 1,600 species beautify waters around the world. Murex shells jut royal spires, ornate spines, and what can only be described as outstretched wings, often frilled, like the offspring of a flamenco dancer and a fabulous bat.

The most sought-after murex in the world, Loebbecke's Murex or *Chicoreus loebbeckei*, evokes mythology rather than taxonomy: a fairy-dragon poised to fly from the deepwater corals of the western Pacific. Seeing one on a boyhood trip to the British Museum cinched Peter Dance's future as a shell curator in a single glance. He remembered the pinkish-white murex as "the loveliest, most exquisite natural object" he had ever seen.

Other murexes secrete their evolutionary genius into spines rather than wings. Coveted from the earliest days of shell collecting, the Venus Comb Murex grows a hundred impossibly thin spines that Tucker Abbott described as "the ribs of a fish picked clean." The murexes that made the royal purple, *Hexaplex trunculus* and *Bolinus brandaris*, build shells less elaborate yet still striking, the former burlier and the latter spindlier, both suggestive of ancient iron weapons that also originated in molluscan shell defense.

The purple hue—more accurately called murexine—is also thought to be a kind of defense. Mollusk mucus has all sorts of important jobs. It might work as a lubricant to move a snail along, an adhesive to help it stick, or a rope that allows a small creature to dangle. Mucus can protect mollusks from pollution, from extreme cold or heat, or, with a touch of toxin, from predators. Scientists hypothesize the purple mucus, released from the murex's hypobranchial gland, is some type of biochemical block against harmful bacteria or UV rays; it only darkens when exposed to the light.

Scientists find the precursor to purple in the spectacular egg masses where these murexes begin life; they think it has a role in protecting the spawn. Mother murexes congregate in the shallows in springtime to lay the cushiony masses. Each is made up of a hundred or more capsules, each capsule holding three hundred or more eggs. Inside the capsules over four to six weeks, the stronger eggs grow to embryos that cannibalize the weaker; develop eyes, foot, and shell; and hatch into miniature murexes that swim or crawl into their world and ours.

Protecting the offspring in life, murex pigment in death adorned Assyrian warriors and festooned Pharaohs. Murex made the sacred blue in the Bible and the blue-violet *tekhelet* of Judaism that inspired the flag of Israel. They secreted a hue so lasting that archaeologists excavated the two-thousand-year-old tomb of Macedonian King Philip II found traces of Tyrian purple from the mask placed around his face before his cremation in a funeral pyre.

The mollusk-made purple, often tinged with gold, also survives on textiles dug from the ruins of Pompeii; evidence of luxury on the eve of disaster.

PEOPLE HAVE WRUNG oil from the earth since darkness fell on caves. The first shell lamps were not the glass cylinders my mom painstakingly arranged with found seashells in the 1970s, but cupped shells filled with melted animal fat or plant oils, a fiber wick threaded through a bivalve bore hole or a gastropod siphon.

Pacific peoples used large sea snails including turban shells and giant frog shells to burn oil well into the nineteenth century. Greasy deposits burned into shells of a large Nile bivalve, *Chambardia rubens*, in Egypt mark how rapeseed and other oils lit the ancient night in shells.

Fossil oil, too, had been tapped for thousands of years by Indigenous people, who knew about its seeps. Fire worshippers tended eternal flames over natural gas seeps in what is now Baku, the largest city in Azerbaijan, before the Zoroastrians built the handsome fire temple still preserved there. In China, the eminent first-century scientist Shen Kuo wrote of the rock oil "born from the sand and rocks on the water's edge . . . and mysteriously gushed forth."

Yet oil had low trade value for most of recorded history; the Phoenicians exchanged it for common items like tin that were not valued by their producers but could fetch high prices elsewhere. The story only began to change in the early 1700s, with the frenzied pursuit of sperm whales by Nantucket hunters. Sperm oil burned much cleaner and brighter than that from other whales or animals rendered to tallow. The great-brained mammal became light, lubrication, and the first commodity oil.

In western Pennsylvania in 1859, in hills risen from what was once an inland sea, prospector Edwin Drake drilled into a petroleum reservoir along an Allegheny River tributary called Oil Creek and hit a gusher so abundant he couldn't contain it. Drake scrambled for empty whiskey barrels to capture all he could.

The rush that followed created boom and bust towns and fortunes and failures in western Pennsylvania, a cycle that continues with shale gas fracking. It also helped create the richest man in America, and for a time, the largest oil corporation in the world. John D. Rockefeller of Cleveland had started in the business with a lone refinery on a tributary of Ohio's Cuyahoga River. He methodically undercut the oilmen of Oil Creek until he had consolidated nearly all of them into Standard Oil. A fourteen-year-old girl named Ida Tarbell watched in anguish as Rockefeller's secret agreement with the railroads destroyed her father's business—Tarbell's Tank Shops, one of the last holdouts. Plotting mergers, eliminating competitors, and colluding with the railroads, Rockefeller's Standard Oil by the 1880s controlled 90 percent of America's oil industry. The company, nicknamed "the Octopus," then looked to spread its tentacles around the globe.

Rockefeller was selling the "new light" of kerosene abroad, but transporting it was a huge challenge. The first bulk oils arrived in Europe in wooden barrels. The barrels wasted space; and worse, they were prone to explode. Americans turned to packing oil in 5-gallon rectangular tins, handy for buyers but inefficient for packing and loading onto a ship.

The other complication for Rockefeller was Russian oil. Once home to the fire worshippers, Baku on the western shore of the Caspian Sea fronted the most productive known oil reserves in the world, yet it was cut off from Europe by the Caucasus Mountains and the Black Sea. Two sets of brothers from two of Europe's best-known industrial families—Ludwig and Robert Nobel of Sweden and Alphonse and Edmond Rothschild of France—were building refineries and cutting railroad tunnels through the mountains. The Nobels had commissioned the world's first oil tanker, the *Zoroaster*, with cylindrical tanks inside its hull.

When their efforts opened the flow of Baku oil over Russia's borders, Rockefeller, taking a page from his U.S. conquests, undercut prices in Europe, employing his famous strategy. Russian oilmen began to suffer, as had those in Philadelphia. The Rothschilds quickly agreed to Standard Oil's "terms of peace" in Europe, accepting a small slice of the market in exchange for Rockefeller's backing off.

But here's what Rockefeller didn't know: The Rothschild brothers were negotiating with British merchant brothers, Marcus Samuel Jr. and Sam Samuel, on a much bigger deal—one that would cut Standard Oil out of the largest emerging market in the world. They would export the Russian oil to Asia, and block his ability to undercut them. This bid for global oil domination was being mapped out in a modest office off a narrow alley in the East End, crammed to the ceiling with seashells, then still a crucial part of the family's trade.

Marcus and Sam, operating respectively out of London and Japan, had acquired the business from their older brother, Joseph, and now ran it on behalf of the family. They'd spent the twenty years since their father's death building on his connections in the Far East, mindful to "keep the good name of Marcus Samuel from reproach." Those decades-long relationships had helped them become the leading British business concern in Japan. The Samuels, along with their trade in tropical shells, brokered half the country's annual rice exports, most of its sugar, and all its foreign coal sales. They also shipped oil in tin cases to Asia through the Suez Canal, though their supply was not nearly enough to meet demand.

Canal authorities permitted this "case oil," but not newfangled ships like the Nobel brothers' *Zoroaster*, nor Standard Oil's tank ships, nor any other tanker; the danger of explosion was too great. Anyone who tried to export bulk oil to Asia through Europe would have to travel the length of Africa and around the Cape of Good Hope, which made it exorbitantly expensive. If someone could design a bulk tanker safe enough to satisfy the canal authority, the transportation costs would be slashed. Even Standard Oil could not possibly compete.

Marcus Samuel Jr., not yet forty, was the idea man at the front of the company. He was short and impeccably dressed; his eyebrows arched over his pince-nez and a neat walrus mustache that plumped through the years with the rest of him, his weight commensurate with his success. Samuel agreed to partner with the Rothschilds only after traveling to Baku in 1890 to see for himself what the British journalist Charles Marvin described then as the city of "all-pervading" oil. Oil soaked Baku's forest of wooden derricks, hung in the clouds, ignited clothing, filled every breath, infused the taste of food.

Samuel looked beyond the black clouds and saw a route and market for the Rothschilds' oil now accumulating in surplus—via the Nobels' railroads, the Suez Canal, and a series of storage facilities across Asia. The plan avoided Standard Oil's tentacles at every point. But it had a great risk. It relied on a ship the likes of which the world had never seen. While the Rothschilds stocked up the oil in Russia, and Sam Samuel set up the trade deals and storage tanks in Asia, Marcus Samuel Jr. worked with a marvelously named marine engineer in London—Fortescue Flannery—to design the ship that would become the modern oil tanker. The Samuel family christened it the *Murex*.

WHEN YOU WATCH a live murex creep along the ocean floor, the carnivorous snail looks more like a military land tank than a seagoing oil tanker. As the soft underbody glides across the sand, the turret-like shell slowly advances, its siphon canal jutting forward like a gun barrel bracing the soft proboscis that noses out bivalve prey. As beloved as they are to shell collectors, murexes are despised by shellfish farmers. They are the "drills" that sniff out oysters, bore into the shell with their radula, and deliver an enzyme that paralyzes the morsel; all the better to devour and digest it. The spiky predators can wipe out an entire abalone or oyster farm.

For humans, the murexes were the morsels since prehistoric times. Paleolithic hunter-gatherers ate *H. trunculus*, also known as the Banded Dye Murex, across the Mediterranean; some tucked them into burials to fortify the dead. It was at least five thousand years ago when the first cultures—likely the Minoans—discovered that the abundant sea snails could also be crushed into striking colors.

Mythology told that the Greek hero Heracles (Hercules, to the Romans), was walking along the shore with his dog on the way to court a sea nymph, Tyros, when the dog gnawed on a spiny shellfish that turned his mouth and lips a deep purple-red. What first appeared to be blood was an indelible dye. When hero and hound arrived at their destination, Tyros was stunned by the color. She told Heracles she wouldn't see him again until he brought her a robe of the same hue. He managed to make her one by smashing "a great number of shells."

No doubt, producing royal purple took a great number of mollusks. In impressive deposits around the Mediterranean, archaeologists still unearth hordes of broken murex shells, evidence of dye workshops that bustled from early Minoan to late Roman times. Archaeologists discovered the earliest-known dye factory at the Minoan harbor town of Kommos on the island of Crete. The palatial beachfront complex is paved with slabs and drainage channels and buried in murex shells crushed nearly four thousand years ago.

No one in modern times had even a rough idea of how many murexes it took to produce the dye until archaeologists began experimenting with an ancient recipe recorded by Pliny the Elder in his *Natural History*. The Canadian archaeologist Deborah Ruscillo had spent many years researching murex debris at Kommos. She grew weary of seeing, in academic papers and in tourist materials around Crete, what she considered a dubious fact: that it took 10,000 snails to produce 1 gram of purple dye. Twenty years ago, she set out into Kommos harbor with a baited basket and pot, just as Pliny had described, to see for herself.

It took Ruscillo six weeks to figure out how to collect live murexes; more trial and error to break them open with tools at hand to the Minoans—a sharp awl and rock turned out to be the trick; and still more hit and miss to master slicing out the purple-producing hypobranchial gland. As she eviscerated the sea snails under the Mediterranean sun, her operation drew large flies and wasps to the butchered remains—the former to lay their larvae and the latter to feast. Soon, maggots were born. But all that was nothing compared with the stench, which grew stronger as she brewed the dye. The putrid smell infused her skin along with the beautiful cloth she was teaching herself to color—with Pliny's recipe proving misguided at every step.

Crushing murex and concocting dye through the summer, Ruscillo closed in on some of the facts not found in Pliny's chapter, "The Nature of the Murex and the Purple." Murexes produce a wide range of blue-violet color, from royal purple to the biblical blue described in the Old Testament. A few hundred of the snails could make enough dye to trim a garment. A little saltwater added to the glands produces the deepest colors; the dye is colorfast even without additives; three days is the perfect time for steeping; and wool absorbs the deepest shades. She also definitively concluded: "Pliny never made dye himself."

Her most important findings involved the human toil behind the color purple, underscoring its elite status and expense. From the dangerous diving and baiting to the maggots and the terrible stains and odor that would have plagued the dye-makers—likely enslaved people—conspicuous consumption always had a flip side: human suffering and ecological calamity.

THE MOVING, MULTINATIONAL parts in the plan to export Russian kerosene to Asian markets through the Suez Canal had to be carried out in secret to keep Rockefeller and the rest of the men of Standard Oil in the dark. The plan included contracts for oil from Baku; construction of storage tanks in Singapore, Thailand, Hong Kong, Saigon, Shanghai, and the port of Kobe in Japan; and many transportation puzzles, none trickier than construction of the *Murex*.

The marine engineer Fortescue Flannery had to "divine the minds of the Canal authorities," according to a company history of Shell's tanker fleet, though it seemed the authorities themselves couldn't imagine a ship safe enough to carry bulk oil through until they saw it in the *Murex*. Flannery placed its oil tanks amidships to isolate them from the boiler and engine rooms and then designed the rest of the ship's features to protect them: a water-ballast system to keep the ship from grounding and easy to drain when the *Murex* was heavy with oil; special expansion trunks atop each tank to let the oil expand and contract as temperature changed; a steam-cleaning system to flush the tanks; and electric lighting throughout the ship.

Canal authorities released their safety specifications in early 1892, along with the requirement that oil tankers would also have to earn a first-class rating from Lloyd's of London to pass through the canal. The rest of the oil world learned about the *Murex* only when it met the specs and earned the rating. By that time, Samuel had the oil contracts signed, the Asian storage facilities built, and ten other tankers near completion.

Standard Oil hired London solicitors to sow doubt about the new bulk shipments in Parliament and in the press. Samuel's opponents riled up the tin makers who would be put out of work once case oil disappeared. (Even though his opponents, too, were working to eliminate tin cases for bulk.) The propaganda was fierce, and often anti-Semitic: An *Economist* article tried to tamp the persistent innuendo that the scheme was "purely of Hebrew inspiration."

Protracted efforts to cancel the canal authorities' permission failed, largely because the British government refused to be drawn into the controversy. Marcus Samuel Jr.'s connections at home and in Asia helped insulate the firm from the attacks. In the same years he'd traveled to Baku, negotiated with the Rothschilds, and commissioned the first modern oil tanker, Samuel also ran for alderman of his ward in the city of London. The press raised doubts that he had the money, status, or clout. But the voters decided. Samuel won the election. His admiring biographer, Henriques, was convinced that, more than business success, Marcus yearned for "status for himself and his family on the long road to acceptance" for Jews in England. His father would not have been able to run for office at the same age, nor send sons to Eton and Oxford, as did Marcus Samuel Jr.

In August 1892, the *Murex* became the first bulk oil carrier to pass through the Suez Canal, hauling 4,000 tons of Russian kerosene. It steamed into the Red Sea and onward to the Indian Ocean and its first port, filling the Samuel brothers' tanks in Singapore. Marcus Samuel Jr. soon launched the ten other ships in the fleet, all named for seashells in his father's honor: the *Conch, Clam, Elax, Bullmouth, Volute, Turbo, Trocas, Spondilus, Nerite,* and *Cowrie.* By the end of 1895, sixty-nine bulk oil shipments had made their way through the canal; sixty-five of those were on ships named for seashells.

The real-seashells part of the story ended in October 1897, when Samuel incorporated the larger trade group including the Rothschilds into the joint-stock Shell Transport and Trading Company.

Steeling the firm to compete with the rival Royal Dutch Company in the Dutch East Indies, Samuel sold the family shell-box business to nephews. But as a brand, few could argue with Henriques's claim that Samuel's symbolic vision for his firm practically changed the common meaning of the word *shell*. The company's first logo was a bland mussel. In 1904, Shell introduced its iconic ribbed scallop, so recognizable that it no longer requires the company name.

Samuel died having earned both great wealth and the civic luster he'd always wanted. He served as sheriff and later lord mayor of London. Queen Victoria knighted him after Shell's tanker the *Pecten* rescued a Royal Navy ship from sinking in the Suez Canal. Samuel earned a full baron's title following World War I, for putting the Shell fleet at the service of the British Admiralty. He died in 1927 with the title First Viscount Bearsted. By most accounts, he had kept the good name of Marcus Samuel above reproach.

THREE DECADES AGO in the Mediterranean, where the Phoenicians had once sailed and steeped their royal dye, scientists began to find that female murexes were growing penises. The species are not hermaphroditic, as are about 40 percent of mollusks. This was endocrine disruption, and clearly tied to the maritime industry. In coming years, nearly all female murexes sampled in some of the Mediterranean's busiest yachting, fishing, and shipping harbors were found sterile—unable to lay their lovely egg cases, the precursors to purple.

The offender was tributyltin, or TBT, a biocide used to keep ship hulls free from the likes of algae, barnacles, and mussels. In the 1950s, the Dutch chemist Gerrit van der Kerk, a pioneer in the study of metal–carbon bonds, discovered the antifouling properties in a group of compounds called organotins. The find was momentous for the shipping industry, where a 1-millimeter layer of algae can slow a ship's speed by 15 percent. TBT killed any sea life trying to colonize a hull. By the early 1960s, it was in standard use in ship-bottom paints. Soon, it was killing far more sea life than that hitching itself to ships.

French scientists warned as early as 1970 that female oyster drills—murexes—in Arcachon Bay were losing their ability to spawn. The drills started to die off. But the loss of an oyster predator was considered acceptable in the oyster-loving Aquitaine region. Through the end of the century and into this one, TBT was implicated in reproductive mayhem in hundreds of mollusks across the world. It prevented spawning in three species of murex, one of them the wide-mouthed, purple-staining *Purpura patula*, in the U.S. Virgin Islands; in dog whelks in England's Plymouth Sound; blue mussels in Canada's Gulf of St. Lawrence; and in flame shells in Ireland's Mulroy Bay. The anemone-like bivalves, named for the flaming orange tentacles that lick from their shells, also build extraordinary stone nests that draw other marine life. TBT poisoning decimated the gravelly beds.

Still, governments around the world were slow to ban TBT. Weighing the benefits to the shipping industry versus the harm to mollusks—in other words, lobbied by all global interests that rely on ships versus those that fight for snails and clams—there was little contest. Only after TBT was found to deform shells and cause reproductive failures in commercial oysters did bans begin, in fits and starts and weak regulations that exempted the largest ships. Not until 2008 did international treaties ban the compound once and for all.

In another decade, researchers would find increasing evidence that organotins may threaten human health, too, notably in disrupting hormone and reproductive systems. Mollusks were again prophetic in their burden.

PURPLE-DYE MUREXES STILL mosey over the rocks and sand-mud bottoms of the Mediterranean. They and many other mollusk populations poisoned by TBT, including the lovely flame shells, are recovering thanks to the ban—though many ports, marinas, and shipyards are still contaminated. The saga suggests that humanity can reverse harm on a global scale despite financial pressure to keep the damaging course. It also speaks to the impressive resilience of the purple-dye murexes, which remain heavily fished, their meat a local staple and an export. Neither the ancient exploitation nor the modern fishery nor the TBT poisoning has permanently endangered the species, marine scientists at the University of Carthage have found. The scientists are less certain about the next pollution scourge, which weighs in at a much heavier 10 million tons swept into the seas *every. single. year.*

The scale of plastics smothering coasts, oceans, and sea life is almost impossible to comprehend. Plastics have become so abundant on Earth that they are expected to mold into geologic strata, an indicator of the Anthropocene. The five great ocean gyres, the huge, slow-moving whirlpools that circulate seawater around the globe, now eddy more than 5 trillion drink bottles, food wrappers, fishing nets, and other plastic buoys.

The gyres carry our plastic waste even to far-flung tropical islands like the Keelings in the Indian Ocean or Pitcairns in the Pacific, where uninhabited Henderson Island is covered in the greatest density of plastic debris recorded anywhere in the world. Hermit crabs, those thrifty recyclers of seashells, frequently mistake plastic for shell homes. In the Keelings alone, more than 500,000 of the crustacean boarders die each year after becoming entrapped in plastic.

The beaches of Bali, once famous for tropical seashells, are heaped in plastic each morning of wet season November to March, now called "garbage season" for the dystopian mire of bottles and bags. Plastics are embedded in the deepest ocean at Mariana Trench and in the highest-soaring seabirds. In remote Hawaiian atolls, albatross chicks die with stomachs full of bottle caps, golf tees, and lighters, all fed to them by doting parents mistaking plastic for prey.

Over years, larger plastic debris degrades into smaller and smaller bits. At sizes down to 5 millimeters, they're called microplastics, snack size for filter feeders. Owing to the copious seawater they filter, bivalves are especially prone to ingest the bits. Scientists find microplastics virtually every time they open a pair of shells. Pacific oysters in the northwestern United States have been found with on average eleven bits of microplastic each—mostly the micro-threads that shed from yoga pants and fleece jackets in the wash. Plastics have contaminated mussels from Canada to China, even the ones living in seas where humans have little direct contact. Blue mussels in Arctic outposts thought immune show some of the greatest concentrations of plastic of any tested along the Norwegian coast; our harm swirling on gyres.

The largest and deepest enclosed sea in the world, the Mediterranean also suffers some of the highest concentrations of plastic in the world. Its limited outflow, along with intensive industry, make it the sixth-largest accumulator of marine litter after the ocean gyres. Plastic debris teems on the surface where the Phoenicians sailed, litters the seafloor where the murexes live, washes ashore the great tourist beaches, and winds up in the stomachs of Mediterranean marine life.

In Tunisia, University of Carthage scientists analyzed six economically important mollusks from the Lagoon of Bizerte: three bivalves, a cuttlefish, and the two murexes—*H. trunculus* and *B. brandaris*— known for imperial purple. They found tiny synthetic fibers, fragments, and films lodged inside all six species, with highest concentrations in the filter feeders. Purple-dye murex likely ingest the plastic when they consume their bivalve prey. In turn, the scientists hypothesized, when murex are eaten by people, the plastic-congested mollusks will become "an additional exposure route for these harmful substances to humans." A month after the Mediterranean study, American biologists published findings in the American Chemical Society journal *Environmental Science & Technology* concluding Americans now ingest more than 70,000 particles of microplastics a year. They have made their way into not only shellfish and other food, but into water and even air.

BACK IN WESTERN Pennsylvania, where the remains of primal marine organisms resurfaced in the first commercial oil wells in 1859, another dark deposit from that former inland sea, rich black shale, has today set off another boom, this time for natural gas. Royal

Dutch Shell, now one of the world's largest corporations, is setting a major course in the Appalachian foothills northwest of Pittsburgh, part of its plan for the large-scale shift away from oil that scientists say is necessary to fight climate change. (Natural gas is better, but still an emissions-generating fossil fuel.)

The company is raising one of largest construction projects in North America in the small borough of Monaca, which happens to have been incorporated by dissidents of the New Harmony utopia. Shell's contractors have relocated state highway and interchanges, built bridges, realigned railways, and hired seven thousand construction workers to build a petrochemical complex at an estimated cost of $10 billion. The industrial behemoth, halfway complete, looks like a 400-acre erector set rising on the south bank of the Ohio River.

Beginning in the 2020s, ethane gas will flow into the factory through pipelines stretching for hundreds of miles across Appalachia, fracked from the deep black shale. Inside, great furnaces will heat the ethane to extremes that break apart its molecules and rearrange them into a portentous gas called ethylene. Scientists have identified ethylene as the famous vapor of ancient Greek history that seeped up through the floor of the sacred temple at Delphi, putting the oracle in her trance and inducing her prophetic visions. Today ethylene is the highest-volume organic compound manufactured in the world, owing to its role in a vast range of synthetic chemicals. In a series of high-pressure reactions, the factory will forge the ethylene into long chains called polyethylene, Shell's end product.

Polyethylene is common plastic. The Pennsylvania Petrochemicals Complex will produce 3.5 billion pounds of it a year in the form of small round pellets called nurdles. The plant is designed to manufacture two types of nurdles, the building blocks of our plastic stuff: high-density that shape into firmer goods such as outdoor chairs and toys, and low-density that form throwaway items like sandwich bags and shampoo bottles. Nurdles begat in Monaca, Pennsylvania, will be heaped onto barges, trucks, or more than three thousand freight cars running through the factory on the company's own rail line.

Shell chose Monaca for its 700-mile proximity to most of the North American plants that make the plastic life we live—and the plastic waste we have no idea how to get rid of.

For better and for worse over the past century, Shell, which merged with its Royal Dutch competitor in 1907, often led the oil industry in developing petrochemicals that radically changed material culture and the world: plastics, resins, detergents, solvents, fertilizers, pesticides, herbicides, and countless other products. Like oil itself, those chemical compounds that seemed as morally praiseworthy as they were profitable—such as aldrin and dieldrin, used as insecticides touted to reduce world hunger—were later shown to do great harm. Shell held exclusive rights to what its chemists called the 'drin family, chlorinated hydrocarbon pesticides that were among its most profitable chemicals in the 1950s and '60s. When Rachel Carson published *Silent Spring* in 1962, sparking calls to ban these chemicals, Shell, while careful to stay out of the limelight, joined the rest of the pesticide industry in tackling the public relations problem rather than the pollution. Governments ultimately did ban aldrin and dieldrin, along with other chlorinated hydrocarbon pesticides such as DDT, but only after awaiting solutions from industry and seeing none. While the company's scientists insisted in the 1960s that poisons killing wildlife posed no danger to human life, its historians half a century later concluded the fundamental reason Shell wasn't willing to change course on the 'drins was that there was no financial incentive to do so.

The same is true for the fossil fuels now altering the climate, and for the plastics harming the sea and its life. Marcus Samuel Jr. once explained that "the mere production of oil is almost its least valuable and least interesting state. Markets have to be found." Sometimes, Samuel added, markets have to be created. These trade fundamentals—known to the Phoenicians when they sailed past the Rock of Gibraltar, risking the end of the world for new markets—drive the global plastics crisis. The more plastics are manufactured, the more markets must be found. That simple math foils efforts to recycle, clean up beaches, and deploy technological solutions. A landmark study on all the plastic ever manufactured—upward of 9 billion U.S. tons since production took off following World War II—found that more than 90 percent of it had never been recycled, not even once.

DURING A WINTER storm off the southwestern British coast in 1997, an "extraordinary" rogue wave hit the container ship *Tokio Express* as it traveled from Rotterdam to New York. Sixty-two cargo containers lurched overboard, one packed with nearly 5 million plastic Legos. The colorful toys spilled into the sea, among them 26,600 Lego life preservers—no help to the Lego men and women who were neither wearing them nor fastened into the 28,000 Lego life rafts that also pitched over. By fate's humor, many of the toys were in nautical-themed packs. Soon after the accident, plastic scuba tanks, flippers, spearguns, sea dragons, and octopuses began washing onto the north and south coasts of Cornwall, England's southwestern county that juts into the sea.

Tracey Williams had scavenged the Cornish beaches since childhood, when her parents made her lists of naturalia for treasure hunts: dog whelks, cockle shells, periwinkles, top shells, razor clams. Pebbles, mermaid's purses, sea glass. Her own beachcombing children were six and four when the Lego pieces began to wash ashore by the thousands near her parents' home, perched on a cliff in South Devon. They collected bucketsful. Her son still remembers finding a cutlass in a sea anemone that looked for all the world like it was defending itself from marauders.

Over the years, as her son and daughter grew up and she spent most of her time inland, Williams forgot about the Legos. Then, a decade ago, she relocated to Newquay on the northern Cornish coast. As she began to walk the beaches again, she was astonished to find the bright plastic bricks still washing ashore. To this day they come in, a quarter century after the spill, along with other synthetic species: Plastic Monopoly houses and plastic cars. Plastic pop-together beads and plastic fishing beads. Plastic army men and plastic fairies. Plastic flip-flops and visors; pacifiers and cigarillo tips; cassette tapes and film reels.

Williams began to curate a collection. What was a lark when the kids were little now struck her as an indictment of disposable culture, and a history in need of an archive. There is too much plastic to deal with it all; she thinks about her task similarly to that of a malacologist organizing type specimens and lots, or a marine biologist who surveys mollusks on a small square of the seabed to understand a larger whole. Williams picked up 427 cable ties on one Cornish beach in one day, 253 lighters on one Friday the 13th. She has collected hundreds of brand-new shoes and printer cartridges from cargo spills. She has picked up 20,000 little blue ale-cask stoppers from one beach. And she has scooped hundreds of thousands of nurdles like those that will be made in Shell's Pennsylvania factory. The plastic pellets, which tend to spill from the hoses, trucks, trains, and ships that move them from place to place, have become the second-largest source of microplastic pollution after the tiny fibers shed from our clothes.

Williams catalogues her finds in a taxonomy of ocean plastic, organizing them in grids around colors or themes and photographing and posting the montages on social media under the name Lego Lost at Sea. The medleys are oddly compelling, like museum drawers or cabinets of curiosities for the industrial age. They are artful, but Williams says not meant as art. "It's more about creating order out of chaos," she says, "a museum of plastic artifacts."

From Cornwall to California, Williams is among the new breed of beachcombers who make art, collect, or form cleanup clubs around beach litter. They are as inspired by the problem of human-made flotsam as were previous generations by the wonder of shells. Their numbers have overcome those of shellers by urgent call. Southern California, once home to about a dozen conchological clubs, is down to two, including the oldest in the nation, the Pacific Conchological Club, founded as the Tuesday Shell Club in 1902. Yet Los Angeles County alone has more than two dozen local beach cleanup groups.

A countywide marine debris database contains the same number of localized trash data points—4 million—as the number of mollusk specimens in the Natural History Museum of L.A. County.

The Victorian-style seashell boxes brought to market by Marcus Samuel Sr. are still sold in the British beach shops where they originated, and around the world. But on store shelves and on too many beaches, the scale of plastic far overwhelms the beauty of shells. Scientists at the University of Plymouth used Williams's Legos in a study that found they may persist in the sea for more than a thousand years. She has found century-old blocks from a west Java rubber plantation made from gutta-percha, the gum of a Malay tree. Williams traced them to a Japanese cargo ship sunk during World War I. The oldest plastic she's tracked to the source are American cowboy and Indian figures from 1957. They came free in cereal boxes, testament to the cost of free.

Not infrequently, she finds plastic seashells washed ashore. Many are the discarded cases of a lick-able German candy called *Schleck-muscheln*. They look so realistic that when Williams showed me her elegant grid of white cockles on white background, I couldn't tell the artificial from the molluscan. I could, however, imagine the plastic half-shells fused with marine strata long in the future, fossil layers of the Anthropocene.

Nine

~~~~~~~~~~~~~~~~~~~~

SHELL SHOCK

THE JUNONIA
Scaphella junonia

If anyone could capture the depth of obsession possible for a single shell, it might be a famously obsessed poet. Edna St. Vincent Millay collected valentines when she was young and lovers in adulthood. Less known, from her youth on the coast of Maine to the troubled last years of her life, she also collected seashells.

"The very thought of the words *Conus gloriamaris*," she wrote of the Glory of the Sea Cone in an overlooked scholarly citation, "fills me with an ecstasy of longing and despair."

The beloved American poet is remembered for her intense romantic sonnets. They were inspired as frequently by nature as by men. Her soul, she once wrote, was "Earth-ecstatic." In her poem "Exiled," published during her Greenwich Village heyday in the early 1920s, she wrote of her heart's "true sorrow." She was sick of the city, words, and people, and "wanting the sea."

I have a need to hold and handle Shells . . .

In a poem called "Eel-Grass" in the same collection, she wrote:

No matter what I say,
All that I really love
Is the rain that flattens on the bay,
And the eel-grass in the cove;
The jingle-shells that lie and bleach

At the tide-line, and the trace
Of higher tides along the beach . . .

Millay's yearning for the seashore with its tidal surprises and solitude drew her to remote islands. She and her husband, Eugen Jan
Boissevain, bought the 80-acre Ragged Island off Casco Bay, Maine,
in 1933 as a getaway. In 1936, they took the train to Florida to make
their way to Sanibel Island. Millay had been struggling to complete
her verse play *Conversation at Midnight*, and hoped to finish at a seaside retreat called Palm Lodge. "Fine Shelling," read an advertise-
ment from the era.

The couple checked in at the lodge in the waning daylight of
May 2, 1936. While Boissevain saw to their luggage, Millay headed
to the beach to hunt for shells at sunset. Just a few minutes into
her walk, she turned around to see the hotel engulfed in flames.
Her husband and the other guests escaped. But everything in their
room—including her books and her manuscript for *Conversation at
Midnight*—burned.

Much has been written of the fire, and of Millay's efforts to
reconstruct the versions she had finished, "an exhausting and nerve-
wracking time," she wrote to a friend. But rarely noted is what she
described as her "sweet insanity" for shell collecting—and the rare
shell that helped draw her to Sanibel.

I live in the strong though ebbing hope of finding someday on a
briefly uncovered sandbar a right-handed Left-Handed Whelk;
or even, someday, after propitious foul weather, of digging out of
the beach under the jealous eyes of hundreds who dare not quite
attack me and wrest it from me, a perfect Junonia.

Then and now, no other shell excited Sanibel collectors—or
sparked their envy—like the Junonia. The near mythical volute
stretches 4 inches or more. Its plump beige fusiform shell is covered
in mahogany dots. The intensely private animal that makes it lives

and dies in deep rocky areas offshore, so it's rare for one to tumble to the shallows. Finding a Junonia on the beach takes luck. A big storm increases the odds, as Millay wrote in the 1930s, sharing wisdom she may have found in Julia Ellen Rogers's *Shell Book*.

"When the northwester comes down across the Gulf, churning the sea to its rocky depths, a Junonia may be unexpectedly flung ashore, and buried in sand," Rogers wrote of Sanibel. "The morning after such a storm, the Floridians and the conchological aliens in their midst go forth to gather the spoils of the gale."

A persistent urban legend has a Sanibel chamber of commerce lackey tossing a few Junonia shells onto the beaches every so often to fan the promise of finding one. (Not likely today for shells that fetch $100 or more.) Like most visitors to Sanibel, neither Rogers nor Millay ever experienced the thrill, which to this day lands the lucky sheller in the local newspaper.

Lamarck named the shell *Voluta junonia*—"Juno's Volute"—in 1804, evoking the goddess Juno's intense jealousy of others' beauty. A London dealer in that era called it one of the most coveted shells in the world. Only four specimens were then recorded in European cabinets, and no one knew their provenance. Collectors assumed they lived under the sand in tropical waters near the Philippines, home to so many other showstoppers.

Edmund Ravenel of Edgar Allan Poe lore knew differently. He collected a Junonia sometime before John James Audubon paid a visit to him in South Carolina in the 1820s. Audubon painted Ravenel's shell onto the shoreline in his illustration of a pair of terns. Audubon must have decided the gray-white shorebirds, which appear on plate 409 of *Birds of America*, needed some color and flourish. The huge Junonia pops off the page like the Dutch shell portraits of a century before. The Junonia steals the scene.

TO HEAR THE ocean's softest song, walk the southern beaches of Sanibel Island. Listen closely at the break line. As each wave pulls back to sea, a sparkly tinkle rises from the rumble; the roil of tiny shells. They ring from the quiet end of the aural spectrum, place of fairy-dust notes and first rains.

Lying off Fort Myers Beach in southwest Florida, Sanibel is not a metaphor for seashells, but a synonym. The island's landmass, itself, was formed out of "innumerable millions of shells, their reduction to fragments, and finally to shell sand," as the Field Museum malacologist Fritz Haas wrote in 1940. "The beach represents a vast mortuary."

While most of Florida's 4,500 barrier islands run north and south parallel to the mainland, Sanibel formed sideways—terminus of a band of Gulf of Mexico currents that heaped the shells, fragments, and sand over the past five thousand years. For shell lovers, every tide brings a treasure hunt. At water's edge, the tinkling slurry of mini clams, cockles and coquinas, augers and tritons, drills and murex, opaque jingles, and other miniatures paves the landing for the larger bounty: Horse Conchs and Crown Conchs. Lightning Whelks and their crepe-paper egg casings. Tulips and turbans. Glossy Lettered Olives. Sand dollars that cover your palm and starfish big as dinner plates. Pen shells that look utterly homely or ravishingly nacreous depending on which side lands up. Those are some of the commoners among more than three hundred Gulf and Caribbean species that tumble up at Sanibel.

At low tide, the shoreline is a mosaic of rippling flats, the toe-tickling shell slurry, tidepools, and lagoons. It's also full of life, slight to the eye. Kneel to the wet-sand world of a Florida Fighting Conch, shell polished deep brown and orange. Protruding from two notches at the tip of its shell, two curious eye stalks periscope the scene. Coast apparently clear, the animal stretches its soft pink body onto the squishy sand and uses its nimble foot to tai chi toward the sea.

For more than a century, collectors have scoured the shore before dawn for Junonia and other prize shells, typically less interested in the mollusks within. Headlamps shining in the dark, the methodically sweeping figures can look eerily like an amphibious invasion. But those shell soldiers miss what makes Sanibel Island exceptional. It's not Haas's vast mortuary or the empty shells of dead mollusks.

It's the quiet, slow-moving world of living ones.

～～～～～

THE NAME SANIBEL would seem to evoke Spanish history. Historians consider it the likely area where the Spanish conquistador Juan Ponce de León first sailed into the Calusa's great cities of shell in 1513. The explorers kidnapped four Calusa women during initial skirmishes with the Indians; the women's fate went unrecorded. Juan Ponce returned eight years later to what a Spanish mariners' guide described as *Costa de Caracoles*—the coast of seashells—to claim *La Florida* for Spain. The Calusa were waiting. A warrior armed with bow and arrow, its tip possibly coated with the poison of an apple from the deadly manchineel tree, struck Juan Ponce's thigh. The Spaniards retreated to Cuba, where their leader died of his wound.

Hispanic fishers were next to settle the shell cities, establishing *pescadores ranchos*, "fishermen's ranches," to cure and export fish to Cuba and blending families with surviving Indians. The Spanish fish traders set up on the coast or anchored offshore in smacks—schooners with two sailing masts and deep live wells that could hold thousands of pounds of grouper and snapper. The *pescadores* and their families were made American citizens when the United States acquired Florida as a territory in 1821. But Anglo-Americans interested in the land soon protested against them, claiming they were squatters. Many of the *ranchos* were destroyed during the Seminole Wars, when the Indians who had coalesced in South Florida fought being forced to Oklahoma under the Indian Removal Act. As American soldiers ferreted out mixed-race children and burned *ranchos*, many Spanish fishing families fled. A few persisted in Sanibel and on a small island called Cayo Costa, at the heart of *Costa de Caracoles*.

Legend has it that remote Captiva Island at Sanibel's north tip, severed from the rest of the land by a major hurricane in 1921, was named for women prisoners held captive there by José Gaspar, the Spanish pirate known as Gasparilla. But André-Marcel d'Ans, a French anthropologist of the Caribbean, traced the legend of Gasparilla to its roots in early Florida land sales. He found that the modern names Sanibel and Captiva were likely the products of Anglo land boosters evoking the romance of the Spanish buccaneers.

Sure as the tides, the next people forced out those who had arrived earlier. In 1831, New York investors hoping to build a town, homes, and farms in America's newest territory gained dubious title to a Spanish land grant, set up the Florida Peninsular Land Company, and embarked on a scouting trip to find a dream spot. They also conceived the Florida press junket. A Key West town councilor, physician, and newspaper man named Benjamin Strobel sailed to Sanibel with the survey team and published the first detailed description of the island in a series that ran in the *Charleston Courier*. With verisimilitude, an old PR trick, he described arriving to "immense swarms of flees"—midges now cursed as "no-see-ums"—which he blamed on the remaining Indian and Spanish inhabitants. He complained that on his first night, heavy rains flooded him out of his ground-level palm hut, an event he likewise blamed on Indian and Spanish workers for not raising the structure.

But once he could explore, he was captivated by Sanibel, especially the sloped southern beach and its seashells:

On the south side is a beautiful sea beach, extending the whole length of the Island. Great quantities of elegant shells are washed ashore on this beach, by every unusual swell of the sea. It is a beautiful place for children to run about, in the cool of the evening, or for persons to ride upon. The north side of the island is six or eight feet above the level of the bay; the beach being narrow, and a complete wall formed within fifteen or twenty feet of the water's edge, by banks of shell, which have been piled for centuries by the sea.

Strobel gushed over shell-calcified soils prime for Sea Island Cotton; the ideal climate; and the ready market to Pensacola, Mobile, and other points around the Gulf. He promised would-be settlers vibrant farms, healthy living, and bountiful nature: A seine net could pull more than one hundred sheepshead in a single haul. The oysters and clams were among the biggest he'd seen. The deer and wild ducks "may be had with very little trouble."

Strobel soon lost his position with the company for killing a man in a duel. His Sanibel dreams would come true, if not in his time. Sixty settlers who bought into the Florida Peninsular Land Co. abandoned their farms just two years later to escape the surrounding battles of the Seminole Wars. Census takers made no mention of Sanibel in 1840, 1850, or 1860. In 1870 the island's census taker counted only himself and his son.

Two flashing beacons finally drew investment in Sanibel: the iron tower lighthouse that still stands on the southeastern tip of the island, and the first tarpon caught on rod and reel, hauled in by a New York sportsman from what is now Tarpon Bay on the island's north side.

Sanibel still rose with hulking Calusa mounds when the first archaeologists visited in the late 1800s. Before the turn of the century, homesteaders were using the mounds as rock quarries, hauling the shell away to mix tabby building materials, erect structures, and lay shell roads across sand. The homesteaders farmed in the calcified soils as Strobel had imagined, growing plump tomatoes that were famous in northeastern markets by the turn of the century. Many farmers doubled as fishing guides and innkeepers to nurture the burgeoning tourism industry. Tourism would prove both more lasting and more lucrative than tomatoes; only the very wealthiest Americans could make their way to the island. Women walked the beach in long skirts, parasols in one hand and shell baskets in the other, their hands covered in gloves to protect them from the sun and mosquitoes.

One afternoon as locals battled a brushfire near the seashore, three anonymous visitors wading near the lighthouse ran up to help. They turned out to be Henry Ford, Harvey Firestone, and Thomas Edison, who shared a love of seashells with his wife, Mina. Edison's infamous industrial research labs in West Orange, New Jersey, were crammed with countless thousands of articles mechanical and natural, from gears to gastropods. But the great inventor's gravestone in New Jersey features only one iconic image. It is not a light bulb or a phonograph, but a large, scalloped shell.

JULIA ELLEN ROGERS stepped onto this island of scrappy homesteaders and wealthy holiday-goers in 1905, during her summer on the Dimocks' houseboat. Judging by her attire in other photographs and what other women then wore on Sanibel's shores, she likely wore dresses with high collars, long sleeves, and sand-sweeping skirts as she walked the beach for seashells. Marveling over the island's "richest finds" in *The Shell Book*, Rogers noted that Sanibel "seems to be the meeting ground of the Atlantic and Panama faunas, suggesting that far-off time when no intervening land separated these now dissevered regions."

Yet even then, Rogers and others were beginning to lament the inevitable descent of too many people suffocating a fragile spit of sand and shell, and their daily hoarding of all the best specimens. "Sanibel is too popular," she wrote in *The Shell Book*. "Too faithfully are her beaches scanned."

Late in the summer of 1921, William J. "Bill" Clench, then a young field biologist, took his second shell-collecting trip to Sanibel and found a bounty of miniature specimens such as pointy augers and coquinas along the beach, outer sandbars, and in front of Sanibel's iron tower lighthouse. The haul was disappointing. "Many shells formerly common on Sanibel have disappeared, while others have become quite rare," he wrote. "The abundance of shells on the island, especially the larger and more showy species, attracts many tourist-collectors during the winter season . . . this might in part explain the paucity of many of these forms that were abundant a few years ago."

In October that year, and again in September 1926, major hurricanes pummeled Sanibel, and the Gulf of Mexico swept over the island's fruit and vegetable farms. They never fully recovered. Islanders increasingly turned to making a living from fishing and shells, and the visitors drawn by them. A ferry called *Best* carried tourists from the mainland to Sanibel, still home to fewer than a hundred permanent residents.

Since 1906, Sanibel's best hotels had hosted an annual shell show, a competition for the finest collections among their guests. The town put on the first island-wide shell show in 1927 as a financial and psychological lift while Florida spiraled in its first real estate bust, beating the rest of the nation to the Great Depression. The Gulf washed up gold in dark times. Even in the Depression, the famous and hermetic flocked to Sanibel and its more isolated sister enclave, Captiva. They rented out boats and barges, hotel rooms and local guides.

Born into a Cuban fishing family on Cayo Costa in 1901, Esperanza Woodring married the son of a Sanibel homesteader and took over her husband's guide service when he died, becoming a rare woman guide and earning a reputation as the best fisher and sheller on Sanibel.

Former President Theodore Roosevelt encamped on an elaborate fishing barge in what is now Roosevelt Channel on the east side of Captiva. Pulitzer Prize–winning editorial cartoonist Jay Norwood Darling, better known as "Ding," first visited the islands in 1935, the year after President Franklin Roosevelt appointed him to head what's now the U.S. Fish and Wildlife Service. Darling soon bought land and buildings on Captiva for his winter art studio overlooking Pine Island Sound. He put in a drawbridge to raise while working at his wooden drafting table.

The first malacologist to settle on Sanibel figured out how to find a Junonia—and so did commercial shellers. It didn't involve walking the beach. Dr. Louise Merrimon Perry, an ophthalmologist from Asheville, North Carolina, ventured to Sanibel with her husband in 1918 for vacation. They found the island "so lacking in conveniences," writes Sanibel historian Betty Anholt, "they turned around to leave." But the steamer wasn't returning to the mainland until the next day. Surviving the night, the Perrys saw the beach in the pink light of dawn and decided to stay.

After wintering in Sanibel for a decade, the Perrys moved down full-time and built a home near the southern bend of the island. Dr. Perry erected a marine lab, filled it with tanks, and outfitted a special dredge with a local boat captain. Among other sea creatures, she found Junonias living beneath rocky reefs a few miles offshore, and brought live specimens back to the lab.

Publishing the first guide to the marine shells of southwest Florida in 1940, she was also the first scientist to describe the denizen that made the envy-rousing volute: "The animal is strikingly marked with velvety-black spots and blotches on an ivory-pink ground color." Still, for a long time, most of the Junonia's admirers cared only for its striking shell.

IN WINTER 1939, the popular biblical writer Albert Field Gilmore filed a travel feature, "The Spell of 'Shell Shock,'" for the *Christian Science Monitor*. Gilmore had traveled to Sanibel Island "for rest and peace," and to bask in the warm sun on vacation. Instead, he caught a syndrome that was spreading, and barely slept. "Things fell out somewhat differently than we had planned. Immediately upon our arrival, it became evident that the visitors in the various cottages were hopelessly shell-shocked. Shell-talk was in the atmosphere! At the table, when we met a guest in our walks, in the quiet tête-à-tête in the evening, always the conversation, no matter upon what subject it started, reverted to the dominant topic—'shells.'"

Gilmore described the knee-deep windrows that piled up on the high-tide line. Professional harvesters and collectors alike crawled the shell heaps before dawn to pick up "anything new." By this time, commercial shellers were sending workers to sweep the beaches onshore, and to dredge for live specimens out in the Gulf.

In 1941, the novelist Theodore Pratt wrote with derision about Russel T. "Bing" Miller, the island's principal commercial shell shipper: He was "a wiry little man who, as a gesture, put on a shirt and tie" for the Sanibel Shell Show. "Living in a shack at the remote end of the beach, he has nearly every Negro on the island collecting shells for him," Pratt wrote, "He ships out three to four million yearly to shell art novelty makers and supplies choicer specimens to some seventy-five dealers in rare shells."

"Shell shock" was spreading across the nation. The first American shell club—the Tuesday Shell Club—had been founded in Southern California in 1902. Brooklyn shell aficionados launched their Conchological Club in 1907. Large shell clubs launched in New York City and Philadelphia; they would be followed by those in Chicago, Pittsburgh, and dozens of local clubs across the country, particularly in Florida and the Gulf states.

By 1940, conchologists had set up in thirty-eight states to buy and sell specimens by mail, corresponding with tens of thousands of amateurs who, if they couldn't find shells on their own, could order most species for between a nickel and fifty cents and have them delivered to their front door. Shell auctions such as those at Philadelphia's Buttonwood Farm in the 1950s drew collectors from across the northeast and coverage in the *New York Times*. "Miss Elizabeth W. Wistar, who collects volute shells, puts together and catalogues the auction items from experienced conchologists from all over the world," the *Times* breathlessly reported. "From the Aleutians to Africa, they snorkel and dredge for mollusks, often at great risk."

But finding one's own seashells was the dream. American collectors didn't have to journey to the Aleutians or Africa for volutes to fill a poet with longing. They could make their way to Sanibel Island. One amateur drove down from Chicago to load up on shells. On the return trip home, he called for his car at the hotel garage where he'd left it overnight and was seized by the police, who'd been called to investigate the stench coming from the trunk.

The police found not one dead body as the hotel employees had suspected—but thousands.

ON MARCH 27, 1955, a slim book called *Gift from the Sea* appeared on the *New York Times* bestseller list. Author and aviator Anne Morrow Lindbergh's marriage to Charles Lindbergh had put her in a grueling spotlight of celebrity—and tragedy when their first-born child was kidnapped and murdered in 1932. But she also had a soaring career of her own, having twice won the National Book Award for her first and second books. Now, "all the outward explorations she had joined were replaced by an inward journey," wrote her youngest daughter, Reeve Lindbergh, "one she described later as a 'journey toward insight.'"

Charles and Anne Lindbergh had first seen Sanibel Island in twilight on a clear evening in January 1940. They planned the getaway as an escape from the growing anger surrounding Charles Lindbergh's crusade to keep the United States out of World War II. Where Edna St. Vincent Millay used her celebrity to warn Americans against neutrality—"And he whose soul is flat—the sky," she wrote in "Renascence," "Will cave in on him by and by"—Charles Lindbergh pushed for a neutrality pact with Hitler.

The couple practiced their own isolationism on the remote islands of southwest Florida. In his diary, Charles described their approach on the Sanibel ferry that first year: the flash of the lighthouse, the palms on the beach rustling in dark silhouette against the evening sky. They drove shell roads through palm and pine forest to a small frame cottage on Captiva. The couple were restored by the primal beauty and privacy. Nine months later, their first daughter was born; a day before the release of Anne Lindbergh's book, *The Wave of the Future*, on the inevitability of fascism. The short manifesto was intensely criticized for its defeatism; Interior Secretary Harold Ickes called it "the bible of every American Nazi, Fascist, Bundist and appeaser."

Anne Lindbergh felt the book was misunderstood, and regretted that she'd rushed her ideas into print. She was also struggling to find an identity and voice separate from her husband's. Being the most important person in her children's lives, she lamented, was restricting her freedom to write with the depth and clarity of a man. As she and Charles returned to the islands the following winter, she began to write what she called her "feminist essay" on married women and creativity.

Lindbergh took her time with this book. In January 1950, ten years after she first traveled the shell roads, and after returning for memorable vacations with the family, she came to the islands alone. She rented a cottage at the bend of the road that linked Sanibel and Captiva, walked the beaches in solitude, and collected seashells as she thought through the manuscript she called *The Shells*.

In *Gift from the Sea*, Lindbergh used seashells to meditate on marriage, motherhood, and middle age. The Double Sunrise Shell reflected the early stages of marriage; two flawless halves bound together with a single hinge, an ideal but fleeting time. Oyster beds were the middle years, intricate and enduring bonds built by two people, "not primarily beautiful, but functional."

The moon shell represented the need to be alone, an idea that found an enthusiastic audience in the wives and mothers who flocked to the book. Lindbergh seemed to speak directly to mid-twentieth-century women dissatisfied with their confinement as homemakers. There were a lot of dissatisfied women out there. *Gift from the Sea* sold 320,000 copies its first year. The book remained on the bestseller list for eighty weeks—forty-seven of those in the No. 1 spot. It continued to sell millions of copies in a twentieth-anniversary edition, and later in a fifty-year edition.

Lindbergh's feminism for the 1950s homemaker seems quaint in the #MeToo era. The wealth and other privileges that allowed her to hibernate in remote places to work on *Gift from the Sea* surely were not available to most of her readers. The irony of the liberation that gave her time alone emerged two years after her death in 2001. Charles Lindbergh had led a double life. He'd fathered seven children by three German women, two of them sisters more than twenty years his junior, and one the private secretary who helped him with his business affairs in Germany.

Still, Anne Lindbergh's insights about the natural world imperiled by the amassing culture taking hold in the 1950s have stood the test of time. She succumbed to shell madness on Captiva, only to realize that her shell-crammed window ledges and bookcases obscured her enjoyment of any one seashell: "The acquisitive instinct is incompatible with true appreciation of beauty," she wrote.

"One cannot collect all the beautiful shells on the beach. One can collect only a few, and they are more beautiful if they are few. One moon shell is more impressive than three. There is only one moon in the sky. One double-sunrise is an event; six are a succession, like a week of schooldays."

Lindbergh warned against the excesses of consumption and development that would define the rest of the twentieth century and ordain the crises of the twenty-first. "It is only framed in space that beauty blooms," she wrote in a line that could apply to a condo-crammed beach as easily as a shell-crammed display.

Like the great conchologist Rumphius on Ambon three hundred years before, Lindbergh also felt a potency in found seashells, but frowned upon those dredged up for commercial gain or greed. Wisdom, like seashells, "must not be sought for or—heaven forbid!—dug for. No, no dredging of the sea-bottom here. . . . The sea does not reward those who are too anxious, too greedy, or too impatient. To dig for treasures shows not only impatience and greed, but lack of faith."

Yet the ethos of filling car trunks and entire rooms with seashells collected alive, their doomed animal huddled inside, would not change on Sanibel for another half century. Shells again revealed the human impulses around them, foretelling a new rise of materialism that would outrival even the Dutch golden age that had launched the first global corporations.

As Lindbergh helped her readers understand their own oppression, neither she nor they saw the real gift from the sea. The gift was not the shell. It was the life within.

BY MID-CENTURY, SANIBEL motels offered an amenity even more ubiquitous than the shuffleboard court: special boiling stations to encourage tourists to kill and clean their mollusks down by the beach instead of in their rooms. Writing in the *Saturday Evening Post*, the novelist Pratt complained that so many pots boiling at once, "so that the mollusks may be removed from their shells without a tug-of-war," created a hellacious stench in the Sanibel night air.

The many articles about live shelling from the era stressed that empty shells were for amateurs. "Beach-worn empty shells are likely to have lost their true color, luster and shape," the *New York Times* lectured.

"Real shell collectors must take a shell alive in order to have it count," the *Washington Post* reported, "just the way bird watchers may not score birds observed in captivity, heaven forfend."

Shell books and guides, too, advised that, in the words of the 1961 *How and Why Wonder Book of Sea Shells* for children: "To get the best shells, you must bring 'em back alive."

Shell lovers not only brought back conchs, whelks, and olives alive, but plunged them into bleach alive, froze them alive, and boiled them alive. Not to consume the meat—a smelly burden to discard—but to hoard the shells.

Some guides recommended burying live shells in ant piles for a few days. Let ants be helpful for a change, observed A.P.H. Oliver in the 1975 *Hamlyn Guide to Shells of the World*. "A bent piece of wire is often the final answer," Oliver also wrote, after enthusing that all mollusks have a brain, lay eggs, and sometimes even sit on their eggs. (Rather than a brain in the strict sense, they have sensitive nerve cell clusters known as ganglia.)

A popular volume called *How to Clean Seashells* might have been better titled *How to Kill Mollusks*. The author, a Canadian-born radioman named Eugene Bergeron, had caught the shell-collecting fever down in Florida when he and the century were both in their late teens. He joined the Navy and collected throughout the Pacific including at Pearl Harbor, where he found the best shells on the leper colony of Molokai. His guidebook was novel for its techniques honed to different species: Refrigeration for conchs, whose body could then be extracted with a strong pointed tool, and for cowries, which could be picked out with a dissecting needle. A nut pick for tun shells. Alcohol for whelks and murex. Chitons should be mummified in sturdy tape to keep them from rolling up, their way of dying. Hermit crabs in a desired shell should be plunged in freshwater with a little Clorox: "The little crab will crawl out of his shell and die in a very short time."

Bergeron did not recommend muriatic acid, but so many collectors insisted on it that he included instructions on how to tie the shell to a string and lower it into the acid bath.

AMERICA'S POSTWAR SEASHELL craze was partly tied to the Pacific Theater and all the young soldiers who pocketed tropical beauty to bring home to sweethearts—or perhaps cradle in shaky hands. R. Tucker Abbott had loved seashells long before, since creating his teenage shell museum in the basement of his parents' house in Montreal. When he left for college at eighteen, he donated the seashells to McGill University's natural history museum. He never again curated his own collection. But he devoted a lifetime to helping other shell lovers with theirs.

At Harvard, Abbott found a "second father" in William Clench, the professor to whom he had written his "dear curator" letter. Clench was a warm mentor and peerless field leader—he made his 2,500th field station in Florida at the age of eighty-five. He estimated he wrote thousands of letters to students and former students fighting in World War II, including to Abbott. The young scientist served as a bomber pilot for two years before the Navy put his snail savvy to use. Abbott was assigned to the Navy's Medical Research Unit in Guam to help battle schistosomiasis, a fatal tropical disease passed to humans from freshwater snails. Trudging through rice fields in China, and working from a makeshift lab in an ambulance strapped to a railroad flatcar, he was credited with discovering and describing a tiny flatworm killing thousands of people.

Abbott went on to found research and scientific journals devoted to mollusks, and to curate some of the most important scientific collections, including the world's largest at the Smithsonian and America's oldest at the academy in Philadelphia. But he seems to have been made for his more-public role, as an approachable expert eager to answer anyone's questions about shells. He published the first of his many popular books, his tome *American Seashells*, in 1954 while he was still at the Smithsonian. In 1962 he came out with *Seashells of the World*, just the right size to tuck in a beach bag. Carole Marshall, a retired truck driver who is president of the Broward Shell Club, told me a typical story of how she'd idolized Abbott ever since her mom gave her a copy of his *American Seashells* for her twenty-third birthday. "I pored over it and over it and at first I never imagined I would join a shell club, or own a golden cowrie or a slit shell," Marshall says. "It never crossed my mind that I would meet the author of this amazing book. Much less that someday he would know my name and say, 'Hi Carole!'" when he saw her at shell events.

Over the years, Abbott increasingly doubted that research science could move the world. The urgency as he saw it was to change hearts and minds. Before he turned sixty, public demand for his seashell books became so great that he resigned his scientific malacology position and moved to Florida to form his own publishing company. The move was a point of tension with some of his former colleagues. "The ivory-tower boys look askance if you get out of the tower," Abbott told a reporter. "They feel like you've sold your soul to the publishing devil."

In 1977, Abbott embraced his "Mr. Seashell" role full-time, walking the beaches in his Panama hat and working toward "the diffusion of knowledge more than the increase of it." He agreed to help local collectors pursue a dream to build the first American museum devoted entirely to mollusks on Sanibel. Seashells would be on display for the public, rather than stored away in drawers as they were in so many museums.

BY THE MID-1970s, visitors to Sanibel Island were collecting so many live mollusks that some locals feared the island's signature attractions would be wiped out. A popular joke had it that the best shells in Florida were to be found at the Georgia border. That's about where cars headed north from Sanibel started to stink, and families had to pull over to dump their shells on the side of the road.

Some residents and business leaders began to suggest the unthinkable: a ban on live shelling. Mollusks needed protection from the very people who loved them most. Shell shop owners and several noted malacologists were aghast. Abbott was among the scientists who weighed in to say that collectors couldn't possibly make a dent in the populations of such fecund animals, which would tumble to shore each day in the tides and storms regardless of how many were pocketed as keepsakes.

Some of Abbott's biggest fans disagreed with him. The new environmental ethos taking hold in the United States had trickled down to the modest mollusk. In the summer of 1977, dozens of Girl Scouts from around the nation met on the island for a weeklong shelling extravaganza. Abbott was generous in giving his time to young people, and keynoted the event as he had even when he was duPont Chair of Malacology at the Delaware Museum of Natural History. The girls treated him like a rock star; their flashbulbs popped as soon as they spotted him. Weighing in on the local controversy during his speech, Abbott told them they shouldn't feel guilty about collecting live shells, because no local species were endangered, "certainly not from over-collecting, though the whole west coast of Florida is endangered from overdevelopment."

"A conservationist," he told the girls, "is someone whose shell collection is already big enough."

The girls chose not to heed his advice. That year, the Scouts and their leaders voted that each girl could collect a maximum of two live specimens during their week on Sanibel. Several girls wanted to go further, and pledged to collect only empty shells. They were two decades ahead of the grown-ups.

The city council appointed a Live Shelling Committee that included shell people of all walks. Its eloquent chairman, a retired entomologist named K. C. Emerson, had fallen in love with seashells as a young soldier serving in the Philippines during World War II before walking the Bataan Death March and being held prisoner of war in Japan. He made the case that Sanibel had an obligation to protect its soft mascots not from the ordinary collector, but from "flagrant abusers who hoard live shells by the bucketsful and pay for their island vacations by selling the shells they gather."

The panel heard testimony about annual visitors who came at Horse Conch breeding season to gather Florida's gargantuan state shell by the dozens. Esperanza Woodring, the longtime fishing and shelling guide who had been on the islands since her birth on a Cayo Costa fishing rancho in 1901, implored the panel to limit live shelling to two specimens a day, with special protections—one per person— to Horse Conchs, Angel Wings, and True Tulips.

Anne Joffe, owner of the She Sells Sea Shells tourist shops hawking specimens from around the world, was among the shell experts who helped delay such limits for years. A tall redhead with Junonias embroidered on her pants, she led the Sanibel Shell Show and the local shell club on and off for half a century. Joffe made an argument also being stressed by malacologists: Live shells washed up on the beach were doomed to die, anyway. "Once a shell hits the beach," she said, "the chances of it surviving are nil." (To the contrary, mollusks can survive when they are gently placed back into the sea, or drawn back by the next tide.)

It was 1987 before Emerson and fellow conservationists were able to convince the city council that "time is going to run out for our shells if we don't take some action." That fall, the city limited live shelling to two mollusks per person per day. Nearly ten years later, in 1995, the Florida Legislature agreed to let the city ban all live shelling; it's now law in all of surrounding Lee and one other Florida county. California, already facing extinction of its prismatic delicacy the abalone, also came to ban live collection of tidal invertebrates of all kinds, except for shellfish collected by licensed sport-fishers.

By that time, Abbott's dream of a "monument to shells-for-people, not just a museum full of shells," was nearly complete on the highway running between Sanibel and Captiva. Abbott died of a stroke two

weeks before the grand opening of the Bailey-Matthews National
Shell Museum. In a retrospective on his contributions, the retired
Smithsonian malacologist M. G. "Jerry" Harasewych called Abbott
"undoubtedly the most widely known malacologist in the world."
But Abbott probably would have been proudest of the museum,
which has led a new ethic in the island's relationship with mollusks.
In the years since, it has opened the world's first aquarium devoted
solely to living mollusks.

The motels have kept their mid-century motor court charm, but
not the killing stations. Tide charts are stamped with reminders to
take only empty shells. At the Island Inn's shell-cleaning station—
now just a big sink—a cartoon gastropod in flip-flops reminds visitors: "Don't take us home if we are still alive!"

PART OF THE new generation of malacologists born after the
first Earth Day in 1970, Gregory Herbert grew up in Louisiana, but
his earliest memories—picking up seashells with his grandfather—
were etched on the beaches of southwest Florida. He had a shell
collection by age three. As soon as he could read, his grandfather
bought him Abbott's *Seashells of the World.* Herbert read shell books
voraciously, but no one at school noticed that strength; an auditory
learning disorder kept science and math out of reach. In college he
earned a philosophy degree. He went to work building and reno-
vating houses in major cities including Philadelphia, where a shell
lover couldn't spend much time without discovering the Academy of
Natural Sciences. Herbert went to meetings of the Philadelphia Shell
Club at night after hanging drywall during the day. A lecture by the
academy's curator of mollusks led to a conversation, which led Herbert to volunteer on a project, which led academy scientists to notice
his aptitude for mollusk research.

Four years later, at age twenty-six, Herbert began a PhD program
at the University of California in Davis with Geerat Vermeij. The
shell prodigy with a hearing disorder found his mentor in the shell
prodigy who couldn't see. For part of his doctoral research, Herbert

analyzed drill holes in shells 2 million years old, fossilized at the bottom of the Gulf of Mexico. Vermeij's work had shown how the rise of predators led to an evolutionary arms race in which mollusks developed increasingly elaborate shell defenses. Herbert's work now foreshadowed how the *disappearance* of predators could remove the drive for natural selection, bringing evolution to a halt.

Herbert's results raised alarms about modern biodiversity losses, most obvious among top predators like the Horse Conchs that once flocked like gulls on the barrier islands. (Groups of mollusks are often described as "congregations," though conchs herd, scallops school, oysters might be found in a clutch, clams in a bed.) Now a professor of paleobiology at the University of South Florida, Herbert began to sample modern shells in the Gulf, comparing numbers of mollusks that lived on the seafloor before 1950 to numbers living there now. "We've forgotten what's normal, what's supposed to be there, how ecosystems are supposed to operate," Herbert told me. "By the time we began to take our first ecological surveys in the 1960s, that was already the height of change."

Plying the Gulf on a 115-foot research ship called the *Weatherbird*, Herbert and his crew of students and scientists scoop up mollusks and shells at more than two hundred spots across the seafloor. Back at Herbert's lab in Tampa, they identify shells and analyze the diversity of life at each spot to see how things are changing.

In 2018, the marine biologists José Leal and Rebecca Mensch of the Bailey-Matthews museum joined Herbert on the *Weatherbird* with hopes that its dredge might pull up a live Junonia. Of the trillions of images stored on the internet, no one could find a photograph or video of the elusive creature (though one turned up in an old copy of *National Geographic*), making it that much harder to share the life history of the most beloved shell on Sanibel.

On a clear night in February, Mensch was working the dredge at 2 a.m. when she saw a mahogany-dotted form spill to the deck in the vessel's bright spotlight. She was too tired to yell. With quiet elation, she picked up a perfect live Junonia. By the end of the expedition, she would have three. As shrimpers and commercial shell dredgers had known for half a century, the animals were not really rare in their rocky habitats offshore. It was just rare for one to tumble all the way to the beach undamaged.

Mensch packed her charges in a glorified cooler, drove them carefully back to Sanibel, and set them up in their aquarium home with a couple of Lettered Olives to eat. The Junonia liked to burrow under the sand. But sometimes they'd reveal their billowy mantles, mottled in bright yellow and velvety black as Dr. Perry had described them in 1940.

Scientists long assumed Junonia, like other volutes, would envelop their prey with that lovely foot to trap and devour it. But they had missed a shocking step, which Mensch captured in the first live-action video featuring the star of Sanibel. In the clip, a Lettered Olive creeps up the Junonia's right side from behind. Creamy beige body matching its polished projectile, the olive begins to wrap itself around the Junonia's black and yellow flesh. In a flash, the Junonia draws a long, sword-like proboscis from the folds of its mantle and stabs it clean through the olive's soft center. *Oliva sayana*, named for the nation's first conchologist, recoils violently into its shell, which slides to the bottom of the tank, unmoving.

The paralysis suggests a fast-acting toxin, Leal and Mensch concluded in a paper on the Junonia's strike, until then unknown by the humans who craved the shell for beauty alone.

THE *WEATHERBIRD* EXPEDITIONS in the Gulf of Mexico are revealing patterns that scientists also see in many other parts of the world: Even coveted mollusks living well away from humans and their harvesting, coastal development, and polluted runoff seemed to be maintaining their populations. The closer they live to people, the harder the animals have it. The Iberian Peninsula where our Neanderthal cousins once gathered seashells has now seen a precipitous drop in the number of shells on the beach, the paleoecologist Michal Kowalewski has found, a signal that the ecosystem is also in trouble.

In Gulf of Mexico habitats, Herbert's preliminary results show some species have disappeared where they once flourished. Many that remain are clumped in hotspots offshore, largely out of human reach. At least that's true for now; a federal moratorium on oil and gas exploration in the eastern Gulf of Mexico expires in 2022.

Beyond harm from drilling, says Leal, the malacologist who came to Sanibel to lead the shell museum after Abbott died, the ocean warming and chemical changes wrought by burning the oil and gas will be "many orders of magnitude larger" than all the collections of all the shell aficionados in all the history of the world.

Part III

~~~~~~~~~~~~~~~~~~~~~~~~~~~~~~~~~~~~

ORACLE

Ten

~~~~~~~~~~~~

THE END OF ABUNDANCE

THE BAY SCALLOP
Argopecten irradians

Along the coastal curve of Florida, where the peninsula begins its westward turn to the panhandle, seagrass shimmies like laughter beneath the shallows of the Gulf of Mexico. My husband and I and our two children have zipped out to the grass flats by way of the Steinhatchee River in our 16-foot skiff; little outboard too loud to talk above, speed too fast to admire the run past an unpretentious town and palm and pine forests that open to sweeping salt marshes where the river meets the sea. Such is our hurry to reach the laughing grass.

At the undersea meadows, we slow to a putter. The morning is calm and cloudless. The water shines in the mirage travel writers call glass; a miss on metaphor and place. Even with a slack tide and no wind, the Gulf is always animated. Rhythmic little waves cut the surface and thump our flat-bottomed boat. The children are still small enough to lie side by side on their bellies on the bow to look down into the water while Aaron motors slowly across the seagrass. At 6 feet of depth, the view becomes so clear that we can see individual grass blades undulating below.

Turtle grass reigns in this sweep of the Gulf of Mexico known as the Big Bend, home to the most extensive seagrass meadows surviving in North America. Flat green ribbons with tawny tips sway in thick, foot-high fields grazed by the sea turtles that give the grass its name. Also named for its gentle grazer, manatee grass grows in thin, round shoots. The meadows pass below as if we're in a glass-bottomed boat. The kids peer down with keen eyes. When they see a promising patchwork of sunlit blades and sand, shallow enough that they can easily dive to the bottom, they call out for their dad to cut the motor. We toss the anchor to give it a try.

The first scallop is always a celebration. Our daughter gets it with rigged odds, snorkeling in the brace of my arm. I spot one and point. With a kick of her legs, she dives down to snag the inaugural prize and drop it in her mesh bag.

The Bay Scallop, *Argopecten irradians*, rests on the grasses with the dark, mottled side of its shell at the top. It's like many other bivalves in that the two shell halves are not mirror images; the lower is rounder and lighter colored—often snowy white. For the most part, though, scallops aren't like many other bivalves—or any other creatures on Earth: blue-eyed, hopping, skipping, spitting, finger-pinching, jet-propelled, zigzagging, shell-clapping, free spirits of the laughing grass.

These Bay Scallops see with twenty-two electric-blue eyes, one at each ridge of their rippling shells. Seen through our dive masks, the eye rows glow battery-charge blue, like tiny flying saucers have landed in the seagrass. Peering through powerful microscopes, scientists find that each eye contains a tiled mirror rather than a lens, bearing a striking resemblance to the segmented mirrors on reflecting telescopes such as the Hubble. The mirrors reflect images back to two retinas in each eye. An upper retina allows the scallop to see a dark, moving figure like a girl with a mesh bag or a crab with a hungry pincer. A peripheral retina appears to help the creature navigate while swimming in its zigs and zags.

Scallops dart in the water column by clapping their shells together with their adductor muscles—the sweet white delicacy that Aaron will later sauté briefly in butter. The clap forces water out of the shell, sending the creature darting like a cartoon clam. Using its mantle to push the water out in different directions, a scallop can zip about strategically, running a lateral play to trip up crabs or other predators giving chase.

On this day, the scallops are big as my daughter's hand, and prolific in our spot, some congregating in pairs. Gathering them is a zen experience, like picking berries. The kids get caught up filling their bags with the snapping shells. Well before lunchtime, we've hauled up all the scallops we're allowed to keep for the day—two gallons, in their shells, for each person. That's more than enough for two meals for a family of four, and it means we'll spend hours carving white muscle away from dark viscera. The scallops' zigs and zags couldn't outrun us humans and our dive bags, now heavy with our bounty— and our burden.

FOR EVERY TEN species of gastropods shambling about the world on their big, soft foot, there is roughly one species of bivalve. But given the huge numbers of what are collectively known as shellfish, and owing to their importance as food, bivalves get most of the attention. They make up the bulk of ancient shell mounds and molluscan catch historic and modern. Oysters and clams support and define many of America's last-surviving fishing villages. They are the subjects of books all their own, not to mention sandwiches: po'boys in New Orleans, clam rolls in New England. Among all mollusks, bivalves draw the greatest research interest, by far the most government funding, and almost all the regulatory concern.

They are less popular among shell lovers. Most oysters, mussels, and clams (save the giants) are hardly even considered seashells. Scallops are a regal exception. Once ambrosia for the gods, scallops are still a delicacy for seafood lovers, and their shells are also collected, especially the bright yellow genetic anomalies among them. Scallop shells became religious icons, inspired artists and architects for thousands of years, and "truly stood among the most favored of shell collectors' trophies," Tucker Abbott wrote in his *Kingdom of the Seashell*.

The scallop, for many people, defines a shell. It is the sloop of Aphrodite. It is a favorite icon in coats of arms. Its stylized logo is so recognizable for Shell Oil that the company need not include its name. The words *scallop* and *shell* come from the same common ancestor: *skal*. Germanic tribes in Central Europe in the centuries before Christ used *skal* to describe hard coverings. The word made its way into several dialects; we get *shell*, *shirt*, and *skirt* from the same forebear. The modern Dutch word for shell, *schelp*, became *escalope* in French and *scallop* in English.

To scientists and shell people they are the pectens, Pectinidae. Their shape and ribs reminded the Roman naturalist Pliny of the round hair combs of his day; he called them "comb shells"—in Latin, *pecten*. After Linnaeus lumped scallops with oysters, the Danish biologist Otto Friedrich Müller restored their combed distinction. Müller is better remembered for inventing the scientific dredge.

The recurring depiction of scallops in art and architecture is part aesthetics and usefulness of shape: as an arch, a niche, a marble fountain, a buttress for a lounging goddess, or a Queen Anne chair. A scallop niche is carved into the red cliffs of Syria's Banias (Arabic adaptation of Paneas, meaning "place of Pan") above the ancient grotto dedicated to the hooved god of the wild. Scallops became such common architectural motifs that the Romans seemed to raise the shells on every sort of public space. From ancient niches, scallop domes rose in grand Christian churches, mosques, and synagogues like the former Ibn Shushan of Spain. Inside the Great Mosque of Córdoba, a great scallop-shell dome with gold and blue mosaic ribs was inspired by the shell motifs prevalent in the ancient Umayyad dynasty that ruled the Islamic world from Damascus after the death of Muhammad.

Scallops were among the marine animals associated with a life-giving goddess who came before the patriarchal gods; an earth mother dating to Paleolithic times and defined by her generative powers rather than her dalliances. Painted on a private garden wall in the ruins of Pompeii, Venus appears as this older goddess reclining on a huge scallop-edged shell. Around her neck hangs a delicate cowrie.

In an ancient celebration of Demeter—the Greek goddess of agriculture—women were said to march while holding scallop shells aloft. On special occasions, celebrants feasted on scallops and also served them up to the gods. Archaeologists researching archaic feasts at the sanctuary of Apollo in Thrace, Greece, find more scallops than any other sorts of shells. Light fire marks reveal they were roasted, briefly, over the coals.

Symbolically, the round-ribbed birthplace of Aphrodite suggested the fecundity of the sea. Scallops signaled birth and resurrection long before Botticelli painted *The Birth of Venus* (1485), the Roman version of Aphrodite rising from a scallop shell. The Maya buried scallops and spiny oysters with the dead—a ritual connected with a watery underworld and dream of rebirth.

At the turn of the twentieth century, Julia Ellen Rogers would note their aphroditic abundance:

> To see hundreds of scallops the size of a silver dime flitting through the shallows on a bright summer day will certainly convince you that even mollusks can express the joy of living, as plainly as a flock of blackbirds or a troop of boys bound for "the old swimming hole."

Rogers didn't see the need to devote a scientific entry to the Bay Scallop in *The Shell Book*. Flitting through the shallows and dredged up in heaps from the seagrass for flash-frying in boiling fat in the pushcarts of New York, "our" scallop was too common.

SEAGRASSES GROW LIKE native prairie grasses, roots digging into sand and tangling sideways, shoots springing toward the sun in coastal shallows from the tropics to the Arctic Circle. They flower with the seasons. They fruit and seed. They make the largest pollen grains in the world, then send them into the currents to fertilize a bloom. The Spanish call them marine meadows—*praderas marinas*. Marine scientists say they are some of the most overlooked habitats in the sea.

Coral reefs get the glory. But the world's seagrasses are also vital to marine life, supporting as many as 50 million organisms in each acre. The undersea meadows rank close to estuaries and wetlands in all they give, including keeping great stores of carbon dioxide out of the atmosphere. Scientists who calculate nature's value say the grasses contribute $1.9 trillion a year to the global economy in just cleaning coastal waters by cycling nutrients.

Bay Scallops are among the huge range of marine creatures, from the tiniest gastropods and starfish to some of the biggest fish in the sea, that begin life sheltered in the grass. More than a hundred types of sun-loving algae live on the turtle grass alone, clinging to its tips. Eggs and larvae glom on to the stalks. Sea anemones and urchins tuck themselves in the roots. Small fish hide and big fish feed among the blades of grass. All those lives lure sea turtles, sharks, and marine mammals to cruise through, and sea birds to wade or plunge from above.

A. irradians is born in the grass and dies there only a year or two later. Hermaphrodites with female and male sex organs, Bay Scallops spawn en masse, spraying both eggs and sperm into the grass. Fertilized eggs squirm into larvae known as veligers. Only a few will make it to baby scallops. Called spat, the babies sport the perfect classic scallop shell in miniature. Placed atop a penny, a spat's shell covers just the tip of President Lincoln's pointy chin.

To make it to adulthood, the mini-scallop holds on to the grass for dear life. Spat weave a silky anchor rope, the byssus, to tether themselves to a blade. Like a belly button, a small, distinctive opening at the base of most scallop shells and their ancient fossils marks the spot where the lifelines once unfurled.

The spat outgrows Lincoln's chin to reach the size of a dime in a few months. That's when a scallop will let go of its byssus to swim free.

"BAY SCALLOPS POSSESS an indefinable lusciousness," the American zoologist and journalist Ernest Ingersoll wrote in 1887, "not possessed by any fish or fruit, yet approximating a combination of them all."

Ingersoll, a Michigan native, was twenty-seven years old when the U.S. Fish Commission hired him to conduct the first survey of the nation's shellfish industry. He had a dark mustache then that pitched over his lips like a field tent. He'd helped survey the Rocky Mountains with special attention to mollusks, already naming a half dozen and having one named for him. He'd also proven his gift for interviewing locals and describing their experiences with wildlife and sea life, publishing science articles in *Scribner's Monthly* and other popular magazines and newspapers.

His first *Scribner's* article made clear his fascination for mollusks— "little snails with slippery tails," he wrote in verse, "who noiselessly travel across my gravel." He celebrated them as creatures of "vast multitude and variety, ancient race, graceful form, dignified manners, industrious habits, and gustatory excellence."

Ingersoll's government monographs, uncommonly vivid, are the earliest official accounts of America's commercial shellfish harvest. He began with the great shell middens. Native Americans harvested fewer scallops than oysters, mussels, clams, cockles, and whelks, though Ingersoll knew of the soaring scallop-shell heaps on Florida's Gulf coast and described how some northeastern tribes fashioned the shells of the largest species into shamanic medicine rattles.

For Natives and European settlers along the Atlantic, serious scalloping followed other types of shellfish harvesting by centuries. Gathering scallops in commercial numbers required mechanical dredges for the seagrass beds. In middens along the northeastern U.S. coast, Bay Scallop shells were scarce in the deepest—oldest—strata, and more common in the younger layers. Likewise, when Ingersoll visited America's commercial scalloping industry in 1879, it was just taking off—while oystermen had been plunging their wooden tongs into the great bays of the Eastern Seaboard and reseeding the beds for a century.

From Cape Cod through Rhode Island, Long Island, New Jersey, and points south, Ingersoll found a family industry that employed hundreds. Men set off into the bays in rowboats or sailboats towing small, triangular dredges through the seagrass to scoop up scallops. Onshore, they hauled the bounty to long wooden scallop houses, where women stood at waist-high counters prying shells open and cleaving white muscle from the dark guts they pushed into holes with refuse barrels below. Young wives worked "with cradles behind them containing less than year-old babies, opening scallops with their hands, singing merrily some baby song to quiet the young ones."

Like Rogers, Ingersoll took note of the Bay Scallop's abundance. In Sag Harbor, a boat captain named Pidgeon told Ingersoll he'd seen scallops swimming in schools 10 feet deep. In Oyster Bay on Long Island, the eelgrass blades grew so heavy with baby anchored scallops that the stalks eventually broke from the added weight. Come fall, when the grass floated about the bay in huge clumps, it spread the young scallops, creating a bountiful crop.

Still, some of the most heavily dredged scallop grounds were already showing signs of depletion. Long Island Sound, New York Harbor, and much of the New Jersey coast, Ingersoll reported, had started to see a "depopulation" of scallops. Dredgers that hauled up to a hundred bushels a day at Greenwich, Connecticut, a decade before were now landing about ten. One Cape Cod fisher complained of "the ruin which was being perpetrated by the too-greedy pursuit of scallops in the waters south of Barnstable."

Ingersoll reported a peculiarity of scallops that made it difficult to suss out human-caused declines from natural cycles: Scallops can be wildly abundant one year, and scarce the next. In the Long Island ports, fishers told him their numbers fluctuated on five-year cycles: "The second year following the season of plenty would produce a few, the third year a scattering one or two, the fourth year absolutely nothing. Then would come a sudden accession from some unknown source."

But he was rankled by a theory that scallopers shared everywhere he went: The more they raked up scallops from the seagrass, they assured him, the more abundant future harvests would become. "I heard this from many dredgers myself, and the reports of others contain the same assertion," Ingersoll wrote. "Raking, they say, scatters the young and keeps them from crowding one another; in short, it lets them grow. Yet in each locality they will tell you that the yield there does not compare in quantity with ten, fifteen, or twenty years ago."

He could not then see how American notions of progress, particularly coastal development, were beginning to alter the future of scallops and other marine life more profoundly than the triangular dredges and rakes ever could. Bay Scallops had evolved with seagrasses; their lives were enmeshed with the underwater meadows. Like the filling of wetlands and felling of mangroves on land, the shallow meadows were being dredged for harbors and deeper channels, and scooped up in fill for waterside development.

Progress also seeped into the water from septic systems under the houses being erected on the coasts.

Progress poured out of factories then pumping toxic waste into rivers that fed the bays.

Progress spewed into the atmosphere from smokestacks. Its harm to the scallops, and the seagrasses, did not become clear for another century.

FROM ITS EARLY ties to abundance and harvest, in Christianity the scallop came to symbolize the Apostle James—and the pilgrimage to his legendary burial place at Santiago de Compostela, a town in northwestern Spain.

As a young man growing up by the Sea of Galilee, James left his fishing family to become one of Jesus's first disciples. After Jesus's death, he brought God's word to Galicia, lighting a flame that continued to burn across Spain after his return to Jerusalem—and his execution there in 44 CE.

The legend goes that after witnessing his martyrdom, the remaining apostles whisked James's body to Jaffa, where they found a ship, guided by an angel, to carry him back to Galicia. That's where the scallops come in. In one version, a horse and rider who've fallen into the sea at Galicia surface at the ship, encrusted with scallop shells. In another, James's own body rolls into the sea and surfaces, covered with the shells.

The real origin might have been more simply the beloved saint's fishing heritage. Regardless, the scallop shell became the emblem of St. James. And in 814, the discovery in Galicia of his supposed tomb set off a stampede of devotees to Santiago de Compostela. Scallop shells became the abiding symbols of the pilgrimage.

During medieval times, thousands upon thousands of Christians from across Europe set out on the long, self-punishing journey called the Camino de Santiago ("Way of St. James"). Many had convinced themselves they carried such grievous sins that no priest or confessor could absolve them. Only a mission to Santiago would do. From the ninth century onward, they would sew a scallop shell onto their hat or clothes to identify themselves. The pilgrims drew both fascination and great sympathy. Those not trying to scam them felt obliged to help, often contributing halfpennies toward the journey.

The twelfth-century *Pilgrim's Guide* is considered one of the world's first tourist guidebooks. It detailed four primary routes through France, including an arduous trek up and down the Pyrenees, all coming together at Spain's Puente la Reina for the final leg to Santiago. The guide offers practical advice like which rivers were safe to drink from and which would kill, eels to avoid eating, or where to wash one's private parts. It's also full of stereotypes about the local people pilgrims would encounter, detailing different groups' shabby clothes, piggish manners, or perverted sexual habits.

The first-person account is believed to have been written by the French monk Aymeric Picaud. Once he gets to the city and basilica, Picaud describes the glory and miracles that make the pilgrimage worthwhile: "For health is given to the sick, vision to the blind, the tongue of the speechless is released, hearing is revealed to the deaf . . . the chains of sin are loosened, heaven is closed to those who assail it, solace is given to the sad, and people of all nations, from every part of the world, come trooping together bearing gifts of praise to the Lord."

And everywhere, scallop shells: for sale to the pilgrims at the entrance to the city and built into the basilica itself, including a "miraculous fountain . . . upon which is a beautiful stone shell, like a dish or cup, round and enclosed, so large, I think, that fifteen men could bathe in it."

Compostela is inland; regulated by the church, vendors brought the scallop shells in from the coast to sell to pilgrims as keepsakes and as proof they'd completed the journey. On the way back home, travelers could present their shell at homes, churches, and abbeys for as much food or drink as they could scoop up in it. The small dish meant even the poorest households could give without burden.

In the first half of the thirteenth century, more than a hundred English ships sailed to Spain carrying thousands of seekers. Many more came by Picaud's land routes. The poor limped and walked. Royals were carried by slaves and horses. Fearing bands of robbers and other threats warned of by Picaud, pilgrims went in bands for safety, and groups of knights organized to protect them. The pilgrimage was such an honor that high-ranking families who could claim it added scallop shells to their coats of arms. Soon, scallops adorned family crests throughout Europe, the shields of soldiers in the Crusades, and abbeys and churches along the pilgrimage and well beyond.

In his *Church History of Britain*, Thomas Fuller wrote of the pilgrims who returned from Compostela *"obsiti conchis,* 'all beshelled about.'" Returned pilgrims so cherished their shell keepsakes that they often passed them down as heirlooms from father to son, and on to grandchildren. Many pilgrims were buried with their shells.

Archaeologists still find natural and crafted scallop shells associated with Santiago. Nearly a thousand years old, they lie along the pilgrimage routes and far away. At the site of an old British leprosarium at Winchester dating to the Norman Conquest, researchers recently excavated the grave of a young man who died in the twelfth century, around the height of the pilgrimage. He was between eighteen and twenty-five years old, and DNA evidence showed that he suffered from leprosy. Still on his torso was a large, reddish scallop shell—pierced with two holes for attaching to pilgrim's garb.

Perhaps he'd read Picaud's words along the way: "Health is given to the sick," and never lost faith in his scalloped charm from St. James.

The beautiful Mediterranean scallop came to be known as the St. James Scallop. In Spanish, St. James translates to Santiago, and in French to St. Jacques. Thus the name of the creamy scallop dish broiled and served in the elegant shells: coquilles St. Jacques. More popularly, people call it the Pilgrim's Shell.

Today, hikers set out for a month to walk the 450 miles from the French border town of Saint-Jean-Pied-de-Port. The route is lined with all manner of scallop shells, set sideways with the top pointing the way. Some local towns have mounted elegant brass scallops into stone walls along the trail; some are stylized yellow shells carved into cement markers; others, bright orange scallop shells fashioned into an arrow pointing onward. Residents along the route festoon their gardens and houses with scallops in solidarity with the pilgrims.

With scallop shells now strung on necklaces or fastened to backpacks, the modern hikers and bicyclists still follow the route trod by long-ago pilgrims.

BEFORE THE CENTURY turned to 1900, Ingersoll was lamenting the decline of Bay Scallops in the nation's bays. "An effective reply," he wrote, "to those men who told me that they thought the more the scallop beds were raked the more plentiful the mollusks became." He also described the extirpation of the Giant Scallop, "which formerly abounded on the coast of Maine . . . now become so rare as to be a prize in the cabinet of the conchologist rather than an edible commodity—a result unquestionably due to over-greedy catching."

Overharvesting was part of the story. In 1930, Virginia's eelgrass beds nurtured the most productive Bay Scallop fishery in the United

States. In 1933, the scallops vanished and never returned. The collapse was long blamed on a wasting disease that wiped out the eelgrass, followed by an August 1933 hurricane that swept away the last grass and straggler scallops. Later, researchers trying to restore the scallops found that grass-scraping dredges also contributed to their demise. The numbers of scallops reported harvested in 1930 likely exceeded the adult population, meaning commercial fishers took young scallops that never had the chance to spawn.

But while Ingersoll was right to cast greed as an existential threat, the dredgers were only the first link in a long chain of coastal development, destruction, and dumping that ignored the consequences amassing in the bays.

Dredge-and-fill projects, ports, and bridges were another link in the chain. Though Ingersoll never made it down to Florida, museum collections show that Bay Scallops once thrived in disparate populations on the southeastern coast, up the Gulf side, and out to Texas. In the mid-1950s, a University of Miami researcher described two healthy commercial scallop fisheries in Florida, in Pine Island Sound near Sanibel and Bay County in the Panhandle. The labor force was mostly fishers and crabbers who scalloped in the summer, or "shell draggers" who dredged the sea bottom for seashells and seahorses to sell to tourists, turning to scallops when they became more profitable. The state's Bay Scallops, he reported, were "safe from economic or biological depletion caused by man . . . so long as . . . he does not pollute or otherwise adversely alter the environment." Within decades, both fisheries had collapsed. Pine Island Sound had escaped the worst pollution. But every alteration to land altered the water, too, changing the sound: The replumbing of the Everglades; dredge and fill to build cities in swamps; construction of the Sanibel Causeway; deep cuts in the Intracoastal Waterway. They were all deep cuts.

Water pollution was another deadly link in the chain. For much of the twentieth century, Americans flushed every sort of waste to local waters, which ultimately drained to the grass-filled bays. Sewage, farm, and factory waste plumed into Long Island Sound and the

Chesapeake Bay, into North Carolina's Pamlico Sound, Miami's Biscayne Bay, Tampa Bay, and most every other major estuary. Scallops and other mollusks, crabs, fishes, and marine mammals had flourished in the seagrasses of Tampa Bay. Aerial photographs from the late 1930s show the meadows covered 76,500 acres before dozens of bayside cities and four ports dredged their harbors, deepened their channels, paved their shorelines, and flushed sewage into the water. By the early 1970s, only 14,200 acres of seagrass remained. Algae mats—so putrid the stink wafted into waterfront homes—suffocated the bay and its wildlife. Around the country, algae drawn to nitrogen-laden waste deprived waters of oxygen and seagrasses of light. Shallow seagrass meadows vanished. Clacking schools of scallops 10 feet deep were never seen again.

The Clean Water Act of 1972 helped clean up the sewage and industrial plumes that Americans could see. But the massive infrastructure projects altering the delicate mix of sea and freshwater in the coastal estuaries only accelerated. The act also neglected farm runoff, and the underground septic systems that today percolate the waste of one in every five American households.

In the Northeast, commercial fishers could still land a decent scallop harvest—averaging 300,000 bushels a year nationwide—until 1985. That year, a brown tide invaded the bays. The toxic algae bloom darkened the clear waters, starved the scallops, and killed their larvae. Cutting sunlight, it killed much of the surviving eelgrasses. By the time New York lawmakers designated *A. irradians* the official state shell in 1988, less than 1 percent of the animal's historic seagrass habitat remained.

Scientists had identified nitrate pollution as a serious threat to the region's water as early as the 1970s. Yet, "everyone treated the brown tides as alien encounters," Kevin McDonald of The Nature Conservancy on Long Island recalled. The *New York Times* used the word *mysterious* to describe the brown tide in no less than twenty-four articles in the 1980s, '90s, and 2000s. But around the world, the rise of toxic algae blooms is not such a mystery. The foul brown tides, rust

tides, red tides, and blue-green algae choking the world's waters proliferate on nutrients. They prefer warmer water. We have created their
ideal conditions.

WHERE STINKING ALGAL mats covered the surface of
Tampa Bay in the 1970s, sunlight now streaks through clear water
across more than 40,000 acres of seagrass. Manatees have returned
with turtles and fishes; the shuttered Tampa Tarpon Tournament is
back on.

The turnaround began with tough sewage regulations; Tampa
and other cities could no longer dump barely treated waste into the
bay. Nearby power plants stopped burning coal to cut nitrogen pollution carried from air to water. Programs for farmers and landscap-
ers pared fertilizer runoff. The Tampa Bay Estuary Program rallied
citizens, businesspeople, NGOs, and government around more than
five hundred projects over three decades to restore the region's liquid
heart.

None of it has been enough for the Bay Scallop. State fish and
wildlife officials shuttered Florida's harvest in all but the Big Bend
while scientists spent years growing millions of larvae in a hatchery and transferring them to waters where scallops once thrived. In
Tampa Bay and Pine Island Sound, numbers of zigzagging scallops
would dramatically increase following each round. Then, after a few
generations in the wild, the populations would again collapse.

For twenty-five years, volunteers with Tampa Bay Watch have
snorkeled the mouth of the bay each August, counting every scallop they can find. In 2010 they counted a promising 674 after scientists planted larvae in the bay. In the following years they found
thirty-two, then five, then a dozen. In 2018, red tide made the water
too toxic for snorkeling. In 2019 they found fifty. The citizen hunts
expanded to other communities, but they have faded for lack of scallops. In Pine Island Sound, where earlier counts had seemed promising, the most recent search found one scallop. Scientists said an
ill-boding red tide was likely to blame.

Restoration efforts are likewise hit-and-miss where Bay Scallops once thrived in the Chesapeake Bay and the coastal bays and estuaries of Virginia and North Carolina. In New York's Peconic Bay, the largest Bay Scallop restoration in the United States proved successful enough to support a commercial harvest. At least, before spiking temperatures in the bay led to a massive die-off.

Following the deadly brown tides of the 1980s and '90s, neither the scallops nor the seagrasses returned naturally to the Peconic, the 155,000-acre estuary flowing between Long Island's north and south forks. For decades, scientists have reared Bay Scallops at a Southold hatchery and transferred them to the historic grounds.

Hanging from buoy lines at the surface, hundreds of thousands of scallops spend winter suspended in cylindrical lantern nets, ensuring they'll be close enough together to spawn in the spring. Hundreds of thousands more are tossed into the bays to swim; congregations of dime-sized scallops flit across the surface as they did a century ago, now from a boat rather than on their own. The project of Long Island University and Cornell Cooperative Extension helped increase spat and adult scallops—and commercial landings—by more than 1,000 percent in the Peconic and nearby bays. But New York's Bay Scallops cannot tolerate water hotter than the mid-80s. In summer 2019, temperatures in the Peconic soared to those heights. Fishers set out on the first day of the state's commercial scallop harvest to find a massive die-off—with no toxic algae to blame.

Scientists well knew the old truism noted by Ingersoll a hundred years before: Scallops can be wildly abundant one year and scarce the next. But this die-off seemed different. Bivalves, sensitive enough to be put to work as water-quality monitors, were now warning about climate change, too.

WE ARE PLOWING west along another river toward the Gulf of Mexico and the laughing grass. The kids are bigger now—teenagers—and so is the boat, a 20-foot metal pontoon. These recreational barges, tricked out with thin outdoor carpet and lawn chairs, are institutions on the Homosassa River, one of the byways to what has become the most popular scalloping grounds in Florida.

Our friends also have two teenagers, meaning sixteen long limbs flop and sprawl from a huge round float centered on the gray carpet. The pontoon gets a lot of stares. Melanie and Charlie bought it from a guy who christened it *Poontoon*, affixing the name to the side in big sticky letters. But for years the family read it as "pontoon." When they finally noticed the extra *O*, the kids died a thousand deaths of mortification.

The Homosassa begins at a great turquoise spring and historic town of the same name, and winds for 7 miles to the Gulf. Anthony Dimock called the outpost a realized utopia. His Homosassa flashes today in brief glimpses between speaker-mounted, flag-waving fishing boats: A clear blue spring nearly 100 feet across. A manatee surfacing in the brown river. A perfect arrowhead turned up on a lonely shell island.

For the first few miles we pass multistory marinas and tiki-torch restaurants, mini-mansions and Old Florida fishing shacks, with a Monkey Island thrown in the middle of the river for good measure. The fake island, planted in palms and cedars and touting its own putt-putt-type lighthouse, is an Alcatraz for a small group of monkeys removed from nearby Homosassa Springs Wildlife State Park for hijinks including swiping candy from little kids and breaking into cars.

Since their imprisonment in the 1960s, the spider monkeys have seen increasing numbers of their human relations motor past to Florida's busiest scallop hunt. Midway between Tampa Bay to the south and the Big Bend to the north, Citrus County's harvest was closed, along with most of the state's, while scientists tried growing scallops in wire cages on the seafloor. Tens of thousands of the caged bivalves spawned over several years, helping restore the population well enough to reopen a recreational harvest in 2002.

The mollusk biologist Stephen Geiger says that like the perils that killed them off, the aids that brought back Citrus County's scallops were many: The closure. The restoration. Proximity to Big Bend scallops sending larvae south. Relative lack of red tides—or the coastal hardscaping that altered Tampa Bay and so many others.

Most important: "Hundreds of thousands of acres of relatively healthy seagrass habitat." Still, the seagrass is not a St. James miracle. The scallop population is holding steady. But the human population chasing behind with dive bags is growing. "If you follow that logic," Geiger tells me, "there will come a time when there are more people harvesting than the scallop population can sustain."

THE BOAT TRAFFIC on the Homosassa feels like a Disney queue. Being part of the traffic, we can hardly complain. Despite it, the tea-brown Homosassa is still a drink of wild Florida. Soon the mini-mansions give way to fishing shacks reachable only by water, and those to limestone islets covered in shell rubble. From the jungle-like riverbanks, sabal palms poke here and there like the teenagers' limbs, angled out of the woods or sideways over the water.

The river gradually shifts from its freshwater source at the blue head-spring to its salty estuary at the mouth of the Gulf. Rising seas are pushing the saltwater further inland, aboveground and below. Midway down the river, the tallest palms are shorn to poles or blackened as if in a forest fire. They are dying as the saltwater saturates the coastal aquifer: Florida's official state tree embodying the state's existential threat.

Other life is running wild in the new conditions. Among the dead palms, red and black mangroves twist in tangled labyrinths, soaring to 20 and 30 feet as their range expands thanks to fewer freezes. Salt marsh entombs old fishing shacks and houseboats collapsed in the worsening storms.

Where the river nears its terminus, golden salt marsh unrolls to the north and south like midwestern fields of wheat. Ahead to the west, the Gulf of Mexico opens in a flat panorama of sun-glinting sea and forested islands. We're swept up in a procession of boats that accelerate in the channel and make a beeline for the seagrass meadows. After running for a while with an eye on the GPS, our friends turn out of the channel to find a favorite spot. It is no secret. We join a flotilla of bimini-topped boats and red-and-white diver-down flags, close enough to hear other boaters' music but far enough to avoid impinging on each other's scallops. Our towheaded first mate, Aidan, rises from the big round float to toss the *Poontoon*'s anchor.

Jumping off the platform and into the Gulf, I find the water feels cool for just a moment, then comfortably warm. Beneath the surface, the music and the memory of the dead palms dissolve. The Gulf buoys body and spirit, though the water is turbid and the turtle grass fuzzy with algae.

There seems to be no shortage of scallops. I snorkel over congregations of two and three. But for the first time in my decades of gathering the hard-shelled berries, I can't bring myself to pluck even one from the grass. I toss my blue mesh bag back into the boat. Like Dimock hanging up his gun for his camera, I take aim at the life in the laughing grass with our underwater point-and-shoot.

Seeking pictures rather than scallops, I notice much more of the life in the grass. Dozens of long-spined star shells cling to the blades. They are the shape and size of nickels, with triangular spikes jutting from the periphery like a child's drawing of the sun. I place a few of the small gastropods in the hands of my daughter and her friend, our second mate Phoebe. Later I get to tell her that she shares a name with the animal that tickled our palms: *Lithopoma phoebium*.

I spot red-orange sea stars and dozens of purple sea urchins. I photograph a small black-tipped shark. This I do not tell the teenagers, who have binged on Shark Week to affliction.

Translucent moon jellies pulse by, grace undeserved by a Shark Week world.

THE KIDS ARE also ignoring the scallops. Only my husband and Melanie, chums over British novels and neighborhood gossip, keep harvesting. They collect just enough to flash-fry an appetizer at Melanie and Charlie's cabin halfway back up the river.

The cabin, built as an end-times bunker by a Florida billboard lawyer, is now an electronics-free weekend getaway. Melanie and Charlie have stocked it with mystery novels and used-bookstore gems about the sea. A blue paperback catches my eye: *Stalking the Blue-Eyed Scallop* by Euell Gibbons, the beloved wild-foods guru once described by John McPhee as "a man who knew the wild in a way that no one else in this time has even marginally approached."

I'm a fan of Gibbons's wild-food writing and had been meaning to read his guide to harvesting and cooking seafood, published in 1964. I flip to Chapter 6 to see what he has to say about the blue-eyed scallops. My brain latches onto a familiar phrase:

> No one who has watched scallops flitting about the tide pools and shallows will deny that they have an instinct for play as well as for survival. They seem to express the joy of life.

I'm struck by the similarity to a favorite line by Julia Ellen Rogers:

> To see hundreds of scallops the size of a silver dime flitting through the shallows on a bright summer day will certainly convince you that even mollusks can express the joy of living.

I borrow my friends' well-worn copy of *Blue-Eyed Scallop*, and later compare Gibbons's and Rogers's chapters on the pecten.

On the shells' legacy of pilgrimage:

> Rogers: A member of the First Crusade starting home picked up a pretty shell.
> Gibbons: Once, a soldier-pilgrim picked up a pretty shell.

On their locomotion:

> Rogers: The scallop does not crawl or burrow.
> Gibbons: The scallop never crawls or burrows.

Whether the famous wild-foods man borrowed too much from the turn-of-the-century shell-writer seemed debatable until I read their descriptions of the scallops' blue eyes:

> Rogers: A row of bright eyes heads the fringe. Each eye is an iridescent green spot, encircled by a rim of turquoise blue. . . . They have the cornea, lens, choroid coat, and optic nerve. Dr. Cooke calls them bona fide eyes, approximating more closely to vertebrate eyes than any other found among bivalve mollusks.
>
> Gibbons: Around the edge of the mantle there is a row of as many as 50 bright, shining eyes, dots of iridescent green encircled by rings of turquoise blue. Biologists tell us that these are real eyes, having cornea, lens, choroid coat, and optic nerve, and they more closely approximate vertebrate eyes than any found among bivalve mollusks.

I don't doubt Gibbons' foraging and culinary credentials. He fed McPhee for a solid week one freezing November on what he found and cooked on the Susquehanna River and Appalachian Trail. He told McPhee his childhood story of foraging to save his mother and three siblings from starving to death in their New Mexico dugout during the Dust Bowl.

I read his acknowledgements, including to Ed Ricketts and Tucker Abbott—two scientists who credited Rogers as an inspiration. Gibbons never mentions her. "There are dozens of others," he wrote, "from which I have taken what I needed at the time and then, ungratefully, forgotten."

Something else strikes me about Gibbons on seafood. His perspective has little of the economy for which he is remembered—foraging acorns for grits or watercress for a simple salad. Reflecting his time, he holds forth in his seafood book on countless ways of killing "that very peak of molluscan evolution, the Octopus," gaffing the creatures in their holes at low tide or spearing them on the sea bottom using their prized cowries as lures. He gushes over extravagant dishes like Newburg, clam croquettes, seafood crepes, and bouillabaisse. The secret of the latter, he explains, is the rule of threes. No self-respecting bouillabaisse chef uses only one fish, one crustacean, and one mollusk, but three of each: Start with a seabass, a red snapper, *and* a flounder. Then then add crab, shrimp, *and* a lobster. Then add clams, mussels, *and* oysters—or any three mollusks, shells on if you're cooking for someone you want to impress.

Having gotten to know Rogers through her writing, and reading a hundred little stories about her in the historic Long Beach newspapers, I suspect she would have been crushed more than anything by the popularity of Gibbons's marine-life scavenging—his lessons to gaff octopuses in their holes, rather than study them quietly at low tide.

No one thing is broken. It is rarely only the harvest. It is, rather, the exponential growth in harvesters *and* the coastal development *and* the destruction of seagrass *and* the rising carbon emissions that have made the past five years the five warmest for the oceans in human history.

It is not the seafood, itself, but the bouillabaisse—the rule of threes, and that of thousands.

Eleven

~~~~~~~~~~~~~~~~~~~~~~~~

SAVING THE QUEENS

THE QUEEN CONCH
Aliger gigas

When the tide tugs the Atlantic Ocean from the southern tip of Andros Island, a lustrous sand plain emerges across the horizon, barely covered in shallows that make a splashy turf for the Bahamian children racing across the flats on a Sunday afternoon. The retreating sea leaves a perfectly round swimming hole banked by the sand, one of the glorious "blue holes" of Andros, the largest island in the Bahamas. Usually hidden by the ocean, the sapphire pool exposed at low tide at the small settlement of Mars Bay pops from the plain like an oasis in a white desert. While their parents chat under mangroves grown big as shade trees, the boys and girls back up for a running start to dive, somersault, flip, or otherwise hurl themselves into the blue.

The kids are giddy and jumpy because—Conch Fest. The annual festival in South Andros celebrates the Bahamas' favorite animal and food. The Queen Conch also makes the coral-colored shell that so many tourists want for the mantel or the garden path, a thick, heavy spiral with a flared lip that reveals a cavernous interior, glossed in pink. From the whorls to the polished entryway, the shell is a palace, fit for an impressive queen. *Aliger gigas* are strong, agile, prolific, determined. They are acrobats among mollusks despite their rock-heavy shells. Like the kids at Mars Bay, Queen Conchs skip and flip. They make somersaults look easy. Using their muscular foot and claw-shaped operculum like a pole vault, they hop and jump across the sea bottom to elude predators and break up their scent trail. Should a shark or turtle turn one over in search of meat, the conch pitches itself back. It is not a maneuver that works on humans.

While the kids flip and dive, a couple more grown-ups, far out on the sand flats, are hauling heavy sacks and planting objects in the shallows for one of the festival contests. Conch Fest is a homecoming for South Andros, which, as a remote island within a remote island, is both untouched and uncompensated by cruise ship commercialism. Natives living in Miami or the Bahamian capital of Nassau fly into the small airport at Congo Town or book passage on one of the private ferries that turn into a conch party boat for the weekend. Mars Bay is 20 forested miles south of the airport on the Queen's Highway. The only paved road through Andros is lined with conch shells.

At Mars Bay Community Park near the blue hole, a rake-and-scrape band trades the sound stage with a pop DJ. The food stalls are painted hot pink or draped in pink fabric. Piles of live Queen Conch dwindle behind the stalls as piles of empty shells grow. Local cooks are slicing and selling cracked conch, scorched conch, curried conch, steamed conch, BBQ conch, two kinds of conch salad, and fried conch fritters. At a pair of messy tables near the stage, knives clank and conch guts fly during competitions for conch cracking and conch skinning.

Back at the beach, the flats are soon studded with the chunky forms, some flashing pink, some hidden at a mangrove island or in one of the wrecked fishing sloops interred in the sand. When it's time, the kids line up in teams in front of the blue hole, facing the horizon. On "Go!" a child from each team dashes across the flats, collecting the spoils in small batches before racing back to pile them in front of their friends and send another kid out for more. By the time the splashy relay is finished, each team has amassed a small heap of conch shells. They suggest the copious mounds that spill over at the nearby community docks and beside boat ramps here and across the Bahamas.

The game is an effort to pass on a tradition in lieu of a trade. The festival itself, like the dozens of salmon festivals in the Pacific Northwest, the abalone festivals of California, and Oyster Week in New York, reenacts abundance lost. The pink flashes in the shell piles are like the blue hole. They are there and not there. They are real as a somersault and as illusory as a mirage.

QUEEN CONCHS LIVE across the Bahamas, up to Florida and Bermuda, and all around the Caribbean Sea, meaning their lives are entwined with humans and their unsettling ways in no less than twenty-six countries. In ancient Mayan ruins, archaeologists found imagery of Queen Conchs as melee weapons for hand-to-hand combat—spiked five-pound boxing gloves, the inner cavity a perfectly smooth, knuckle-protecting grip. In modern times, the queens were at the center of a CIA plot to kill Cuban President Fidel Castro with an exploding shell placed on a coral reef. In 1963, agents considered booby-trapping a "spectacular shell which would be submerged in an area where Castro often skin-dived," according to John F. Kennedy assassination files declassified in 2017. "The seashell would be loaded with explosives to blow apart when the shell was lifted." Castro loved diving and spearfishing offshore from the still pristine Zapata Peninsula—just where the United States had been humiliated in the Bay of Pigs fiasco. The agents ultimately decided no native Caribbean shell was both large enough to hold the number of explosives that could kill Castro and spectacular enough to inspire him to lift it from the reef. Had they asked a conch fisher or conch scientist, they might have changed history.

No human reimagining of a shell comes close to the Queen Conch's own transformative life cycle. The animals ride the currents as larvae, hide in seagrasses when they are little conchs, hang out in grainy sand and rubble in middle age, and hop and leap to deep-sand channels when they reach old age. They are not all queens. They are female or male and must join up to mate, unlike the bivalves that send their eggs and sperm into the sea to meet in the currents.

In springtime, mature conchs gather in large herds and graze on algae, plowing the nourishment into eggs and sperm. The herds are crucial to their survival; scientists say it takes at least ninety Queen Conchs in a hectare to successfully reproduce. Each female will develop millions of eggs. A male scoots over to stretch its long, spade-tipped penis underneath her shell. Within a day after her eggs are fertilized, the mother makes a little trench in the sand and piles up half a million or so in a gelatinous strand that, if extended, would stretch longer than a basketball court. She uses her all-purpose foot to camouflage the strand with sand as she goes, coating and heaping until it could pass for a hunk of white coral. She lays about nine of these egg masses each season, bringing nearly 5 million larval conchs a year into the world. Fewer than 1 percent—50,000 or so—may survive to become adult queens.

The larvae's metamorphosis is every bit as royal as that from chrysalis to monarch: Inside the hideaway their mother has made, larval conchs start right away on the shell, which begins as a delicate gossamer bubble. Within days, the soft-shelled embryos begin to spin inside their eggs—as if practicing the somersaults to come. The spinning signals they are ready to hatch and drift into the water as veligers, small as this period. The weeks ahead mark the only time in its life a Queen Conch swims freely, floating for many miles on the ocean currents. The atom-shaped veliger sprouts petal-like lobes; first two, then four, then six. By the time the veliger is three weeks old, its diaphanous shell is a perfect spiral, and the six lobes have stretched out to limbs that brace its landing on the sea bottom and set it crawling on the grass.

It crawls, then swims again, then crawls, then swims, looking for just the right spot to settle. Its freewheeling infancy comes to an end as the conch resorbs its wiggling limb-fins. It grows the clawed foot and other molluscan features; its snout-like proboscis to eat, gills to breathe. Now complete with a perfect tiny shell that can sit on the tip of your finger, the juvenile Queen Conch buries itself in the sand for the first year of life, invisible but for the periscope eyes. For that year and about the next four, young conchs devote their energy to evading predators; first marine worms and small crabs, then larger stingrays, lobsters, octopuses, sharks. By the time they are about five, they have bulked their palace into a fortress so secure that most of their natural predators no longer try to break in. There is only one menace left to worry about.

AS A BUDDING teen naturalist in the 1860s during his father's stint at Fort Jefferson in the Dry Tortugas, Charles Frederick Holder visited Key West and made his way to Conchtown. He hired out a boat and some sandy-haired "Conch boys" to show him the only barrier reef in the United States—the great corals spreading south-west from the Keys. Holder described the branching reef as "a world beneath the sea—a coral city populated by uncounted millions," crowded in bright colored fishes, crabs, and innumerable mollusks. Large cowries, which the Conch boys called micramocks, hunkered low on the reef, mantles billowing up their shells. Queen Conchs levered along in the sand and grass. The boys took Holder to their conching grounds, where "the conchs were feeding on the weed in such numbers that the boys turned them into the boat in dozens."

"I now learned why the name Conch had been applied to these people," wrote Holder, who became a big-game fisherman and one of the nation's most popular outdoors writers. He watched the boys carve out the animals, divide the meats for soup and bait, and add the shells "to a vast pile, to be sold by the ton as curiosities and made into jewelry and other articles."

"Conch" originated as a disparaging term for working-class whites in the Bahamas, who migrated to the Keys in various waves, the first after the Revolutionary War. Thousands of loyalists, fleeing the United States with thousands more enslaved Africans, had flooded the British-ruled Bahamas with a vision to turn the fishing archipel-ago into an empire of cotton. The wealthy newcomers clashed with the Conch old-timers, a number of whom set off across the Florida Straits to the Keys. In the following century, white and Black Baha-mians alike continued to migrate to the Keys to work in wrecking, fishing, sponging, and farming. In the 1880s, the *New York Times* reported that the "Conch" nickname was "usually applied in derision, but the Conchs themselves do not seem at all ashamed of it." Reporters sent to the southernmost city for a travel feature often filled their dispatches with bigoted details on the city's Black, Conch, and Cuban people and neighborhoods, even while raving about the seafood they caught and prepared. The Conchs, wrote one, "are a large, rough class of men, and apparently very ignorant," dropping their aspirate like Londoners.

The Conchs still held Cockney accents and traditions when the WPA Federal Writers Project described them in the 1930s, "although the term," advised the WPA's *Guide to Florida*, "is now applied to anyone living on the Florida Keys." And that is how the Keys became the only place in the world where an entire population is named for, and identifies with, a mollusk. Babies born in Key West are issued novelty Conch birth certificates. At Key West High, home of the Conchs, the buildings are painted pale pink, and a huge metalwork Queen Conch rises on the lawn. And every evening as the sun sets at Mallory Square in Key West, the southernmost point in the continental United States, a Queen Conch trumpet bellows its mournful good-night.

Conch identity reached its apex in 1982, when the U.S. Border Patrol set up a roadblock at U.S. 1—the only land access on and off the islands—to search all cars leaving the Keys for illegal drugs and undocumented immigrants. The foreign-border-scale searches caused hours-long traffic jams, chased off tourists, and angered locals. Political and business leaders tried every remedy to shut down the checkpoint, including the federal courts. The day after a Miami judge denied their injunction, Key West Mayor Dennis Wardlow and supporters orchestrated a mock secession from the United States, declaring Key West a sovereign nation—the Conch Republic. The spectacle was not as lighthearted as it is often remembered; FBI agents converged on the city, and Naval Air Station Key West was on high alert for attack. Wardlow read a proclamation evoking the original Conchs' rejection of British taxes and tyranny. The rebels raised the blue-and-yellow flag of the Conch Republic, emblazoned with a large Queen Conch.

The rebellion drew so much publicity that the feds dismantled their roadblock. The micronation had won—and the human Conchs had finally earned respect. "We Seceded Where Others Failed," declared the new Keys T-shirts and bumper stickers.

Yet as they spent a century building up their Conch image, the collective conch fishers, seafood wholesalers, chowder chefs, curio dealers, and shopkeepers selling gloss-pink souvenirs in the Keys steadily depleted the very symbol of their worth. Scientific reports began to note the excess in the early 1920s, when food vendors strung choice meats on a stick to peddle around the city for 5 cents a conch, and novelty shops hawked the shells for as little as a quarter. The U.S. Fish Commission noted in 1923 how easy it was to catch such large, slow-moving animals. "The supply could easily be depleted by over-fishing," the commission stressed in its annual report that year.

By 1940, virtually everything nearshore to Florida's tip was up for grabs and up for sale. Commercial shellers steered barges over reefs and did not stop at the conchs, but broke off coral branches with crowbars and hoists to sell them for décor. Old-timers in the Keys complained they could no longer gather an easy mess of conchs for chowder. Marine scientist Gilbert Voss warned that if Americans did not preserve their great undersea monuments as they had the nation's iconic mountain peaks and canyons, their coral reefs and seagrass beds would be lost.

Voss and other marine biologists petitioned the state and U.S. Department of the Interior to preserve the reef. It took ten years from the time President Eisenhower set aside part of the ocean bottom until the grand opening of Pennekamp Coral Reef State Park. America had its first undersea preserve. But while the living reef was saved, the Queen Conch remained imperiled. Rachel Carson wrote that even as the queen was "becoming rare in the Florida Keys," its pink shell "is displayed by the hundred at every Florida roadside stand selling tourist souvenirs."

People don't take to warnings of excess when they are surrounded by abundance. By the 1960s, commercial fishers in the Keys were taking a quarter-million Queen Conchs a year. The state banned commercial harvesting in 1976. But with so many tourists and locals chasing so few conchs, the recreational harvest continued to devastate the queens. At the state natural history museum, the malacologist Fred Thompson compared their extirpation to the slaughter of plume birds a century before. "There are no ethical limits," he lamented. Florida finally banned all Queen Conch harvesting in 1986, a move the state Marine Fisheries Commission said "will enable the species to replenish itself so that harvest can eventually again be allowed."

Going on four decades later, that promise has not materialized. Conch is still the beloved Keys cuisine. The shining pink palaces are still a favorite keepsake from a Keys vacation. And the locals are still proud Conchs. But the meat and shells come from the Bahamas.

ON A MAP of the Caribbean Sea, Andros is the huge landmass that dominates the Bahamian archipelago, bigger than all the other seven hundred islands combined. The island's shape evokes the shell that has been part of its people's livelihood for two millennia. It begins as a pointy tip in the north, widens to a broad flare at the shoulders, then contracts to a thick trunk that opens at its southern end to miles of sand flats and leggy mangrove estuaries.

Itself an archipelago 100 miles long, Andros is only a twelve-minute flight from Nassau, but the contrast is deep as the plunging trench known as the Tongue of the Ocean between them. Andros is one of the "Family" or "Out" islands of the Bahamas, modest communities beyond Nassau or Freeport/Grand Bahama where tourism has not wholly overrun fishing, farming, and other traditional livelihoods. The island's three enclaves—North Andros, Mangrove Cay, and South Andros—are cleaved by shallow, serpentine bights that connect the inhabited east to the national park wilds that cover the west. Its 7,500 residents live mostly along the Queen's Highway that follows the eastern coast, many working in tourism that caters to bonefish enthusiasts and divers drawn by the blue holes and barrier reef. The U.S. Navy has a major underwater research center here, and other industries center on agriculture or crafts. But fishing is a mainstay for both income and sustenance. Despite long-promised development of ecotourism around its rich natural and cultural histories, Andros's poverty, potholes, and frequent power outages reflect a weary neglect.

Queen Conchs have been at the center of those histories since the Lucayan Taínos, who had lived in the Bahamas for more than a thousand years by the time Columbus "discovered" the islands in 1492. A huge conch lies at the foreground of the idealized scene illustrating Columbus's "Letter Announcing the Discovery" that was widely circulated in Italy the following year.

The Lucayans were tremendous fishers and deepwater divers; scientists examining their skulls found bone built up around their ears in response to the pressure. They harvested infinite piles of Queen Conch, honed the shell into tools and jewelry, and roasted the conch meat on *barbacoa*—leaving us barbecue in word and gift. History has interpreted them as friendly and generous; they also showed the Spanish the hanging beds they called *hamaca* and their impressive wooden boats, *canoa*. The single Lucayan canoe that survives today was preserved in one of the blue holes of Andros.

They called the conchs *cobo*—*co* for "outer" and *bo* for "house"— an animal that carries its house.

The Spanish wiped out all the Lucayans in less than twenty years. They forced some 40,000 into slavery at Hispaniola, including making them dive for conchs and pearls. By the time Juan Ponce de León came through the Bahamas in 1513, no Native people remained. The historical geographer Carl Ortwin Sauer famously described Juan Ponce's discovery of Florida as merely "an extension of slave hunting beyond the empty islands."

The Bahamas, including Andros, were deserted for a century before the British took possession; the loyalists and their enslaved laborers added to the population in the wake of the American Revolutionary War. Then, in the 1820s, Black Seminoles from the Florida Everglades, fleeing Andrew Jackson's Indian wars on hired wreckers and huge canoes, crossed the Straits of Florida and founded a community they called Red Bays on the northwest shore of Andros. There they lived "peacefully and quietly, and having supported themselves upon fish, conchs, and crabs which are to be met in abundance," reads a governor's dispatch from when they were rounded up by British customs in 1828. The abolitionist governor freed them after a year's detention in Nassau, and eventually some two hundred Black Seminoles settled at Red Bays. The next settler wave arrived in the 1840s, when newly freed mariners relocated to Andros from other Bahamian islands to work the huge sponge beds.

THE NAME QUEEN CONCH comes from Queen Victoria's time. The Queen's Staircase in Nassau and much else in the Bahamas was named for the young queen who ascended in 1837 and oversaw emancipation of the empire's slaves. Colonialism left its huge stamp on the Bahamas and on the Queen Conch, literally; the first adhesive postage that replaced the islands' local stamps in 1859 featured Queen Victoria's face. The formerly dominant conch shell and pineapple were relegated to the two lower corners.

The fashionable queen loved the pink shell. She employed her own cameo cutter to make her brooches and numbers of commemorative keepsakes for occasions including her wedding. The ornaments were more commonly carved from helmet shells, but Victoria's were usually made from the Caribbean conch. Her delicate cameos inspired a craze and a major import market.

"The profit when converted into cameos and other *objets d' art* is enormous," Sir Augustus J. Adderley, Bahamas fisheries commissioner to Britain, wrote of Queen Conch imports in 1883. Demand for Queen Conch in Europe was so intense then, Adderley warned that "I am under the impression that this fish is not so plentiful as it used to be, and that its protection is desirable." Adderley wanted to advise a closed season for the queens, "but I fear it is not practicable."

Andros supplied the crown with conchs, sponges, and timbers that formed "the wooden walls of old England," according to Adderley. The emancipated seafarers built their own island culture and economy around the ocean's abundance. They relied on conch meat especially in hard times. Across the Bahamas, conchs and their protein helped islanders weather economic downturns and disasters—"when the islands were seemingly forgotten by the outside world," wrote a conch researcher in the 1960s. Dried Queen Conch, nicknamed "hurricane ham," was a staple when other meats and fish were scarce in the wake of big storms.

The shells heaped across Andros are from Lucayan times, from colonial times, and from last week; the animal is bound with the island's food and work, play and lore. In one folktale recorded on the island in the early twentieth century, Conch and Lobster decide to race to see who will get to marry the king's daughter. Conch knows he is slower with only one foot, so he sets off early. At points all along his route, he leaves a young conch in a shoal. As Lobster comes along later, he stops at each shoal to feed, at one point getting his big head stuck. Lobster can't help himself even when he realizes Conch is beating him. As Lobster stops to feed on the last young conch, he hears singing and dancing: With its one big foot, Conch has stayed one step ahead of gluttonous Lobster. Conch is already marrying the king's daughter.

AFTER FLORIDA BANNED all Queen Conch fishing, scientists in the Keys kept expecting signs of a comeback. Instead, the depressing years stretched on. The herds did not come. The conchs' numbers were so low that males and females couldn't find one another to mate. "It was like after the Zombie Apocalypse," says conch biologist Gabriel Delgado at the Florida Fish and Wildlife Conservation Commission's lab in Marathon, halfway down Florida's 125-mile finale of islands. "If the last man is in Canada and the last woman is in Australia, it's going to take them a while to find each other."

Delgado's love for marine life developed when caring for tropical fish in the aquarium his parents bought him for his bedroom in New York City, where he was born and raised. When it was time for college, the University of Miami's Rosenstiel School of Marine and

Atmospheric Science met his Cuban parents' wishes that he live near extended family—and his, to study the sea. In 1997, Delgado was presenting his master's research on Queen Conch density at a conference when he caught the attention of marine ecologist Bob Glazer, who had spent a decade working on the mystery of why the queens were not returning to the Keys.

Delgado joined the state research lab and Glazer's trials to rear conchs in a hatchery and release them to the wild. The conch biologist Megan Davis, now a research professor at Florida Atlantic University's Harbor Branch Oceanographic Institute, had grown queens for six years at the commercial farm she helped found on the Turks and Caicos Islands. She and other scientist-entrepreneurs hatched eggs in tanks on land and moved millions of juveniles to huge, round corrals resembling crop circles in the shallow turquoise sea. The farm succeeded in raising conchs to supply restaurants and wholesalers, taking some pressure off the wild animals.

But efforts to release young conchs to repopulate their habitats seemed to fail everywhere scientists tried. In the Caribbean and Keys alike, little conchs that had evolved to grow on the soft seabed didn't take well to tanks. The hard surfaces wore away their spires and spikes. Once released, having never learned to bury themselves, they were often gobbled up in no time. Delgado and Glazer figured out that by exposing them to a predator, the juvenile conchs would learn to hide under the sand. But it was costing $9 to $12 a survivor to rear a conch—an investment government wasn't willing to make. Megan Davis points out that, given what each conch can contribute to a population, it might have been a small price to pay to bring back such an iconic animal.

About fifteen years after the ban, the last male and female finally seemed to find one another in the Florida sea. Conchs began to herd in greater numbers again, in their reef habitats offshore and in the shallow grasses nearer the Keys. But the repopulating conchs gave scientists something new to worry about. While those congregating offshore began to mate, the herds in the nearshore waters showed no signs of reproducing. To figure out what was wrong, the Keys scientists collected some males and females from each area, dissected them, and FedExed their gonads to biologist Nancy Brown-Peterson at the University of Southern Mississippi's Gulf Coast Research Laboratory in Ocean Springs.

Brown-Peterson specializes in marine reproductive biology at the lab overlooking Mississippi Sound on the Gulf of Mexico. When she slid the tissues from the offshore conchs under her microscope, she noted that the females and males had healthy eggs and sperm. But when she went to examine those from nearshore, "I've never seen anything like it," she told me. "The inshore females just had nothing there." Almost all of the inshore females and most of the inshore males had no healthy gonadal tissue to speak of.

The Keys scientists next tried moving the sterile nearshore conchs to the deeper beds. Within six months, the conchs developed eggs and sperm. They mated. The female queens began to make their clever egg cases. The scientists figured that some human peril from the land must be harming those close to shore. The scientists studied all the causes they could think of: The mosquito-killing pesticides used in South Florida. The septic tank pollution then fouling the Keys. Fertilizers running off green lawns. Metals or other compounds. Delgado, Glazer, and colleagues tested for all of that and more. But while the pesticides lowered the conchs' fecundity, and while metals mixed up their sex organs just as they had in the purple-dye murex and other mollusks, no one threat could explain the devastating absence of eggs and sperm in Brown-Peterson's microscope. Was it a compounding of human harm? Or something yet unseen?

IN THE REMOTE settlement of Mars Bay at the southern tip of Andros, fisherman Bertram Taylor and other organizers of Conch Fest are working to commemorate the proud two-thousand-year history of conch fishing on an island that has suddenly become known to the outside world for something else entirely. The year I visited Conch Fest in Mars Bay, a reality TV family was fretting over pool tile colors twenty-five miles to the north as they turned a run-down hotel into a resort for the HGTV show *Renovation Island*. Some locals took the fame as a sign that the Bahamas' "sleeping giant" of Andros was finally awakening. Whether that proves true—and whether Andros could preserve its cultural traditions amid luxury tourism that has Disney-fied islands from the Bahamas to the Maldives—remains to be seen.

Taylor's grandfather fished for conchs in a small wooden sailboat, leaning over the side with a glass-bottomed wooden bucket to spot the shells and sliding a long-handled hook underneath to raise them hand over hand into the boat. Bahamian conch fishers still plied the Carib-bean in wooden sloops until the close of the twentieth century, some of the last commercial sailing vessels in the Western Hemisphere. The largest "smacks" towed two-man dinghies to offshore conch grounds to access the shallow banks. "A team working in a shallow water area with a large conch population may boat 1,000 or more of the mollusks in a single day," reported the American ichthyologist John E. Randall in 1963. The dinghies returned over several days to fill the smacks, low-riding with their live wells back to the city docks at Nassau.

American scientists and environmental groups have a delicate relationship with the Bahamas. They lament what they describe as lax conch fishing regulations even as the United States fished out its own conchs and now consumes more of the islands' conch than any nation besides the Bahamas itself. Queen Conchs are among the few marine mollusks regulated under the Convention on International Trade in Endangered Species (CITES). But the United States has so far declined to protect the queens under its own Endangered Species Act, which would allow for U.S. regulation of import, trade, and other activities.

The Bahamas prohibited the industrial-style fishing practiced in Honduras, Nicaragua, and some other countries where as many as sixty divers pack large vessels and stay out weeks or months at a time. They send motorized canoes to the banks to fill up with conch meat, leaving shells on the sea bottom before heading back to ice the conchs on the mother ship.

The Bahamas outlawed scuba to gather conch; set export quotas and rules that permit fishers to take only adult conchs; and established twenty-one marine protected areas where conchs cannot be fished. But the island nation is also home to the greatest number of artisanal Queen Conch fishers in the world—nearly ten thousand. As other countries such as Jamaica imperiled their conchs and enacted bans, the small Bahamian fishers picked up even greater demand. They also had to travel farther and farther to find the queens. And according to the pink-tinged shell archives that rise across the islands, they took smaller and smaller conchs that fishers historically left to grow to breeding age. By 2010, scientists studying conch densities in the historic grounds at Andros suggested they were fished out. They recommended closing some areas to fishing, or at least a closed season. Both ideas went unheeded.

In 2018, researchers analyzing decades' worth of data on the conch herds and shell piles across the Bahamas reached a grim conclusion: The once-massive herds at Andros and Abaco, the Berry Islands, Eleuthera and Exuma—all the great Bahamian conch grounds—have thinned below the minimum number the animals need to reproduce. "The data are showing a serial depletion," says research biologist Andrew Kough of the Shedd Aquarium in Chicago. "The Bahamas is repeating the story in Florida."

When he leaves the museum on Lake Michigan's shores to work in warm latitudes, Kough takes stock of the queens with the help of researchers from Bahamas National Trust and local volunteers. They glide across the turquoise waters on handheld tow boards. The researchers cruise to the remote conch grounds on the Shedd's 80-foot research vessel *Coral Reef II*, then motor out in smaller powerboats that pull divers along a transect on the camera-mounted boards. The researchers peer down through their dive masks to find and count conchs; when they find a herd, they return later to measure each individual. The queens are like human teenagers in that some grow in great spurts before they mature. They are easily mistaken for adults. Bahamas regulations allow fishers to take any conch whose shell lip has curved into its famous flare, long considered the sign of maturity. But scientists now believe it's the shell's thickness that counts, and that too many juveniles are being harvested before they've had a chance to reproduce.

Kough is an expert on the larval journeys of Queen Conchs and Spiny Lobsters. Both begin life as translucent free-swimmers that float for miles on the ocean currents before settling at the sea bottom. Queen Conch larvae look otherworldly, alien-flowers spinning through the water column. Spiny Lobster, too, are alien-like as larvae—flat crystalline spiders with ET eyes, small enough that Kough photographs them on his fingertips, their antennae jutting out front as long as the animals themselves.

Reflected in the old Andros folktale about Lobster and Conch, the two species' lives intertwine at the human scale. Spiny Lobster is the largest fishery in the Bahamas, and Queen Conch an important second that makes up fishing families' income when lobster season is closed. If both had closed seasons, the fishers say, their families would suffer in springtime. On Andros, some fishers say they simply don't trust that regulation can ever protect marine life. For one, poachers target their remote cays, taking lobsters out of season and conchs smaller than they should.

Kough, not yet born when the Florida Keys lost its Queen Conchs, has come to see the Bahamian collapse as a test of whether we can save the seas in a world where most fisheries are in decline. In 2017 he led a sweeping study that combined biological surveys with fisher interviews across the islands. The Bahamas National Trust deployed young Bahamians to interview the conch fishers. They roundly agreed the queens were in trouble, but felt dubious about regulations. Still, the conch fishers, scientists, and NGOs did coalesce around one solution. The fishers and the scientific evidence alike support creation of Marine Protected Areas—large sanctuaries in the ocean set aside to give marine life a chance to breed.

While the Bahamas has twenty-one underwater sanctuaries where conch cannot be fished, only one of those is considered well protected and patrolled, the Exuma Cays Land and Sea Park. Kough's team found three times as many adult conchs there as in unprotected areas. Back at his computer in Chicago, analyzing drift patterns of larvae from the robust queens laying eggs in Exuma Cays, Kough found that the juveniles are settling in unprotected areas outside the park's borders, including fishing sites with densities otherwise too low to preserve the species. Kough's research suggests that modeling the larvae's drift could help the Bahamian government make strategic decisions on siting and patrolling sanctuaries to protect the breeding conchs whose offspring replenish the historic grounds. Marine scientists and conservationists around the world imagine such sanctuaries linking together someday in a global chain large enough to save and restore the oceans.

But despite four decades' strict protection for the queens back in the Florida Keys, the beautiful grazers had yet to make a comeback.

AS A DEADLY procession of hurricanes churned across the Atlantic Ocean in September 2017, those of us not scrambling to evacuate could not turn away from the satellite images rotating on our screens in logarithmic spirals and certain tragedy. Hurricanes Harvey, Irma, and Maria were outliers in behavior and in devastation—especially to Puerto Rico, the Virgin Islands, Florida, and Texas. Harvey brought record rainfall, Irma record winds. Maria wrought the worst natural disaster in Puerto Rico's modern history, its indirect death toll nearly three thousand.

Amid the human crises, we don't often consider what happens to marine animals as hurricanes pass over the sea; their triumph and turmoil are unseen by satellites and storm chasers. Fishes, including sharks and snook, can read signs including the drop in barometric pressure that precedes big storms. Some will evacuate dangerous conditions. With their carbonate strength and rough textures, oyster and coral reefs are known for their ability to withstand storm surges and disperse wave energy; they can perform better than human-made bulkheads at protecting life and property in hurricanes. Yet the next set of animals—those that inhabit the reefs—can take a beating. When Hurricane Hugo hit the U.S. Virgin Islands in 1989, its 25-foot waves and 200-mile-an-hour winds lashed Buck Island Reef National Monument for fourteen hours straight, destroying part of the enormous southern reef and flinging the surviving section 90 feet landward. For more than a decade following the hurricane, park employees could hear the displaced reef groan and creak like a lost soul. It finally quieted down, recalcified, and reattached to its new sea bottom. Not something a human-built seawall will do.

Gastropods and bivalves that live in the sand or in seagrasses that can be ripped from their roots and hurled to shore in hurricanes have a more harrowing time. Mollusks—and the urchins, starfish, and other creatures thrown together with them—can be picked up by currents, winds, and breaking waves and tossed to the tops of dunes or rocks where they become stranded. Such was the fate of hordes of juvenile Queen Conchs when Hurricane Michael hit the Dominican Republic.

It's at least as common for mollusks and other slow-moving marine animals to be buried alive by the tons of sand that shift in hurricanes. When the Florida Keys scientists surveyed the still-sparse conch herd after Hurricane Georges passed over in 1998, the queens had been entombed in a bank of sand. Delgado describes seeing only a few sets of conch eyes wriggling from the tips of their tightly buried shells. He and his colleagues dug up all the shells they could find, freeing a hundred or so queens.

Twenty years later, Hurricane Irma drew a bead on the Keys just as the queens had finally begun their comeback in the offshore beds. In the summer before the storm, Delgado and his colleagues had counted 600,000 conchs across the Keys, a record since the ban of 1986. After the hurricane, the numbers dropped to 350,000. Surveys also showed copious new amounts of sand in the conch habitat. Hundreds of thousands of queens may have been buried alive. Their numbers have not yet returned.

The Lucayan Taíno people who gave us words for canoes, ham-

mocks, and barbecue also named the fierce revolving storms—
huracan. The "centers of great wind," *hura* (wind) + *ca'n* (center), were
depicted on Taíno pottery in a simple sideways *S*—with a woman's
face at the center of two clockwise-spinning arms. The form looks
remarkably like today's satellite images of hurricanes hurtling across
the sea.

Harvey, Irma, and Maria strengthened the evidence that climate
change is making hurricanes more severe. Warmer oceans generate more energy, turbocharging tropical cyclones. Warmer air holds
more moisture, filling the storms with more rain. Little more than
a week after Irma destroyed lives and property across the Caribbean
and Florida, another Category 5 storm, Maria, sunk Puerto Rico
into humanitarian crisis, leaving thousands homeless. One disaster triggered others, including 40,000 landslides. The hurricane
destroyed hospitals, roads, and infrastructure for energy, water, and
communications.

Fishers said Maria turned the sea bottom "upside down." The hurricane ripped up seagrass beds. It shattered shallow coral reefs. It
smothered conchs and lobsters with sand and silt. "This has changed
the life of Puerto Rico," said Antonio Torres, president of the local
fishing association. "This ruined families, separated families. Many
people closed their businesses. It brings out the tears."

ON THE EASTERN tip of Puerto Rico at the port city of
Naguabo, local fishers have repaired Hurricane Maria's damage to
their fishing association headquarters. There's fresh yellow and green
paint outside and a new mission within. The fishers are installing
tanks to raise Queen Conchs from embryos to 3-inch juveniles
ready for release in historic grass beds or grow-out pens in the ocean.
Inside, one set of tanks will incubate conch eggs, another set is for
larvae, a third is for metamorphosis. Under a roof outside the hatchery, another set of tanks will hold the juveniles until they're ready for
the sea.

The Naguabo conch nursery embodies every hard lesson that Megan Davis, the Harbor Branch Oceanographic Institute researcher, has earned over four decades of studying and farming the queens. The tanks are made of a soft material that won't rub down the animals' spines. She knows their favorite food at every age: diatoms when they are tiny, seaweed chow when they are juveniles.

The pilot project, supported by the U.S. National Oceanic and Atmospheric Administration, aims to train local fishers to hatch and rear the conchs. Davis hopes to nurture an aquaculture operation run entirely by local people, from the conch farmers to the marketers. Juveniles would be hand-planted in marine protected areas to reestablish wild herds. Farmed queens could be sold directly from grow-out corrals on the sea bottom. The techniques can be taught, Davis says, and replicated on islands across the Caribbean. She spent her Covid-quarantine months creating conch-farming how-to manuals and videos.

"It's not only the science but it's also the art of growing the conch," Davis told me, and it was easy to imagine the long-legged twenty-one-year-old who moved to the Turks and Caicos Islands just out of college to farm conchs. "You need to watch them every day. You need a sixth sense. You have to hear that the pumps are working, and smell that the water smells fresh, you have to touch the conch and feel them and have this intuition where you see what they need—not only that it's time to feed them but what to feed them and how much. It's really knowing them."

In other words, saving the animals is going to require all the instinct, experience, creativity, generational wisdom—and perhaps luck—that it has taken to harvest and eat them; make art, tools, and structures with them; and revere them for thousands of years.

"THE QUEEN CONCH seems an alert and sentient creature," Rachel Carson wrote in *The Edge of the Sea*. "Perhaps this effect is heightened by the eyes borne on the tips of two long tubular tentacles. The way the eyes are moved and directed leaves little doubt that they receive impressions of the animal's surroundings and transmit them to the nerve centers that serve in place of a brain."

Gastropods have bundled cords of nerves, known as ganglia, that send signals to their big foot muscles and mantles; the cerebral ganglia are as close as they come to a brain, controlling their curious eyes, tentacles, and other sensory organs. Cerebral ganglia also control reproduction and shell-building. In the Florida Keys, as scientists continued to search for the cause of reproductive failure in the nearshore queens, they also began to find them building significantly lighter shells than those of their relations offshore. Brown-Peterson, the marine reproductive biologist in Mississippi, examined the ganglia of the nearshore conchs. She confirmed that they were abnormal compared with those offshore.

The queens in the nearshore waters of the Keys are losing their vital nerve centers. While those waters "are hardly pristine," the scientists concluded in an EPA-funded study on the mystery, they could find no fertilizer, pesticide, or other contaminant to explain the troubling turn.

Since the culprit wasn't reaching the offshore conchs, the scientists figured it had to be coming off the land. Delgado started thinking differently when he went to check on a nearshore herd and found their water "sauna-warm."

The oceans have absorbed vast amounts of the heat amassed from greenhouse gases in the atmosphere; more than 90 percent of Earth's warming in the past half-century has been borne by the sea. In Florida, sea-temperature records going back to the lighthouse keepers of the nineteenth century show that corals and marine life today regularly experience summer temperatures seldom reached 120 years ago. The shallows near land are subject to higher temperatures than the deep. Such hot spots can lower egg and sperm counts and fecundity in some gastropods—and decrease calcification in others.

Delgado began to pull temperature data from sites around southern Florida where the Queen Conchs herd. At some sites, not only were water temperatures quite warm, but extremely variable. The nearshore conchs, in particular, were regularly exposed to temperatures above their suspected upper tolerance, about 31 degrees Celsius. In the same way coral reefs can bleach suddenly when water warms just a degree, Delgado suspects the spikes may be pushing the inshore queens to new levels of physiological stress.

It would help explain why Queen Conch abundance hasn't returned to historic levels in the Keys despite a four-decade fishing ban. It would help explain why it will take more than any one country's efforts to save the queens, and all the other lives that depend on the oceans.

Twelve

GLOWING FUTURE

THE GIANT CLAM
Tridacna gigas

Snorkeling amid the tree-tangled rock islands of Ngermid Bay in the western Pacific nation of Palau, Alison Sweeney lingers at a plunging coral ledge, photographing every giant clam she sees along a 50-meter transect. In Palau, as in few other places in the world, this means she is going to be underwater for a skin-wrinkling long time.

At least the clams are making it easy for Sweeney, now a biophysicist at Yale University. The animals plump from their shells like painted lips, shimmering in blues, purples, greens, golds, and electric browns. The largest, *Tridacna derasa*, are hunkered at the seafloor. But most are 5-inch *Tridacna crocea* living higher up on the reef, the smallest of the giant clams. Their fleshy Technicolor smiles beam in all directions from the corals and rocks of Ngermid Bay.*

Having been cooped up in their lab at the Palau International Coral Reef Center, Sweeney and three of the scientists on her team have decided to swim out to these clam grounds one afternoon, about a mile from shore. I stroke through blue-green water beside them as we pass mushroom-shaped islands. We see fruit bats sleeping upside down in the trees above and extensive corals below. Some of the corals are bleached from the conditions in Ngermid Bay, where naturally high temperatures and acidity mirror the expected effects of climate change on the global oceans.

---

* Ngermid Bay is more commonly known as "Nikko Bay," but traditional leaders and government officials are working to revive the Indigenous name of Ngermid.

Even those clams living on bleached corals are pulsing color. Sweeney's ponytail flows out behind her as she nears them with her camera. They startle back into their fluted shells. Like bashful fairytale creatures cursed with irresistible beauty, they cannot help but draw attention with their sparkly glow.

It's the glow that drew Sweeney's attention to giant clams, and to Palau, a republic of more than three hundred islands between the Philippines and Guam. Its sunlit waters are home to seven of the world's dozen giant-clam species, from the storied *Tridacna gigas*—which can grow to 550 pounds and measure over 4 feet across—to the elegantly fluted *Tridacna squamosa*. Sweeney first came to the archipelago in 2009 while working on animal iridescence as a postdoctoral fellow at the University of California–Santa Barbara. Whether shimmering from a blue morpho butterfly's wings or a squid's skin, iridescence is almost always associated with a visual signal—often to attract mates or confuse predators. Giant clams' luminosity is not such a signal. So, what is it?

In the years since, Sweeney and her colleagues have discovered that the clams' iridescence is essentially the outer glow of a solar transformer, running on sunlight and algal biofuel. Giant clams reach their cartoonish sizes thanks to an exceptional ability, optimized over 50 million years, to grow their own photosynthetic algae in vertical farms spread throughout their flesh. The glow is otherworldly, as if from the future and the past. Sweeney and other scientists think it may shed light on alternative fuel technologies and other industrial solutions for a warming world.

I float on my back in Ngermid Bay, waiting for Sweeney and her team to finish their transect. The intense Palauan sun is dropping behind the ancient rock islands, and the fruit bats are beginning to wake, stretch their wings, and take off into the twilight.

FRUIT BATS ARE again above me and giant clams below the very next day, this time on land. The two are among the animals painted on the oldest *bai*—men's meeting house—in Palau. The historic governing houses with brightly painted wood beams and steep thatched roofs once dominated every village. But this one, in the state of Airai on Palau's scantly developed largest island, Babeldaob, is one of the last traditional *bai* still standing.

Fruit bats, symbols of humility in Palauan culture, decorate the entryways at each end of the building, reminders that even the most important chiefs should bow down and show respect as they enter. In a bold black-and-white strip below the bats, stylized giant clams stretch across the front and back of the *bai*, a pattern that also often borders Palauan lintels and carved wooden storyboards. On the walls of the *bai* and in the legends of Palau, "the clam signifies power and it signifies persistence," says Elsa Sugar of Airai's Office of Historic and Cultural Preservation, who is giving me a clam tour of the village where she was born.

Palau's islands have been inhabited for at least 3,400 years, and from the start, giant clams were a staple of diet, daily life, and even deity. Many of the islands' oldest surviving tools are crafted of thick giant-clam shell: arched-blade adzes, fishhooks, gougers, heavy taro-root pounders. Giant-clam shell makes up more than three-fourths of some of the oldest shell middens in Palau—a percentage that decreases through the centuries.

Archaeologists suggest that the earliest islanders depleted the giant clams that crowded the crystalline shallows, then may have self-corrected. Ancient Palauan conservation law, known as *bul*, prohibited fishing during critical spawning periods, or when a species showed signs of overharvesting.

Before the Christianity that now dominates Palauan religion sailed in on eighteenth-century mission ships, the culture's creation lore began with a giant clam called to life in an empty sea. The clam grew bigger and bigger until it sired *Latmikaik*, the mother of human children, who birthed them with the help of storms and ocean currents.

The legend evokes giant clams in their larval phase, moving with the currents for their first two weeks of life. Before they can settle, the swimming larvae must find and ingest one or two photosynthetic alga, which later multiply; imagine self-replicating fuel cells. After the larvae down the alga and develop a tiny shell and a foot, they kick around like undersea farmers looking for a sunny spot for their crop. When they've chosen a well-lit home in a shallow lagoon or reef, they affix to the rock, shell gaping to the sky. After the sun hits and photosynthesis begins, the microalgae will multiply to millions—or in the case of *T. gigas*, billions—and clam and algae will live in symbiosis for life.

The microalgae, called zooxanthellae, keep their clams fat and happy with the sugars they need to survive. In exchange, their hosts provide a safe home and a daily dose of sunlight, delivered with remarkable precision. These intertwining life cycles raise compelling scientific questions for the times: How do the clams collect such intense equatorial sunlight without overheating? How do they distribute the light evenly to millions of microalgae, including those in the darkest depths of the clam? And most urgently, how are these animals proving so resilient in the warming, acidifying tropical seas?

PACIFIC ISLANDERS PUT giant clams to practical use while also seeing in them a certain spirit. The shells served on some islands as ceremonial containers for ancestral skulls or ritual washing vessels. When *T. gigas* were still little known in Europe, the first shells brought home by explorers were revered by collectors and kings, and made their way to Christian ritual as Baptismal fonts.

In the early sixteenth century, the Republic of Venice presented King Francois I with what would become the most famous pair of giant-clam shells in church history. Two hundred years after the gift, the sculptor Jean-Baptiste Pigalle mounted the bowls on oceanic marble bases for holy water fonts in the Church of Saint-Sulpice in Paris, where they are still admired. Victor Hugo, who married in the church in 1822 with the help of a fake baptism certificate, would later donate two giant-clam fonts to Saint-Paul-Saint-Louis, the Baroque church in the Marais quarter of Paris, to mark the occasion of his daughter Leopoldine's secret marriage there. Giant-clam fonts grace La Sagrada Familia Basilica in Barcelona, and now Catholic churches around the world.

The vessels may have been inspired by Pacific peoples' infant bathtubs made of giant clams. One history of the Anglican missions in Melanesia gloats that parish priests over time "ensured that the sonorous notes of the conch shell were summoning the islanders to church rather than to war and that small infants were being baptized rather than brutalized in the giant clam."

The French philosopher Gaston Bachelard is said to have exclaimed over the giant clam, "Who would not feel cosmically heartened and strengthened imagining himself bathing in the shell?" Yet somehow, in the nineteenth-century macho mélange of exploration and deep-sea fantasy, a beloved icon of birth and renewal became maligned as a man-killing monster, its steely jaws poised to trap and drown luckless divers.

THE SCIENTIFIC NAME *Tridacna*, from the Greek words *tri* ("three") and *dakno* ("to bite"), was originally meant to describe not clams biting humans, but humans biting clams. Pliny the Elder explained in his *Natural History*: The expedition of Alexander revealed footlong oysters in the Indian Sea, and "some spendthrift and gourmand" named them *tridacna*, "wishing it to be understood thereby, that they are so large as to require three bites in eating them."

But in the English-speaking world, "giant clam" was synonymous with "killer clam" for most of the nineteenth and twentieth centuries. In German, it was the *Mörder Muschel*. All with no evidence—none historic, none popular, none scientific, not even an unverifiable but possible story—of anyone ever being killed in the grasp of a giant clam.

Perhaps long ago in Polynesia, a child got a foot caught while wading in the shallows among abundant giant clams, and parental warnings spun into lore. Or maybe a spear-fisher told a whopper to his pals, and it grew as fish tales do. In the Maori legend *Rata's Voyage*, a stowaway on canoe journey, Nganaoa, gets to stay aboard after proving his worth by killing first a giant clam that threatens to chomp down on the canoe, next an octopus that tries to drag the boat under the sea, and finally a whale about to swallow the whole party. In a macabre trickster tale told in various versions around Polynesia, a mouse-deer or a tortoise gets back at an ape by taking him clam-fishing and directing him to a gargantuan clam for its delicious meat. The ape reaches in and the mollusk cuts his hand clean off.

The legends reflect the human propensity to conjure grotesqueries of real animals. In the age of sea exploration in the nineteenth century, writers planted existential dread in dark grottos or at the ocean bottom—Verne's fantasy of a supernatural sea. Verne set *Twenty Thousand Leagues Under the Sea* in a time when "there appeared in papers caricatures of every gigantic and imaginary creature, from the white whale, the terrible 'Moby Dick,' of hyperborean regions, to the immense kraken, whose tentacles could entangle a ship of five hundred tons and hurry it into the abyss of the ocean."

The naturalists who observed giant clams firsthand in the tropical Pacific vividly described them with admiration as the eye-catchers of the reefs. Passing over a rainbow of clams in the Philippines, the English conchologist Hugh Cuming reported they were as colorful as "a beautiful bed of tulips." But popular writers adding sellable zing to the science texts of the day hyped up the image of the man-eating—even shark-eating—giant clam.

"So powerful are they that large sharks and rays that have accidentally crossed them have been seized and held," warned the outdoors writer Charles Frederick Holder in his 1885 *Elements of Zoology*, a collaboration with his zoologist-physician father Joseph Holder.

In the 1920s, *Popular Mechanics* reported that in Papua, "Divers who accidentally step into the open lips of the monsters are not infrequently held with such force that they cannot release themselves and are drowned. The shells close with such force that they serve as gigantic traps." In the '30s, the *New York Times* Sunday travel section weighed in on the "nightmare bivalve of the Australian seas." Exaggerating its size—"often described" as 14 feet long—and danger, the newspaper of record reported that its snapping shell "has caused the deaths of many natives and divers who have been caught by the foot and found themselves unable to wrench free."

The stories captured the imagination of the U.S. Navy during World War II, when soldiers fighting in the Pacific were briefed on the "man-eating clams" and large sharks known to inhabit the reefs. The man-eater myth was so persistent that decades later, Navy diving manuals still advised frogmen how to free themselves if caught in the "vise-like" grip of a giant clam: by inserting a knife between the valves and severing the animal's adductor muscle.

Even the pioneering marine biologist Eugenie Clark, who did perhaps more than any other scientist to dispel myths about sharks, was gripped by the trope. During a trip to Palau in the early 1950s, Clark profiled one of her guides, the tall and muscular spear-fisher Siakong, who, when he took off his work clothes to dive into the sea in his red loincloth, "suddenly metamorphosed from a bum into a statue of a Greek god."

The two were spear-fishing in the rain one afternoon when Siakong called, "*Nechan* ('big sister'), come see!" Clark dove down and saw a *T. gigas* so huge that it "could have held all of Siakong with ease." Unable to lift the clam, Siakong dove to extract the meat for a reef picnic. When he failed to surface, Clark dove again to find a sight "that sickened me with horror. *Siakong was caught in the clam!*"

The jaws of the gigantic mollusk were clamped tight, with Siakong's arm inside up to the elbow. He wasn't moving. As Clark panicked, rose for air, and tried to figure out how to save Siakong from drowning, he surfaced with a grin and held up the largest clam muscle she'd ever seen. Saikong had been deep in the clam, sawing through its adductor muscle with his knife. Clark recovered from the joke, and they shared a raw hunk of clam the size of a man's thigh.

Even as scientists began to discover the giant clam's role in coral-reef ecosystems in the 1960s, its popular image as killer clam endured, from television shows to roadside attractions. In the 1975 *Doctor Who* serial *Genesis of the Daleks*, the doctor is attacked by the cheesiest giant clams of all time while trying to stop mad scientist Davros from doing horrific experiments. To this day along Interstate

75 through Michigan, billboards remind tourists bound for Mackinac country to take Exit 326 to see the "Man-Killing Giant Clam" at Sea Shell City in Cheboygan.

Like the bloodthirsty shark, the conniving wolf, and the evil snake, the mythos of the killer clam was born out of fear of the unknown and grew incongruously more menacing the more humanity came to know, and then dominate, the sea. All the while the giant clams, like every pair of shells in the ocean, held their truths about life and survival.

IN THE BATHTUB-WARM waters of Ngermid Bay, I swim with Amanda Holt and Jing Cai, researchers on Sweeney's team, as they search for giant clams. I'd imagined we might find the hundred-pound, blue-glowing *T. gigas* that illustrate every giant-clam article and academic paper. I soon learn that this is like expecting Bengal tigers to pop out of the forests of India. Our most ubiquitous wild icons have become our rarest. *T. gigas* are no exception.

Luxury demand for their ivory-like shells and their adductor muscle—a sashimi delicacy—has driven *T. gigas* extinct in China, Taiwan, and other parts of their native habitat. Some of the toughest marine protection laws in the world, along with giant-clam aquaculture pioneered here, have helped Palau's wild clams survive. The Palau Mariculture Demonstration Center raises hundreds of thousands of giant clams a year, supplying local clam farmers who sell to restaurants and the aquarium trade and keeping pressure off the wild population. But as other nations have wiped out their clams, Palau's 230,000-square-mile ocean territory is an increasing target of illegal foreign fishers.

We see only one *T. gigas* all afternoon. The diminutive *T. crocea* are everywhere. I count twenty affixed to a single coral rock, their soft mantles pulsing color. A velvety-black clam is trimmed with electric-blue spots. A dark turquoise clam has a brighter aquamarine band ringed in black spots. They cling next to a yellow-green clam mottled like lizard skin, and a brown one with dark gold running through its mantle like veins in quartz rock. They look like nothing so much as living scrunchies, gleefully discarded by a passing mermaid.

The riotous colors belie the naturally low pH and high temperatures of Ngermid Bay. Like corals, which have a similar bond with zooxanthellae, giant clams are susceptible to bleaching in extreme heat. Under stress, they expel their algae, which drains their color and can kill them. But the wild clams of Ngermid Bay appear to be thriving despite conditions similar to those predicted for the global oceans in the year 2100.

At the Palau International Coral Reef Center on the busy island of Koror, CEO Yimnang Golbuu, a biologist, explains that Ngermid's conditions are a natural result of its placid flow and isolation from outside ecosystems. Those features also confine larvae to the bay, which may allow marine life to select for tolerant traits and adapt to the harsh environment. "Obviously there are differences in resilience that nature has provided us," Golbuu tells me. "We need to understand and listen to nature and learn from those differences."

Listening to nature is at the heart of a growing confluence of biology and materials science, sometimes called biomimicry. When Sweeney took her first physics class, as an undergraduate biology major in her native Illinois, she found that studying life through its physical structures gave her a welcome sense of order amid the amorphous questions of evolutionary biology. Twenty years later, as the first biologist hired by the University of Pennsylvania's Department of Physics and Astronomy, she was still drawn to the elegant efficiency of living creatures: the squid's eye that can see easily in the dark depths of the ocean, the giant clam's ability to turn sunlight into energy.

It's the modern version of the animal moral stories painted on the Palauan *bai*. Just as a new generation of civil engineers has learned that working with ecosystems can make for superior design, materials scientists increasingly look to biology not only for inspiration, but for actual blueprints.

"Evolution," Sweeney says, "is so much more clever than human engineers."

~~~~~~~~

NEARLY 9,000 MILES away on the other side of the globe, a light January snow is falling when I visit Philadelphia, an unlikely epicenter of giant-clam research. I head first to the Academy of Natural Sciences, founded by Thomas Say and the other scientific revolutionaries in 1812. Now part of Drexel University, the academy is a mélange of Victorian-era collections and modern technology. On the top floor, more than 10 million specimens, ranging from mondo *T. gigas* shells to mini 0.01-millimeter sea-snail spirals, are tucked into 13,500 aluminum drawers. Five hundred giant-clam shells fill the bottommost drawers, specially reinforced to withstand their weight.

In the academy's basement, past funky-smelling stacks of stuffed mammals and beastly horns, the biogeochemist Michelle Gannon guides a *T. crocea* shell collected by Sweeney's team into an electric trim saw with a 7-inch blade. The cross section reveals the feathery gray growth rings that, like tree rings, mark time in a clam's life.

Remember how geoscientists used fossil clam shells to reconstruct ancient climate over millions of years? Gannon uses her scanning electron microscope to read *daily* cycles of sunlight and darkness in giant clam shell crystals, just a few micrometers wide. In the academy's stable isotope lab, she zeroes in still further, grinding the crystals into powder and weighing them. The calcium carbonate that mollusks use to build their shells—$CaCO_3$—contains calcium, carbon, and oxygen. The type of oxygen varies depending on the conditions; mollusks build different amounts of the isotopes oxygen-16 and oxygen-18 into their homes depending on water temperature, evaporation rates, and other conditions. By measuring the relative weights of the isotopes, Gannon can describe daily changes in a clam's environment.

When she binds these bioarchives together, Gannon has a detailed history of the clam's living conditions and photosynthetic output—even the amount of sunlight that reached the animal each day. On the brightest days, she's found, the clams grow an order of magnitude faster than when it's cloudy. She hopes the data can inform research on biofuels powered by algae and sunlight. Whether society will be willing to invest in those new kinds of energy sources seems to be a tougher challenge than even the science.

THE SNOW IS falling harder the next day when I meet Sweeney and her University of Pennsylvania collaborator, materials science and engineering professor Shu Yang, at Penn's Laboratory for Research on the Structure of Matter. Hanging in a close-up photo in the lobby, an iconic *T. gigas* glows in tropical blue.

Yang began her career at Bell Labs' Lucent Technologies, researching the use of light in telecommunications. She's spent nearly two decades at Penn, where she and the more than twenty researchers in her lab work to fabricate materials based on natural forms. They've mimicked the self-cleaning ability of lotus leaves; the adhesive talents of gecko foot hairs and burdock seeds; and the water-repellent colors of butterfly wings and beetle scales. Now, they are modeling the photosynthetic efficiency of giant clams.

The research began with the glow. Unlike a pigment, the iridescent color in a giant-clam mantle, a peacock feather, or a blue butterfly wing is a physical effect produced when nanometer-sized lattices within the surface interact with light. Scientists call these lattices photonic crystals. Materials scientists such as Yang are keen to fabricate them to harness light for any number of applications, from speedier fiber optics to more efficient photovoltaic cells.

The sparkling lattices in butterfly wings and peacock feathers appear to have evolved to attract mates. But the clams' glowing cells—called iridocytes—seek to woo the sun. Inside the mantle of a clam, Sweeney's team discovered, microalgae organize into pillars. The iridocytes assemble into solar collection panels over each pillar. The miniscule solar panels draw in concentrated sunlight, then scatter the light waves that best spark photosynthesis in the algae. By directing blue and red light waves to the pillars and reflecting the rest back into the water, the iridocytes keep the algae fueled up without the clam burning up in the intense tropical sun. "I was fascinated because it's self-assembly, and it's cheap," says Yang.

Yang had spent years working to fabricate photonic crystals with costly metallic materials. Working with Sweeney and supported by an interdisciplinary grant from the National Science Foundation, she and her team discovered that silica nanoparticles embedded in gelatin could mimic the shining clam cells' light-scattering properties— and do so inexpensively. The *gigas*-sized obstacle to ramping up algal biofuels is cost. With no price, penalties, or other restraints on fossil fuel emissions, there's little financial incentive to invest.

Sweeney and Yang believe that a giant-clam-inspired bioreactor could be both cheaper and more productive than existing biofuels such as corn and ethanol. Using cultures from Palauan clams, Yang and her team are now at work on the algae side of the model. So far, the scientists haven't been able to coax the algae cells to line up as dutifully in the lab as they do in the clams.

AS I BRAKE for chickens and their scrambling chicks on the two-lane road that winds by small wooden houses and prodigious tropical flowers in Palau's state of Airai, Elsa Sugar shares some of her favorite childhood memories of fishing and clamming with her family from a bamboo raft. The family spent holidays at their fishing cottage on one of the rock islands, always collecting the biggest clam they could find for Christmas dinner. One year when she was ten or so, Elsa found the Christmas clam by throwing handfuls of sand into the lagoon fronting the cottage to make the clams spit. She followed the biggest jet to a supersized specimen that became the centerpiece of their feast.

Giant clams are a beloved staple in Palau and in many parts of the Pacific. They're eaten raw with lemon, simmered in coconut milk for a buttery soup, baked into a savory pancake, or sliced and sautéed.

But like the abalone of the Pacific, the scallops of the Atlantic, and the oysters of the world, an Indigenous staple became a delicacy and soon an expensive trophy, won in an unfair hunt. In the 1960s,

Chinese and Taiwanese clam boats began to range farther and farther into the Pacific Islands, the Indo-Malay region, and Australia's
Great Barrier Reef to collect the coveted adductor muscles that pull
the clams' two shells together. The muscle, favored in sashimi and
an alleged aphrodisiac, accounts for only about 10 percent of the
animal's flesh, but the fishers often left the remaining meat to rot.
Scientists estimate poachers illegally hauled in up to half a million
clams a year for four decades before international pressure and pros-
ecutions slowed them down. During a crackdown in Palau in the
1980s, a police team seized an illegal Taiwanese fishing boat loaded
with 7,500 pounds of pure adductor muscle. While the officers were
unloading the meat, Palau's tribal chief reported, the poachers tried
to bribe him to look the other way.

Giant clams were once spread so thickly along the shallow coasts
and coral reefs of the tropical Pacific that the conchologist Cuming described drifting over a solid mile of them on a collecting trip.
The clams settle in clusters because they need to be close to each
other to reproduce; they spawn in sync. Mature clams generate both
sperm and eggs and release them into the water from their siphon
like smoke from a little stovepipe, setting off a chain of fertilization.

Yet not two centuries after Cuming reported sea bottom carpeted
with giant clams, they are now spread too thinly in much of their
native habitat to fertilize. A single *T. gigas* can send as many as five
hundred eggs into the ocean in one spawning. Alas, fecundity is no
consolation in a lonely sea.

Overharvesting has now driven the largest species locally extinct
in China, Taiwan, Singapore, and numerous smaller islands, as
giant clams' shells become as valuable as their meat. The rippled
bowls once used as infant bathtubs are now in demand for poolside
fountains and bathroom sinks in seaside homes. Giant clamshells
have long been venerated in China, especially by followers of Buddhism, who count them among the seven treasures of nature.

In a fishing port called Tanmen in the resort province of Hainan
Island overlooking the South China Sea, fishers and artisans made

fortunes mining giant-clam shells and turning them into Buddhist prayer beads and elaborate animal sculptures believed to bring good luck and prosperity. White-shining elephants, fat toads, and Arowana fish carved from the ivory-thick shells packed the shelves of some nine hundred retailers in the town by 2015, when fishers could pull in 80,000 yuan, or $12,000, for a single *T. gigas*.

In the end the charms were not so lucky for the fishers, crafters, and retailers who depleted the clams and destroyed their habitat. Demand for the figurines sent crews to smash coral reefs with their boat propellers to free up anchored clams, decimating miles of some of Earth's most biodiverse coral reefs in the South China Sea. International outcry ultimately led Hainan Province to ban trade in giant clams and corals. The Tanmen workshops and retail shops have shuttered en masse. The clams may never recover.

In Palau, some of the toughest marine protection laws in the world, along with the giant-clam aquaculture pioneered here, have given the wild clams a better chance than most. But illegal foreign fishers are taking their toll. At the coral reef center, Golbuu tells me that even in the mid-twentieth century, giant clams were said to be so abundant in the remote Palauan atolls that their spitting would create a spectacle of dancing fountains. Enormous Helen Reef in Palau's far southwestern state of Hatohobei is known locally as *Hocharihie*, "reef of the giant clam." But in recent decades, as other nations have wiped out their own clams, the remote islands are frequent targets of foreign poachers. Palau has created one of the largest areas of protected ocean in the world. Its 193,000-square-mile ocean sanctuary is the size of California, yet it owns just two marine patrol boats. In 2012, two Palauan marine officers and their American pilot crashed while searching by air for the Chinese mother ship of a fishing fleet that had stolen giant clams from *Hocharihie*. The pilot and officers were never seen again. The clams were dumped by fleeing poachers.

The stolen clams must have tumbled like boulders into the open ocean, down into a deeper sea than they had known.

BACK UNDERWATER IN Ngermid Bay, the same stylized giant-clam pattern from the Palauan *bai* circles the upper arm of one of Sweeney's doctoral students in a banded tattoo. Lincoln Rehm grew up in Texas in a Palauan family and traveled to the islands for vacations as a child, often kayaking through the shallows with his aunties to collect clams. They'd bring soy sauce and lemon and enjoy a clam-sashimi picnic on the reef. Rehm says those summers defined his purpose, and after earning his degree in biology, he moved to Palau to work at the coral-reef center. It was there that he became interested in Sweeney's research on giant clams' glow. In 2015 he returned to the United States for his PhD, funded in part by Sweeney and Yang's National Science Foundation grant.

As Holt and Cai swim ahead to scope for the clams, Rehm dives down to set up Sweeney's photographs. He positions a color palette next to each clam to help log its hue into the computer later. Over the past three years, he has built a database of more than eight hundred giant-clam photos. Holt and Cai wrote a code that uses the palette to organize each clam by color; Cai, a machine learning expert, designed an algorithm that analyzes every pixel, allowing the team to see hue and brightness in individual algae and clam cells, including the glowing iridocytes.

Rehm's research aims to trace the source of the clams' mantle colors, to inform the biofuels project as well as the intense local interest in the animals' coloration and glow. Giant clams are more than beloved cultural icons and food staples in Palau and across the Indo-Pacific. They are also valuable draws for tourists, who delight in finding the sparkling clams underwater. Palauans planted a shallow garden of giant clams decades ago at a popular dive-boat stop called Clam City. Many of those giants have been poached in the meantime. But the dozen *T. gigas* that remain, clustered like neon velvet lounge chairs on the ocean bottom, attract boatload upon boatload of tourists living their *National Geographic* underwater photography dreams.

When I visit Clam City, the largest mollusks in the world strike me as the most defenseless animals I've ever seen. Few life stations would seem more vulnerable than being unable to move. They can't run, or even clunk away like most mollusks. Not only are they too big to hide, but they announce themselves with that neon glow.

The aquarium industry is especially interested in color science. The brighter the clam, the higher the price. The country is known for pioneering giant-clam aquaculture at the Palau Mariculture Demonstration Center at the southern tip of Malkal Island, launched in 1973 with funding from the U.S. government and expanded in recent years with support from Japan. (The islands were home to some of the deadliest Pacific battles between the U.S. and Japan during World War II; vine-grown armored tanks and other military remnants still stand in haunting reminders in the forests and countryside.) Researchers learned to spawn giant clams in the laboratory and grow what is now more than a million giant clam seedlings a year. Small as a coquina shell, a baby giant clam is a mini-me, like a wee aquarium ornament to set next to a teensy diver and treasure chest. They're grown in aerated tanks for a year to about the size of a cupped hand, then transferred to ocean cages until they're ready for clam farms or various giant-clam restoration projects underway across their native habitats. Researchers say the industry holds promise for saving giant clams from extinction, though so far, the work has fallen short of providing a steady income for Pacific clam farmers. Scientists in the Philippines recently reported the first sightings of juvenile *T. gigas* spawned naturally from restocked clams, after a nearly thirty-five-year effort. In Palau and around the region from the Philippines to Fiji, fishers and community leaders say the cultured clams are no less vulnerable to overfishing and poaching. A cove-studded Philippine resort called Hamilo Coast that stocked giant clams as a natural amenity hires guards to protect them from being hauled off in the night.

In Palau, I meet a clam farmer and mother of eight named Bernice

Ngirkelau on the anniversary of her husband's death. Her youthful face and easy smile belie the year she's had. She tells me the couple's early years in the business were joyful as they watched the clams grow. But the larger the clams, the more frequently they were targeted by poachers. Between losing her partner and losing most of the biggest clams they raised, the year has been a trial. She can't afford to hire security for her three underwater farms, or buy the fuel it would take to constantly check on them. Ngirkelau is intrigued with Rehm's color research: If she could control for color and glow, she'd grow 1-inch aquarium clams, which sell for double the price of 12-inch food clams.

Back in the laboratory at the coral-reef center, Sweeney, Rehm, Holt, and Cai are finishing up a series of experiments on live clam tissue. They're measuring how light and heat leave the algae and clam cells. Rehm slices a small bit of flesh from a softball-sized *T. crocea*. He places the sample in water under bright light, attaches digital thermometers, and records the temperatures of the water and tissue over the course of 30 minutes.

Across the table, Sweeney whirs an immersion blender she picked up at a nearby shopping center, mixing algae extracted from the same clam. The resulting froth looks like a coconut smoothie, smells like clam juice, and is as dense with algae as a clam's mantle.

Nearby, Holt uses a spectrometer to expose muscle, iridocytes, and algae from another piece of the clam to different types of light, tracking how much is absorbed and through which cells.

Repeated late into the evenings during their three weeks in Palau, the experiments on the clam tissue and the isolated algae reveal a new clue to the workings of the bivalve bioreactor. Photosynthesis can spike a clam's body temperature several degrees higher than the surrounding seawater. The iridocytes appear to not only draw light into the clam and scatter the most useful wavelengths to the algae; they also collect the excess heat generated by photosynthesis and expel it via light.

For giant clams, this ability to shed heat may be a key to their resilience in environments like Ngermid Bay. For humanity, it may point the way toward new cooling technologies—ways to expel heat from power plants, office buildings, or car interiors without fossil fuels. Human-made materials, Yang says, tend to fight nature—making our spaces hotter when it's sunny, for example. Instead, we can take cues from the natural world to learn how animals deflect heat. Yang looks to animal architecture, colors, even the shape of their cells, to design climate-resistant materials for the energy and construction sectors. The universe inside a giant clam seems to run on the "wasting little, harming not" future that Liberty Hyde Bailey envisioned.

Magnified on Rehm's computer screen, a microscopic fleck of iridocytes looks just like the Milky Way on a pitch-black night.

I SPOT GIANT clams and fruit bats together once more before I leave Palau. Both are delicacies on the menu at a local restaurant called Carp. Fruit bat soup is an island favorite that arrives with the winged creature afloat in the bowl. I pass. I heard on my dive boat that Americans have a reputation for ordering fruit bat soup for a selfie or an Instagram post and then leaving the food untouched.

Likewise, I can't bring myself to order the giant-clam soup, the giant-clam pie, or the giant-clam sashimi. In full disclosure, I've eaten no shellfish but small, farm-raised clams since my mistake of ordering plump moules marinèire in a Washington, D.C., restaurant on the same day I saw the wrinkled cowries in vials on Chris Meyer's desk at the Smithsonian. But it's one thing for Palauans to enjoy the staples that have sustained them for thousands of years. It's another for outsiders to exert so much pressure on these icons. It is, after all, the outside world that threatens Palau: the crowds of sunscreen-whitened, fin-flapping tourists on the reefs; the poaching of marine life; and the warming, acidifying, rising seas of a climate altered by the fossil fuel emissions of much larger countries.

Palau is one of the smallest nations in the world, with a population of 20,000, yet it can see more than 160,000 visitors a year. Almost all of them head underwater. More damaging than the invited foreign tourists, who provide more than half the republic's gross domestic product, are the uninvited foreign fleets: the super-trawlers and small-boat poachers that sneak into Palauan waters to nab giant clams, bluefin tuna, sharks, and other marine life for the ravenous global seafood market.

Drawing on the ancient conservation tradition of *bul*, Palauans are leading the world in new protections designed to save the seas. In 2015, President Tommy Remengesau Jr. signed into law the Palau National Marine Sanctuary Act, which prohibits fishing in 80 percent of Palau's Exclusive Economic Zone and creates a domestic fishing area in the remaining 20 percent, set aside for local fishers selling to local markets.

In 2017, Palau amended its immigration policy to require that all visitors sign a pledge to behave in an ecologically responsible manner. The pledge, stamped into passports by an immigration officer who watches you sign, is promised to the island's children:

> Children of Palau, I take this pledge, as your guest, to preserve and protect your beautiful and unique island home. I vow to tread lightly, act kindly and explore mindfully. I shall not take what is not given. I shall not harm what does not harm me. The only footprints I shall leave are those that will wash away.

The pledge is winning hearts and public relations awards. But Palau's existential challenge is at a larger scale; the rising temperatures, sea levels, and destructive storms tied to emissions.

F. Umiich Sengebau, Palau's Minister of Natural Resources, Environment, and Tourism, grew up on Koror Island and is full of giant-clam proverbs and legends from his youth. He tells me a story I also heard from an elder in the state of Airai: that in old times, giant clams were known as "stormy-weather food," the fresh staple that was easy to collect and have on hand when it was too stormy to go out fishing. It reminded me of the smoked Queen Conchs saved up as hurricane ham.

As Palau faces the tempests of climate change, giant clams are still the stormy-weather food, Sengebau says: a secure source of protein; a fishing livelihood; a glowing icon for tourists; maybe even an inspiration for alternative energy and other low-carbon technologies. "In the old days, clams saved us," Sengebau tells me. "I think there's a lot of power in that, a great power and meaning in the history of clams as food, and now clams as science."

Thirteen

~~~~~~~~~~~~~~~~~~~~~~~~

TRUST IN NATURE

THE GEOGRAPHER CONE
Conus geographus

Once a month or so when she was growing up in Brooklyn, New York, Mandë Holford's mother would drop her and her four brothers and sisters off at the American Museum of Natural History at 10 o'clock in the morning with two instructions: Don't lose anyone, and meet back at the Hall of African Mammals at 5:45 in the evening. The museum was a babysitter for their working parents and a heyday for the kids. They came to know every stuffed beast, and to recognize the ancient cultures on the other side of the world.

Then as now, shells held a faded prominence in the cultural dioramas like the Yorùbá dancer costumed entirely in gastropods in the Hall of African Peoples. They were lowly in the marine displays; what small being can compete with a 94-foot-long blue whale? The museum's collection holds more than 6 million mollusk specimens, but most are tucked away in drawers on the fifth floor, where Geerat Vermeij got his life-changing private tour after writing his "Dear Curator" letter at age fourteen.

Holford was never drawn to seashells in childhood like Vermeij and so many others. She didn't even care much about the museum's dinosaurs. But she was fascinated by the halls devoted to animals in their environments and people in their cultures: the African elephants led by the herd's oldest female; the Great Canoe she could imagine Native people rowing through Pacific swells.

Now a research biochemist, Holford credits those museum heydays with sparking her interest in science: "Especially questions of biodiversity," she says, "how nature occurred and why there were so many animals on the planet." Later during graduate school, the American Museum of Natural History was also where she met the mollusk that led her to the specialty she calls venomics—genomics and animal-venom discovery—and a radically different view of seashells.

She first saw the cone snail in a video that opens to a peaceful aquarium scene. Its three-inch shell is barely visible, humped beneath the sand. The only part of the mollusk seen is its long proboscis, stretched across the bottom like a harmless worm. A curious fish nearly the size of the shell swims over the worm. The translucent proboscis pulses with liquid. The mollusk plunges the tip of the appendage—it is armed with a teeny harpoon—into the fish's belly. In a split-second explosion of sand, the cone snail opens a soft, gaping maw at the tip of its shell, draws in the struggling fish head-first, and swallows it whole. The snail mouth closes around the fish tail as it gives a last, reflexive wiggle. "The fish never had a chance," says Holford.

Conidae, builders of the hieroglyph cones that inspired Rembrandt and shell mania in the seventeenth century, also make an arsenal of neurotoxins. They catch their quarry with precision doses that draw from more than eight hundred chemicals. The novel compounds target different receptors in their prey, allowing one of the world's slowest creatures to kill one of the fastest. Cones also deploy venom to defend themselves, which is why they sometimes strike people who pick them up or step on them. The venom of the Geographer Cone, *Conus geographus*, is one of the deadliest to humans of all the animal toxins known. Nicknamed the "cigarette cone," the story goes that a victim stung by it would have just enough time to smoke a cigarette before dying from the venom; the death actually takes a few hours.

What kills in nature can also cure. Cone-snail toxin famously led to development of a chronic pain drug called ziconotide (sold as Prialt), a thousand times stronger than morphine without potential for addiction. But the drug doesn't cross the blood-brain barrier— the protective barrier that prevents compounds in our blood from invading our brain. It must be delivered by spinal tap, which limits

its distribution to certain cancer and HIV/AIDS patients in extreme pain for whom morphine is no longer effective.

Holford is convinced that somewhere out there, along a sea bottom, shore, or coral reef, one of the many venomous animals in the ocean carries an analgesic chemical that could cross the human blood-brain barrier. Beneath the shell of some unknown mollusk hides an alternative to the opioid painkillers that have caused such heartache in the world. She now maps mollusk genomes in search of snail–venom combinations that could lead to that drug, and treatments for cancers and other illness. But DNA sequencing and molecular phylogeny turn out to be the easy parts. The bigger challenge is saving the animal diversity that Holford considers key to bettering all life—in a time when the world is losing many species before they are ever named.

THE FIRST PERSON known to have been killed by a cone snail's little harpoon was an enslaved woman on the Banda Islands southeast of Ambon, now part of Indonesia's Maluku province. The Dutch East India Company had taken the Bandas by massacre earlier in the 1600s; the Dutch enslaved the survivors and nearby islanders to work nutmeg plantations. Writing in his *Ambonese Curiosity Cabinet*, the naturalist Rumphius told the sad tale of one of them: "She had only held this little Whelk in her hand, which she had picked up out of the Sea, while they were pulling in a Seine net; and while she was walking to the beach, she felt a slight itching in her hand, which gradually crept up her arm and through her entire body; and so she died from it instantaneously."

Rumphius's account is first in the official record of more than 140 instances of toxic cone-snail strikes on people, thirty-six of those fatal. The actual numbers are likely far higher; most deaths would have gone unreported in the earlier centuries. The invertebrate biologist Alan J. Kohn at the University of Washington, his name auspicious for a career devoted to venomous cones, has kept the record for more than sixty years. He first observed a Striated Cone paralyze a fish with "apparently a powerful neurotoxin" in a Yale laboratory aquarium during graduate school in the early 1950s. He went on to study the impressive evolution and ecology of cones.

The cone snails evolved to more than eight hundred species, making them among the most successful living mollusks in terms of diversity. Kohn's research helped explain how so many close relatives could live near each another in the tropics—up to thirty-six different species on the same coral reef—without competing for the same food. The answer is that different cones make highly specialized venoms targeted to different prey. Most cones are worm-eaters. Some eat fellow mollusks. About a hundred are piscivores. Some of the fish hunters evolved to paralyze their prey with harpoons, others with the world's most beautiful fishnets.

The Tulip Cone sidles up to a small fish, its mouth-net billowing with tiny tentacles that look like tassels. With none of the drama of its harpooning cousins, the animal gently envelops the fish in its net, delivers its numbing toxin, and draws the stupefied creature inside before it ever senses its doom.

The fish hunters are the only cone snails known to have killed people. Kohn believes it's very likely that the Geographer Cone is the only one that has. The only person known to have injected himself with conotoxin thought so too—or else he was suicidal. In the late 1970s, a Tokyo marine scientist named Shigeo Yoshiba injected milky white biotoxin he'd extracted from Japan's Thunderbolt Cone, *Conus fulmen*, into various marine creatures, amphibians, and mammals. Fishes convulsed and perished. Frogs' sharp-angled back legs straightened stiff before death. A rabbit lost its ability to walk but recovered within an hour. Yoshiba injected a small amount into his own forearm. "No neurological or functional disturbance appeared," he wrote blithely. "Only local findings such as pain, redness, ischemia, edema and itching appeared and lasted for three days."

THE PHILIPPINE ARCHIPELAGO is rich in tropical seashells, home to many of history's best known: Deep-swimming

Chambered Nautilus. Helmets, conchs, and tritons blown as trumpets. The Glory of the Sea Cones that set off greed and gluttony among the Dutch. The large, translucent "windowpane" bivalves known as capiz shells that filtered light in windows across all of Asia.

Growing up on the islands in the 1940s, Baldomero ("Toto") Olivera Jr. was obsessed with shells. He spent countless hours collecting and classifying. He got school friends into shelling, including Donald Dan, the world-renowned Florida shell dealer. At age seventeen, Olivera prepared the collection of Philippine seashells his country's president would present to shell-loving Emperor Hirohito as a gift during a famous goodwill mission to Japan in 1958.

Olivera collected his best specimens not on the white-sand beaches or rock ledges that band the islands, but in construction piles dredged from Manila Bay and heaped next to the tennis courts at the social club where his father played in the afternoons. He prized the large cones especially, for their heft, rich colors, and patterns. Cone shells are visual language: lettered, mapped, zigzagged, spotted, mottled, hennaed, stitched, grooved, tattooed, feathered in wisps. Even as a small boy, Olivera knew some of the beauty could kill.

After graduating from the University of the Philippines, Olivera moved to the United States on a Fulbright scholarship, earned his PhD in biochemistry at the California Institute of Technology, then immersed himself in the new science of DNA replication as a postgraduate at Stanford. When he returned to University of the Philippines as a research professor, his lab "had absolutely no equipment. It was clear I wasn't going to be very competitive in DNA replication." So he settled into a science that didn't require sophisticated equipment: isolating the lethal compounds in the cone snails he had loved as a boy.

During the political and economic upheavals of the Ferdinand Marcos presidency, Olivera worried about his young family and watched support for science erode. He accepted a research position at the University of Utah and moved to landlocked Salt Lake City. With a decent laboratory, he was able to set up the conotoxins study as a side project attended by undergraduates. In 1979, his youngest researcher, Michael McIntosh, eighteen and just out of high school, suggested that instead of injecting isolated conotoxins into mice bodies, they try injecting them directly into mice brains. Olivera didn't like the idea, but he let the inquisitive freshman go ahead. The results would upend drug-discovery science. One toxin made the mice run around in circles. Another made them shake their heads back and forth. Another put them to sleep. Another made them hyperactive. "We realized that we were not dealing with just a few paralytic components in the venom," Olivera recalled. "There was this incredible diversity of different components that all seemed to have different effects on the central nervous system."

Olivera turned his attention back to cone snails. Among promising discoveries, the lab found that one of the peptides that inhibits a junction between fish nerve and muscle also blocks a neurotransmitter in humans that signals pain. The discovery led to the commercialization of Prialt. The company that took it to market spent two years working to improve the drug. But its scientists couldn't beat 55 million years of evolution. The commercial compound is identical to that of the cone snails.

THE PORTUGUESE EXPLORER Ferdinand Magellan never actually made it all the way around the globe as he is credited. In the spring of 1521, on the island of Mactan in what is now the Philippines, a chieftain, Datu Lapulapu, attacked Magellan and his men, who were working to convert the archipelago to Christianity and claim it for Spain. Magellan was killed by the plunge of an arrow, just like Juan Ponce de Leon had been—though local legend has it that Magellan would have escaped if not for a giant clam snapping shut on his foot, as if to help Lapulapu. A muscular monument of the Filipino hero looms at the center of Mactan, a long seashell necklace hanging down his chest. A local cone snail, *Conasprella lapulapui*, is named after him. It is apropos for Mactan, a fulcrum of the global ornamental seashell industry.

Shell-craft manufacturers cover the island's neighborhoods. Exporters buy from Mactan fishermen and women, who, like other fishers in the Philippines, are some of the poorest and most marginalized people in the country. The exporters pack mixed tropical shells by the ton into cargo containers at the international port in Manila. They ship them to China, Europe, and North America for tourists to buy as beach souvenirs in shell shops from Cornish resorts to American beach towns.

The pressure has decimated giant clams, top shells, and once-thick beds of windowpane shell locally and in reefs across the Philippines and Indonesia that early naturalists described as colorful seashell gardens. Even common species such as Money Cowries—known locally as sigay shells and ubiquitous in placemats, baskets, jewelry, and other crafts—are becoming harder to find. A marine biologist who studied the state of the industry early in this century found that the little white cowries now "may have been harvested beyond their sustainable yield."

The classic Horned Helmet, *Cassis cornuta*, popular for carving cameos in Victorian times, is among a dozen marine mollusks protected under both Philippine and Indonesian laws for more than thirty years—yet it's sold openly throughout the tourist markets of the Indo-Pacific. Wildlife researchers from Oxford University found a vigorous trade in the helmets, giant clams, Triton's Trumpets, and Chambered Nautilus, among others.

Having persisted through the dinosaur extinction, and after its forebears' survival of every calamity in the half-billion-year evolution of mollusks, the Chambered Nautilus finally appears likely to succumb. The animal is now protected under the global Convention on International Trade in Endangered Species (CITES) and the U.S. Endangered Species Act, rare attention for mollusks and other invertebrates. But its tiger-striped shell remains a high-demand luxury good. As with so many threats to animals and the environment, imposing regulations without also tackling human dimensions such as poverty can make things worse: The nautilus researcher Peter

Ward says that each protected animal is now worth even more to the often-impoverished fishers who capture them in nighttime traps.

Politics and economics, much more than science, determine humanity's take from the sea. Indonesia had banned the use of seine and trawl nets for their depletion of fish, damage to coral reefs, and bycatch including native mollusks. But as the coronavirus pandemic drove the country into recession in 2020, the government succumbed to pressure to lift the ban. Some species of cones, volutes, and other sublime shell-makers live only in small sections—even specific reefs—of the great wedge of ocean life between Papua New Guinea, Indonesia, and the Philippines known as the Coral Triangle. Almost immediately after the return to trawling, serious shell collectors began to notice what one described as an "avalanche" of endemic Indonesian shells in online markets.

The story is similar in southern India. Living mollusks hand-netted and trawled from the Indian Ocean are piled by the species in great mounds at the same Tamil Nadu shell processors that grind and polish the sacred chanks. The researcher Amey Bansod describes the stench of dead marine animals, the acid baths, and the thick coat of white carbonate dust that plasters the trees in the coastal village of Kanyakumari. "Scenes of the apocalypse come early," he says. In a just-harvested pile of spider conch shells, the mollusks inside slowly die in the heat, spider-leg shell spines unable to walk them back to the nearby sea.

Once bleached and polished, the shells make their way to Western beach shops through a long chain of local fishers, processors, middlemen, exporters, shell-shop owners, and consumers. At one end is the local fisher who might be paid pennies for mollusks which then sell for as little as 15 cents each in villages such as Indonesia's Pangandaran. At the other is the beachgoer across the world who digs through baskets of cheap tropical shells for a souvenir not found on nearby sands, but taken from the depths of a faraway sea.

When the trawl nets haul up what collectors call a specimen shell—a rare, exceptional, or particularly large specimen, or a left-opening wonder, word quickly makes its way to dealers, who sell them at higher prices in online markets or directly to collectors.

For either the serious collector or the tourist on vacation in Sanibel or Santa Barbara, declining to buy one is to take a stand. Even better, you can find your own, high-spired Florida Cone along the Gulf or the Atlantic—or a soft-shouldered California Cone at the Pacific. They are both cream-colored with tawny markings, with no jab of flesh or conscience from an empty shell.

MANDË HOLFORD WAS a doctoral student studying biochemistry at Rockefeller University when she strode across Central Park to the American Museum of Natural History (AMNH) one afternoon to hear a talk on cone snails and their venoms. The speaker was Toto Olivera. Watching the video of a mollusk harpooning, paralyzing, and eating a fish nearly as big as its shell, "I was blown away," says Holford, who has a wide smile and often punctuates her statements with a laugh. "How is this even possible?"

The answer is what Holford describes as a cluster bomb—not one peptide, but a cluster evolved over time to shut down specific receptors in prey. "If we think of diseases as disorders for which switches can be turned on and off," she explains, "then cancers have the switch on to divide all the time, which leads to proliferation." Her research aims to find peptide cocktails in venomous mollusks that can turn off switches for human ailments—such as neurological disorders, chronic pain, and cancer. The work requires both classic evolution—understanding how mollusks are related to each other and which among them deploy venom—and genetic and chemical analyses of the venoms themselves.

Venomics tied Holford's lab research on peptide chemistry to human problems—and to the biodiversity that had fascinated her since her childhood days at AMNH. After finishing her doctorate, she worked at AMNH for a year in public outreach before moving to Salt Lake City to do research in Olivera's lab. She read Geerat

Vermeij's books on mollusk evolution. She went to Paris to train in taxonomy at the Muséum National d'Histoire Naturelle.

There, in the malacology section housed across the street from the Jardin des Plantes, the marine biologist Philippe Bouchet works in a crowded attic stuffed with shell drawers and small, strange beasts floating in ethanol solution. Bouchet's research straddles two centuries and ages: that of discovery and that of extinction. The old-time naturalists' obsessions to know every creature in the sea has shifted to an urgent, global search to identify marine animals before they go extinct.

Bouchet collects from the New Caledonia–based research vessel *Alis*, equipped with an echosounder and miles of steel cable to drop dredges and trawls into the species-rich waters of the Indo-Pacific. His Tropical Deep-Sea Benthos program has been the single largest source of new marine species discoveries in the world for the past fifteen years. Many mollusks and other creatures pulled aboard in that time still await description. The 50,000 species of marine mollusks known, Bouchet estimates, are but a third of those knowable.

To help identify rare and new species, Bouchet relies on a network of serious conchologists; more than half of all new mollusks are identified by expert collectors. On balance, these skilled amateurs—though they can hardly be called amateurs—do far more good than harm for the animals and their conservation, say Bouchet and many other malacologists I asked. But biomedical scientists who research cone snail venoms for human cures often feel ambivalent about collectors. "They thrive on rareness, which means they can collect way too many," says the biochemist and cone scientist Frank Marí at the Hollings Marine Laboratory in Charleston, South Carolina, part of the National Institute of Standards and Technology. The city that was home to America's first museum and Edgar Allan Poe's old haunt at Fort Moultrie is today the site of a federal "cone snail farm," part of Marí's research group that studies cone venom peptides for basic research aimed at human cures.

Marí helped show that a classic toxin targeting the central nervous system can also influence the immune system; conotoxins appear to have anti-inflammatory applications for medicine. He has recently sequenced the Purple Cone's genome. "The venom is far more complex than we thought it would be," Marí told me. "It can be quite a bit different in different animals of the same species, each expressing its own brew."

That makes each animal worth much more alive than dead, Marí says. A cone snail's genetic map, with its billions of letters, repeat patterns, and surprising flourishes, is for him even more captivating than the shell. He considers it incalculably more valuable.

The National Cancer Institute estimates that more than half of all cancer drugs and antibiotics originated from botanical and other natural compounds; Taxol, one of the most commonly used chemotherapy drugs, was derived from the bark of the Pacific yew tree. But animal venoms have come into the drug pipeline only in the past forty years. The pit-viper-inspired blood pressure treatment Captopril, approved by the FDA in 1981, was the first; the cone-derived painkiller Prialt followed. Dozens are now in trials. Scientists who work on them say that tarantulas seem to hold particular promise for blocking pain.

Indigenous cultures on three continents have used the furry spiders as a cure for toothaches, tumors, and other ailments for thousands of years.

ARMED WITH A pair of salad tongs, her hands protected with thick scuba gloves, Holford joined Bouchet on a research trip to Papua New Guinea so that she could collect terebrids, cousins of cone snails. Many of these slim reef-dwellers with pointy shells also evolved toxic venom. Their apertures are so impossibly narrow that no one had tried to extract it. When I visited him at U.C. Davis, the evolutionary biologist Vermeij showed me a Papua New Guinea terebrid that was long and thin as a darning needle, with more than forty beaded whorls. The opening must have been smaller than the eye of a needle; it was invisible to my eye.

Where Vermeij and Bouchet's taxonomy scrutinizes the tiny,

beaded whorls in shell, Holford's generation examines the whorls of molecules in mollusk DNA. The old and the new are both revelatory; Vermeij's fingers told him stories of environment and predation that eyes peering through a microscope couldn't see. Holford used the same new DNA technologies that mapped the genomes of tarantulas and other venomous spiders to sequence terebrids, better known as auger shells.

The new tools helped the Holford Lab become the first to complete a molecular phylogeny of the terebrids—the evolutionary tree that shows how they are related to each other. Following the different branches that contain toxins led Holford's team to a peptide, Tv1, from the Variegated Auger. Computer models suggested Tv1 could attack liver cancer cells. In lab experiments the compound indeed shrinks liver cancer tumors in mice—the second-most-common cause of cancer-related death worldwide. "We might have found a compound that kills cancer cells more than it kills all the normal cells," Holford says.

Her team has a long way to go in figuring out how the compound might target human tumors. But its existence in one, thin auger shell, she says, is a sign that "there is so much more out there." Scientists estimate 15 percent of the world's animals are venomous; Holford suspects that number is higher, given so much tiny life still unidentified. "It means we have to save the oceans," she says. Which has taken her work far beyond the lab bench or the beaches armed with salad tongs.

IT HAS BEEN nearly four hundred years since Rembrandt drew his Marbled Cone, in 1650. Emerging from shadow to light, the dappled shell was his only still life among hundreds of prints. The shell itself likely came from the Indian Ocean aboard a Dutch East India Company ship. Rembrandt's etching glorifies the wealth as much as the beauty—"not an unalloyed good," as the art historian Susan Tallman puts it.

The animal that makes the stunning shell, *Conus marmoreus*, is not a fish-hunter or a harm to humans. It feeds on other mollusks, including a favorite small morsel called the Humpbacked Conch. In a study, Australian scientists placed Marbled Cones and Humpbacked Conchs together in regular seawater, and seawater with the CO_2 levels that are predicted by century's end. The marine biologist Sue-Ann Watson wanted to find out how rising acidity might change cones' behavior or agility. In the current conditions, *C. marmoreus* successfully captured and gobbled up the conchs about 60 percent of the time. In the water modeling future seas, the cones became three times more active, decreasing their time buried under the sand. But the agitation did not help them catch prey. The cones in the more acidic water got the conchs only 10 percent of the time. The chemical change had made them clumsy.

Cone snails evolved over 50 million years to develop chemical weapons that made them some of the stealthiest hunters in the animal world. Chemical changes to the ocean may threaten their own nervous systems. The acidification studies don't account for the mollusks' extraordinary ability to adapt; scientists find increasing evidence that some are adapting their shells in just a few generations to try and survive acidifying seas. But behavioral changes, which scientists have also begun to see in fishes and other marine life, "might not be so easy to overcome," Watson says.

Without the mutual benefit with which wolves became dogs, or the intention of making watermelons seedless, we are altering the evolution of marine animals in all sorts of ways. Gregory Herbert, the malacologist studying how Gulf of Mexico mollusk populations have changed in modern centuries, once handed me an oyster shell that dwarfed my hand; it was dug from the bottom of a Native American mound. American oysters grew as long as 15 inches even after centuries of harvesting by Native people. They shrank in modern centuries in part from our harvesting the biggest oysters, which gave smaller, faster-growing ones more influence on the genetics of the next generation.

"Many species can handle any one stressor," Herbert says. "But few can handle a bunch at once." As oysters, Queen Conchs, and any number of species have become smaller, fewer, or shorter lived—true of nearly every harvested animal in the sea—they are less resilient to ocean warming and acidity. They may stop reproducing, as the Queen Conch scientists see. They may become more vulnerable to diseases like those that have struck oysters in the Chesapeake Bay. Or simply: "They have all died—there is nothing." That was the report from the Peconic Bay fishers for the Bay Scallop season following the historic die-off linked to climate change. The queens, the cones, the scallops, the giant clams, strong animals that persisted over half a billion years of ecological change and predation—could survive one calamity or even two. They are unlikely to survive them all.

HOLFORD PICKED THE best of times and the worst to devote her career to searching for drug compounds in venomous mollusks hidden in the reefs of the southwestern Pacific. The days when men of malacology adventured to every island and coral reef to pluck all the specimens they could—Tucker Abbott once called it "he who dies with the most shells wins"—were coming to an end. Women finally managed museum collections and ran their own labs. Advances in genetic sequencing and mass spectrometry allowed scientists to quickly identify thousands of potential toxins.

But finding the subjects of the research, the living venom-makers themselves, had become a much more complex and fraught challenge. Centuries of colonialism—and exploitation by earlier scientists who extracted artifacts, knowledge, and species from the Global South— left deep scars in some of the countries where venomous mollusks, urchins, jellyfish, and other promising animals live. When Pacific nations gained political independence from colonial powers, some of the islanders' most important historic artifacts remained locked up in museums overseas. Those artifacts were often shell: tools, artwork, jewelry, or, like Native American wampum, records of diplomacy and exchange. In the famous Kula gift ring in the Trobriand archipelago of Papua New Guinea, islanders exchanged long red shell necklaces to the north and white shell armbands to the south over hundreds of miles, an elaborate ritual of social connections and political trust. The armbands, called *mwali*, were cut from large cone shells. Polished memories ended up inaccessible to their cultural heirs, in museums or private collections, though some are now being repatriated.

The extraction carried over to natural history such as fossils, and the animal and plant life on islands and surrounding seas. During the time of empires, "The Southern Hemisphere basically was our supermarket," says David Schindel, the retired evolutionary biologist who founded the Smithsonian Institution's Barcode of Life initiative to share wildlife DNA as a public, worldwide good. "We could go take anything we wanted—anything alive or dead, we just took it."

When Western universities and corporations began to turn marine compounds into drugs and chemicals, species-rich nations stepped in to protect themselves—and their animal and plant life. The Convention on Biological Diversity and its Nagoya Protocol were meant to safeguard species and ecosystems and stop anyone from patenting Indigenous medicines or genetic resources without sharing the benefits. The agreements led one hundred of the most biologically diverse nations to tighten control over their species and the foreign scientists researching them. Yet the treaties did not stem private control of marine resources; to the contrary, they opened a rush of marine-patent applications before going into effect. Transnational corporations have now patented more than 13,000 gene sequences from marine organisms. Half of those are held by one company, Germany's BASF, the world's largest chemical maker. Meanwhile biodiversity researchers like Bouchet can now spend years awaiting permits. The protocol, he told me, "has become the main obstacle to inventorying the biodiversity of the planet."

Such geopolitical pressure—piled onto the exploitation of people and cultures, the extinction of animals, the warming climate—led Holford to the emerging field of science diplomacy. If scientists could navigate political relationships and treaties as expertly as the ecologies of mollusks, the thinking goes, they could elevate the role of science in decisions on global challenges such as climate change.

Working in Papua New Guinea with Bouchet, whose own diplomatic skills and bonds have kept him collecting in the Coral Triangle during a tense time for foreign researchers, Holford also saw how drawing local students into projects strengthens research and ensures a more inclusive future for science. (It's still relatively rare for Western scientists with projects in the region to take on graduate students *from* the region; Peter Ward says it has taken him many years to get his university to support his first Filipino PhD student to work on nautilus in the Philippines.) Holford tells the story of how many of the Papuan children who interacted with her had obviously never seen a person of color—someone who looks like them—among the many scientists who visit their islands. She felt unprepared to become a role model; it was nothing she'd trained for as a biochemist, but she began trying to figure it out. She has since made science diplomacy and educational outreach foundations of her lab.

Holford now teaches science diplomacy at the CUNY Graduate Center and Rockefeller University where she earned her PhD. She also works on diplomatic projects at the American Museum of Natural History where she first encountered the world's great biodiversity as a child. As they transition from collecting the world's species and cultures to global missions to save species and cultures, Holford believes natural history museums may have no more important role than in restoring a bond between society and nature; sparking wonder for the Earth and its life. Many malacologists I met had arrived at the same conclusion.

At the Smithsonian, cowrie expert Chris Meyer had turned his research to the profound biodiversity losses in the oceans. But he spent lots of afternoons with his shining orbs spread across a table in the museum lobby, asking visitors to "vote for beauty" as part of a project to figure out which natural objects move people most deeply.

On Sanibel Island, José Leal and his scientists and volunteers now brought live mollusks, rather than empty shells, to schools and community events. Visiting the century-old Sanibel Shell Show one recent spring, I moved slowly around the displays inside with a group of elderly shell collectors, several of them relying on canes, walkers, or thin oxygen hoses. The aids reminded me of the spires, spines, siphon canals, and other ornaments that had drawn them to a lifetime of seashells.

Outside, people of all ages were thronged around the Bailey-Matthews museum's display of live mollusks, staffed by local school children. A pixie-haired sixth-grader named Crystal Jones effortlessly answered strangers' questions about shells and their builders while a moon snail mooched up her arm. A man tried to stump her with what he thought was a joke: "What's the fastest mollusk on the beach?" She did not miss a beat: "A Lettered Olive."

"Do you eat seafood?" asked another. "I can't eat marine animals," she said. "But chicken is my favorite."

I wondered where her knowledge and empathy for a soft, vulnerable animal would lead. To a lifetime of discovery and compassion, Julia Ellen Rogers and the nature study progressives would have said. Rachel Carson, whose mother had taught her from their books, championed the sense of wonder that would spark care and curiosity and in turn, ingenuity: The resourcefulness that sparked people thousands of years ago to burn shell for the first manufactured chemical, slaked lime. The love for beauty in tropical seashells that led Toto Olivera to find human cures in cones. The compassion for fleeting life that will inspire a new generation to find answers in nature—never the poison, but the cure.

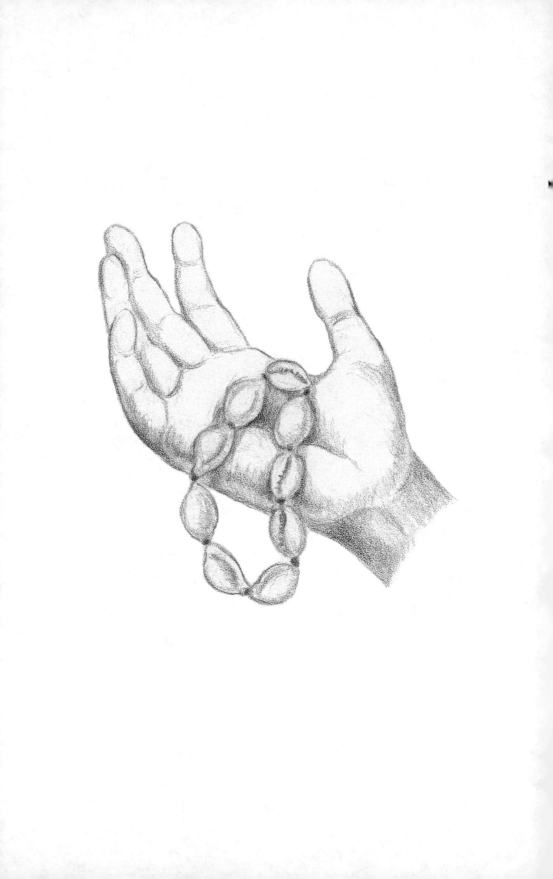

Conclusion

~~~~~~~~~~~~~~~~~~~~~~~~~~~

THE OPEN END

SHELL PEOPLE
Homo sapiens

The Money Cowries seem too beautiful for the evil of this place. A visitor has piled them in a little altar on the stone floor of a dungeon, beside other offerings to the souls kidnapped, chained, and marched hundreds of miles to the Elmina slave castle and the atrocities that waited here and beyond.

Even in bleak disrepair, the whitewashed medieval fortress dominates the beach and fishing port city of Elmina in Ghana on the coast of West Africa. Inside, at the end of a narrow, dark passage, the "door of no return" opens to the sea. The small door, found in many of the dozens of slave forts built along the Atlantic, was the last touch of homeland for the men, women, and children pushed down ladders into wooden boats that carried them to the slave ships. From the sixteenth to the nineteenth centuries, an estimated 12.5 million enslaved people were packed and shackled into the ships' holds. Two million did not survive the Middle Passage to North and South America and the Caribbean.

Piled near the cowrie altar, the other offerings include bottles of water and bottles of gin. A straw hat and a straw fan for torrid heat worsened by crowded, soiled conditions. Flowers from the Mississippi Boulevard Christian Church of Memphis, Tennessee. And dated just a few days before my teenage son and I visit, a handwritten note: "To our Ancestors, You are Us and We are You."

I'm in West Africa to complete the improbable route of the cowries from the tropical reefs of the Maldives across two oceans on sailing ships, ballasts weighted with the shell money that purchased as many as a third of the enslaved people forced into bondage and to the Americas. Will and I are taking time to tour the slave castles of Ghana before visiting the remote southeast coast of the country where the Volta River meets the Gulf of Guinea in a sweeping delta. The mouth of the Volta was one outpost where Maldivian cowries were transferred from ships to canoes for transport to their cruel inland mission.

I did not expect to see cowries at Elmina, where gold rather than shell money dominated exchange. But the shining amulets are ubiquitous in the makeshift altars left by visitors here and at nearby Cape Coast Castle. Cowries are also found stashed in the archaeological layers of these sites and others—hidden in drains and crevices by people once enslaved there. "You are Us and We are You," expressed across centuries by shells.

I HAD SET out to listen to seashells as chroniclers of nature's truth. Just as they'd helped earlier generations understand the age of Earth, and evolution and extinction when those concepts seemed unreal, shells now revealed how humans have altered the climate and the sea—down to its very chemistry. But as much as shells told about oceans, they had more to say about people. In spiraling architecture and art; in natural wonder and lifesaving science; as essential food— the extra protein that may have helped build our bigger brains— shells represented the best of humanity. After they became money, they showed us at our worst.

Linnaeus gave the Money Cowrie its species name, *moneta*, in his *Systema Naturae* of 1758. In the same edition he named the American Quahog, or cherrystone clam, *mercenaria* for what the colonists called the "mercenary transactions" in which they traded wampum with Native people. Wampum was often a gift—discourse and diplomacy exchanged in story belts or strings. Colonists deployed it as money and began to manufacture it themselves. On both sides of the Atlantic, revered cultural objects became mass-produced capital.

For thousands of years and around the world, cowries had promised fertility or warded off the evil eye. Now, cowries were called "slave money," or the "blood money" of the transatlantic trade. The animals, themselves, were said to follow slave ships to feed on the drowned. The archaeologist and cowrie expert Akinwumi Ogundiran told me that fear of the shells, bound up in terrifying historic memory of vanished family members, lasted well into his childhood in southwestern Nigeria. As a young boy of eight or nine, Ogundiran found a cowrie in the street and brought it to his mother's shop. "She went ballistic. She was hysterical. She kept shouting, 'Why did you pick that up? You could disappear! Don't you know you could disappear?'"

Before the infusion of cowrie money into West Africa, Ogundiran's excavations in Nigeria have helped confirm, Yorùbá people had developed glassmaking technology and a manufacturing economy around glass beads. Early in the seventeenth century, human cargo became the primary export, rather than the glass beads, cotton cloths, and ivories of the past. A phenomenal flood of cowries now fueled a market economy that replaced the earlier knowledge economy. The Maldivian shells became currency for every transaction in the Bight of Benin, though the European traders generally would not accept cowries as payment. By the eighteenth century, African chiefs and princes were encrusting rooms and sometimes entire homes in cowries, evocative of the shell-walled rooms of European pomposity. The shells created a grand illusion of wealth in West Africa. All the while, the real wealth was being forced into ship holds and across the sea.

THE SITTING TERRA-COTTA woman with frog-like eyes has a small, deliberate cavity in her head. The little concave appears to be carved in the shape of a cowrie shell. She is among hundreds of clay-fired people and animals discovered with similar cavities, some leading to hidden channels that may have been used to fill them with medicine. Archaeologists began excavating the figurines in Koma

Land in northern Ghana in the 1980s following villagers' reports of finding what they call *kronkronbali* ("Olden Days Children") buried in mounds.

Real cowrie shells are sometimes buried with the Olden Days Children. Many of the figurines are accented with carved cowries as well—earrings, necklaces, helmet trim. The most recent CT scanning of the clay effigies dates the society that made them to between the sixth and fourteenth centuries.

The well-preserved clay woman, dressed in a skirt and heavy necklace, sits under glass at the University of Ghana's Museum of Archaeology. The scholar who unearthed her and leads the Koma Land excavations today, Benjamin Kankpeyeng, estimates she was made sometime between 770 and 900. "The figurines establish a cultural context for cowries" before European traders flooded West Africa with them, Kankpeyeng tells me when we meet in his office across a shady green square from the museum on the university campus in Accra. "We may not be able to recover the meaning," he says, but the cowries' presence suggests an ancient protective or healing role that transcended money.

Revered and infused in human memory for hundreds of thousands of years, the sacred meaning of seashells proved far stronger than their monetary value, says the Nigerian archaeologist Ogundiran. That was true even in the Bight of Benin to the east, where there was little evidence of cowries' use before European traders sent billions into the economy. The cowries rose above their blood-money repute. They overcame vampire myths and mothers' fears. Ultimately, Ogundiran says, the polished charms found their "symbolic, sacred, and metaphorical values as the harbinger of all good things." As they had been in cultures around the world, cowries became protectors in religion and rituals. They became central to divination. To good health. They venerated ancestors and deities and protected against enemies. They helped the deceased transition from this life to the next.

THE VOLTA RIVER meets the Atlantic Ocean at the southeastern tip of Ghana in an unbounded sweep of sea and sandbar. The horizon evokes the legend of a spiritual ferryman paid with cowries and other fees to carry the souls of the dead across the Volta to the afterlife. An abolitionist chaplain who ministered at Danish forts in the early nineteenth century recorded the belief that the ferryman carried deceased loved ones "over the river's various arms to a large, sandy plain which is formed near its mouth," to celebrate with other spirits.

At the town of Ada Foah, a great sand plain still arches in a dramatic gateway to the Atlantic; the beginning of a journey or its end. With the Volta on one side and the Gulf of Guinea on the other, Ada Foah recalls Miami or Copacabana when they were remote fishing outposts with unpaved roads. Large coconut palms sway on the beach side, shading small adobe homes and sand lanes packed in shell. Upriver, waterfront neighborhoods with weekend retreats built by wealthy Ghanaians from Accra give way to the thatched-roof homes of local fishing families.

The larger area, called Ada, was a nation of eight clans before Europeans arrived on the West African coast, its people fishers who also traded salt from their estuary. They were expert navigators of what was then a wild and dangerous river full of rocks and rapids, hippos and crocodiles. The Volta marked the western border of the Bight of Benin, which absorbed most of the cowrie money imported to West Africa between 1650 and 1850; many cowries came in along the Niger Delta as well. Ada was not a notorious slave market, but its traders became middlemen between Europeans at the coast and slavers in the interior. Traders on both sides of the river piled Maldivian cowries onto their canoes alongside the salt and fish. The journeys upriver and onward along the tributaries took months, with frequent unloading to portage. People and donkeys hauled the Indo-Pacific shells still further into the hinterlands.

The name Ada Foah is said to describe "the fort at Ada," Fort Kongensten, built by the Danish in 1783. Today, the ocean has consumed almost every trace of the outpost, taking it slowly and quickly like a sand castle. Kongensten is the first of the colonial forts to wash away in the rising seas and storms accelerating as Earth warms. Cape

Coast and others are also at risk; places of history that transform almost everyone who faces those small doors of no return.

While Ghana lies just north of the equator, its coastal future, like all of ours, yields at the poles. Glacial and ice-sheet melting are now the primary driver of rising seas around the world. Coastlines altered by dams, sand-mining, and sea-defense structures are eroding much more quickly than the rest. The estuary at Ada Foah is subject to all three. Powerful waves have washed away hundreds of feet of coastline since the mid-twentieth century; the village is now eroding at an average of 2 meters a year—double the already rapid rate at which sandy beaches are vanishing globally.

Owing to a hydroelectric dam upriver, the Volta, too, has been transformed since the time of the cowries. Akosombo Dam flooded the river basin to create Lake Volta, the largest reservoir by surface area in the world. The dam forced 80,000 people upstream from their homes, mostly fishers and farmers whose fisheries and lands were drowned. The fishers living closer to Ada Foah and the estuary still ply the river in large wooden boats. They have something in common with their forebears. Their boats are heaped with shells.

WE ARE MOTORING up the Volta in a bright blue wooden skiff with the guide and historian David Ahadzie and an American anthropologist living in Ada Foah, Netty Carey. We pass other brightly colored boats, all piled with African river clams. The burly yellow wedges with black stripes, *Galatea paradoxa*, are at the heart of Ada's artisanal fishery. At camps along the riverfront, thousands of the golden bivalves are spread out to dry in the sun. Women shuck the shells and chat in the shade of huge palms. Half-shells are heaped into small hills. The shells, themselves, are worth as much as the meat being tossed into baskets. Milled into grit, whitewash, or lime, the shells will become the calcium in chicken feed, and a hardener for cement and terrazzo floors.

Before the dam, about two thousand people made their living from clams in the lower Volta, most of them women harvesting in the shallows. The dam reshaped sand and sediment at the river's mouth, shrinking the clam grounds by 90 percent and making shoreline harvesting untenable. Men now had to hookah dive to find the clams—swimming underwater while breathing through a small hose attached to an air compressor in the bottom of the boat.

Between the shells and the meat, the tiger-striped bivalves still bring the small community more than $3 million U.S., or 5 million cedis a year. (Ghana's national currency is the *cedi*, the Akan word for "cowrie shell"; window-sized cowrie likenesses also line the tall West African central bank building in Benin.) Ada and nearby villages are home to several shell mills, their yards piled with shucked shell and bags of milled shell, sold by the truckload. The clam trade also enlivens the riverfront, where tourists enjoy stopping for skewers of clams golden-fried over firepits; many more fried clams are sold along the highway to Accra.

Scientists a decade ago declared extinction imminent for *G. paradoxa*, squeezed between the dam upriver, the shifting sands at the coast, and fishing pressure on the dwindling wild clams. They predicted dire socioeconomic consequences for villagers, especially women clammers. But Daniel Adjei-Boateng, a fisheries professor who specializes in Ghanaian aquaculture, says a sustained shift to clam farming appears to be making a difference. When hookah divers fish the river bottoms, they put aside the small clams. Women plant them in family plots marked by sticks in the shallows—much like the giant clam gardens in Palau, or Megan Davis's vision for local conch farms across the Caribbean. The clams double in size in six to eight months, allowing fishers to harvest from their own plots during an annual closed season that takes pressure off the wild clams.

The survival of the clams and the fishing community, Adjei-Boateng says, will depend on a series of choices from individual to global: from how people harvest to regional policies governing land use to how the world tackles climate change. It is the story of the Queen Conchs and giant clams, and the story of us; it is the story of the world in a shell.

~~~~~~~

LIKE THE SEASHELLS themselves, the sea is too often understood by what we see on the surface: as a beautiful backdrop; a threat churning up a hurricane; a victim of pollution or other human abuse. The oceans, the source of life, are also deep wells of hope for solving climate change and other major challenges of sharing Earth with 8 billion others. Ada Foah and its golden shells are small symbols of big solutions waiting beneath the waves. Farmed clams, oysters, mussels, and scallops clean the sea as they provide food and meaningful work—without hastening extinction or the pollution and disease that spread in fish farms. They do all that while absorbing tons of carbon as they build their shells. Crushed or ground, farmed shells can ease our reliance on quarried limestone. They hold potential for greening cement, the third-largest source of human-caused carbon emissions after fossil fuels and land use changes.

As a kid growing up in Florida and California, I always wished that whoever got to decide these things had set aside the oceanfront—and the dunes and mangroves leading to it—for everyone to share, and had built the cities inland. In the oceans, we can still decide these things; to imagine shellfish farms for food surrounded by the vast marine sanctuaries that scientists consider our best chance for saving imperiled marine life. Protecting and restoring the oceans and their seagrasses and corals will protect people, too; from the conch fishers losing their livelihoods in the Bahamas to the Maldivians who don't want to leave the family islands for the crowds in the elevated capital. The severity of sea rise, storms, warming, and ocean acidity are tied directly to the amount of carbon in the atmosphere. The sea has already buffered us from so much, having absorbed a third of the CO_2 rising since the industrial revolution. Restoring seagrass meadows, salt marshes, and mangrove coasts would absorb more. Scientists estimate that coastal wetlands worldwide can store five times as much carbon as tropical forests.

As it faces the powerful waves, Ada Foah also sits on one of the most promising sources of carbon-free energy in the world. Wave and tidal power lag far behind solar and wind because of the technical challenges and harsh conditions of building undersea. But their potential to generate electricity is significant. In 2015, Ghanaian divers dropped six wave energy converters off a barge and mounted them onto the sea bottom about 10 miles off Ada Foah. Yellow buoys on the surface connect with generators on the seafloor to capture the kinetic energy and send it to a substation onshore. Without blades, the technology seems to pose less harm to sea life than wind power does. Capturing tidal power may also slow beach erosion on the coast. The trial was successful enough that investors are scaling it up: Ada Foah is set to be home to the first commercial wave energy plant in Africa.

A spiral shell, in its infinite repetition, represents nature's economy; a circular economy driven by regeneration rather than waste. Mollusks spend their lives upcycling ocean chemicals for their shells. They fuel up with algae. Their reefs often beat human-made barriers in protecting coasts from erosion and other storm-related miseries. Many traditional homes in Ada Foah are fortified with seashells; people building houses still pay women to wade into the sea at low tide to fetch big bowls of shell to mix into the cement.

But there was a lesson in the cowries beyond their being the "harbinger of all good things" from the sea. No scientific or technical solutions—and no grand gestures to preserve nature—will be adequate without solutions for people, too.

WE HOP OUT of the blue boat onto a long, narrow isle of sand at the mouth of the Volta, where Ahadzie tells us how the estuary has changed in his lifetime: the vanishing of former seashore and entire islands. He apologizes for the plastic scree that covers the uninhabited spit. I've now heard local people apologize for plastic smothering remote islands across three continents; in the Bahamas, the Maldives, and West Africa. The sea-worn bags and distorted bottles have obviously traveled from far away.

Carey, the anthropologist, points out the estuarine island where fishing families have been dealing with a different blight. Nearly a decade ago, the Ghanaian-Italian development company Trasacco Estates gave the families thirty days to leave their island to make room for a luxury resort. Company representatives painted bright yellow numbers on their homes, marking them for demolition. The yellow numbers reminded me of the green dots imposed by planners on a map of the Lower Ninth Ward in New Orleans in the wake of Hurricane Katrina, designating longtime homes as future public parks. Neither the New Orleans residents nor the Volta estuary islanders tolerated being marked; they refused to leave.

Trasacco, which had acquired the land from the government for tourism development, was sanctioned for its failure to work with residents on a resettlement plan or obtain an environmental permit. The company, now partnered with Hilton to remake the fishing island into a high-end hotel-conference center, has finally come up with a resettlement plan. Presenting it at a public meeting, company officers stressed how much better off villagers will be: with hundreds of jobs; concrete houses rather than thatch huts; a flow of tourists and cedis; daily removal of plastic from the beaches. They showed images of plastic-strewn, uninhabited beaches as if to suggest residents don't take good care of their community. The managing director said the resort will give locals "proper jobs" rather than having to fish or farm.

Carey moved to Ada Foah for her research on the double displacement of climate change and tourism development along the Ghanaian coast. People welcome good jobs and concrete houses, she says. But they also have deep ties—ancestral, economic, and ecological—to the seaside. In Ghana and around the world, development continues to forge unbending, divergent roads of winners and losers rather than imagining a stronger, regenerative spiral inclusive of people and nature. Trasacco's relocation plan is unclear on whether fishers will even be able to access the estuary to throw their fishing nets—at least without patronizing the hotel.

~~~~~~~~

THE MALDIVE ISLANDS produced the first worldwide currency, and the West African coast absorbed most of the little white
shells. Centuries later, the two coastlines bear some of the worst burdens of climate change fueled by that new global economy. The great
naturalist Rumphius, "Pliny of the Indies," might have deemed it
fate. Rumphius saw seashells as gifts of nature that lost their power
when bought or sold. Like his Roman namesake, he condemned
greed—often taking the side of the local people on Ambon against
his superiors at the Dutch East India Company—even as he saw no
choice but to work for the company.

Pliny himself, the Roman naturalist, had warned of the rise of
excess in the empire, including the number of murexes that had
to be killed to make the royal purple dye. He expressed sympathy
for the animals that built such handsome shells, without comment
on the enslaved people who had to make the fetid brew.

Pliny's blinders were purple. For those of us who grieve what's happening to Earth and its life, they can be green. It's easy to condemn
greed and injustice—and much harder to change their centuries-long
grip. But that is what a growing movement fighting the exploitation of
people and nature is working to do. "Never should it be forgotten," as
our Elmina Castle guide, Ato Ashun, put it in his memorable parting
words, "that anything that puts money or profit above the suffering of
the people is as terrible as what happened in the past."

The understanding that climate change harms unequally is sinking in across U.S. cities and worldwide, where the Global South faces
greater frequency and intensity of floods, droughts, and heat waves.
The shift in values is evident in the work of young scholars and scientists like Netty Carey, Mandë Holford, and Alison Sweeney, who
treat the people and communities where they do research as partners
and primary sources of wisdom rather than academic grist. It is clear
in the women—community leaders, scientists, writers, and educators
like Julia Ellen Rogers—being added to the history books, or to the

leadership positions where they are forging a better future for people and the environment. It's apparent in the recognition that Indigenous cultures have innovations that overdeveloped countries need to find cures, preserve wildlife, and adapt to ecological change: The traditional Palauan practice of *bul*, now codified in modern law to ban fishing, mining, and other extraction in 80 percent of the nation's waters. The ancient engineers who moved with the rising and receding Florida coasts.

This shift seems palpable in Will's generation, many committed to social justice and climate change, and participating in their first elections in 2020 in record numbers. (After his high-school graduation on a Florida racetrack during the pandemic, Will moved to the Hague to attend Leiden University, where he hopes to study international justice.)

Seashells have often been messages—to scientists, to diviners, to worshippers called together by the voice of a shell. They have often been gifts: Marcus Samuel Sr.'s Victorian shell boxes, the "Gifts of Brighton" that would become one of the largest oil companies in the world. Anne Morrow Lindbergh's *Gift from the Sea* that spoke to a generation of women as the book also warned, perhaps too gently, of our growing excess.

Shells were jewelry and art. They were money and weapons. Their makers, the mollusks, symbolize all of nature in being exploited and brought to the brink of what is bearable—the dissolution of their exquisite homes in the acidifying sea. Now, some of them are proving life's persistence as they begin to adapt to those chemical changes.

Their best and highest role for Earth and humanity is to keep living their lives in full glory. Even more beautiful than in a cowrie bracelet or museum drawer, or beside me on my desk as I write these last words, is the sight of *Monetaria moneta* alive on the reef. Tiny eyes peer sideways from the bottom of its tentacles. Its siphon hose juts ahead to sniff for algae. Its living cloak spreads over its mounded shell, taking carbon from the sea and turning it into beauty. That life is the seashell's one and only truth.

ACKNOWLEDGMENTS

~~~~~~

I owe my inspiration to the barrier islands of the Gulf of Mexico. As a child I gathered seashells on Sanibel, Marco Island, and Boca Grande with two grandmothers who when I showed them even the tiniest coquina would react as if I'd found Blackbeard's treasure. It was in Cedar Key, a more remote island to the north, where I was drawn again to shells. On a sandbar one winter at low tide, Aaron and I found an enormous Lightning Whelk, took it home, and stuck it atop our first Christmas tree. We no longer live on the island, but I write these words here on another low tide, with gratitude for all who helped preserve the Cedar Keys National Wildlife Refuge, its thirteen coastal keys, and Florida's Big Bend estuary where the whelks are born in their miniature shells and scallops clack in the seagrass. Nearly twenty-five years later, the Lightning Whelk still tops our Christmas tree.

My deepest debts are to Aaron Hoover, my husband and first editor, who read every word nearly as many times as I did, while steadying two teenagers during this six-year undertaking. The love and support I felt from Aaron and our kids kept me going. Will and Ilana, thank you for your respect for my work and your many contributions to this book. *The Sound of the Sea* has been a family endeavor.

My editor, Matt Weiland at W. W. Norton, understood and believed in this book from its earliest iterations. I am grateful for his encouragement and grace as I finished it as slowly as a mollusk. I thank the entire Norton team for their enthusiasm for a story of seashells, especially during such a difficult year of production in the pandemic.

My writing partner, the environmental historian Jack E. Davis, was there from the idea to the last word. Writing can be solitary and intimidating. Having a trusted writer friend who is also at his desk by 5 a.m., ready to trade sentences or just listen, is a gift that I cherish. Jacki Levine was also an indomitable ally, and as priceless a seashell as she has been a lifelong friend.

You would not be holding this book if not for the malacologist José H. Leal, curator at the Bailey-Matthews National Shell Museum and editor of *The Nautilus*, the scientific journal of mollusks. José answered my questions for years and with great patience. He read chapters to check my molluscan science and proved just as knowledgeable about arts and languages. Any errors that remain are mine. Also on Sanibel Island, former Bailey-Matthews director Dorrie Hipschman; marine biologist Rebecca Mensch; Sanibel-Captiva Shell Club President Thomas Annesley; and many other shell club members and "shell ambassadors" shared their time, shell books, and lovely specimens.

I thank the many other generous collectors I met on beaches, at shell clubs, and in their living rooms, especially Conchologists of America president Harry Lee, and Shawn Wiedrick of the Natural History Museum of Los Angeles and Pacific Shell Club. I thank each scholar and source named in the chapters and notes, and so many others who talked with me at conferences or suggested paths of inquiry; it is impossible to thank them all here. I owe special thanks to Capt. Karen Chadwick, research scientist Stephen P. Geiger, and paleobiologist Gregory S. Herbert for all the conversations, and comments on early drafts. At the Academy of Natural Sciences in Philadelphia, curator Gary Rosenberg and collections manager Paul Callomon answered many questions over many years. I thank Robert M. Peck, Patricia Stroud, Jim Carlton, and Alan Kabat for historical assistance.

I owe much to the scientists who introduced me to the wonder and importance of mollusks by way of clams and oysters. When this project became unwieldy, I decided to forgo shellfish to concentrate on seashells, but their wisdom still grounds *The Sound of the Sea:* Shellfish aquaculture agent Leslie Sturmer was the first scientist to show me baby clams the size of sand grains in her Cedar Key microscope. Fisheries scientist Bill Pine taught me a lot about oysters, and even more about conservation ethics and trade-offs. Doug Jones, director of the Florida Museum of Natural History, helped me understand sclerochronology, the study of all that's recorded in hard shells. Also at the museum, I thank naturalist Cindy Bear, invertebrate paleontologist Roger Portell, and invertebrate zoologist John Slapcinsky.

I also thank my champion Steve Seibert; Doug Brinkley and Dave Walter for scholarly insights; my agent, Elise Capron at the Sandra Dijkstra Literary Agency; more-than-a-fact-checker Tim Meyer; and scientific illustrator Marla Coppolino, whose animals animate these pages.

I am grateful to the many libraries, newspaper archives, and librarians who assisted me, especially Alice P. Staples, special collections librarian and archivist at Plymouth State University in New Hampshire, who helped me track down Julia Ellen Rogers's surviving relatives. My deepest thanks to those relatives, grandnieces Sarah A. Kinter in New Hampshire and Susan Gillespie in Washington state, who shared their family's stories and photographs.

For their help with the global and justice connections, I thank the historian Ibram X. Kendi for urging me to visit Ghana as I tried to understand how cowries could remain beloved, having been considered the "blood money" of the slave trade. Deepest thanks to anthropologist Brenda Chalfin for her introductions in Ghana, and to Jacob U'Mofe Gordon, Evelyn Adjandeh, Samuel A. Atintono, and Kelvin Addo Boateng for additional assistance there. Journalists J. J. Robinson and Daniel Bosley provided invaluable background and connections in the Maldives. Kaashidhoo dive guide Ambra Federica Dugaria was a live-cowrie whisperer.

Many colleagues at the University of Florida, where I teach environmental journalism, contributed directly and indirectly, including journalism dean Diane McFarlin, the late historian David Colburn, President and Mrs. Kent and Linda Fuchs, Ted Spiker, Matt Jacobs, Carolyn Cox, Joe Delfino, Vasudha Narayanan, Christine Klein, Mike Allen, Ann Christiano, Matt Sheehan, Kenneth Sassaman, Stuart McDaniel, Kristen Stoner, and Thomas K. Frazer. Every student in my classes has bettered my work and perspective; Rachel Damiani, Jennifer Adler, Michael Munroe, Joan Meiners, Madison Jones, and Danielle Chanzes all directly strengthened this book.

A busy journalist with kids at home needs nothing so much as time and space to write. I am deeply grateful to Marsha Dowler for giving me an inspiring place to complete several chapters in Seaside. I thank Leslie Lee for inviting me to write at Pine Hollow in Elk Rapids, Michigan, and Bob and Debbie Knight in Celo, North Carolina. Special thanks to Birgie Miller of the Ding Darling Wildlife Society for helping me find places to stay and write on Sanibel.

My professional organization, the Society of Environmental Journalists, has supported me through every book, no less this one, as have fellow authors Charles Fishman, John Fleck, Craig Pittman, and William Souder. I thank the science journalist Michelle Nijhuis for her deft edits on the giant clam chapter, which originally appeared in *The Atlantic*, and *National Geographic* environmental editor Robert Kunzig, who edited my cover story on marine-protected areas. I also owe thanks to the talented writers on the "seashell beat" for their work, especially Richard Conniff, who has written beautifully about shells and people obsessed by them in *The Species Seekers* and *Smithsonian* magazine; U.K. scientists and prolific seashell writers S. Peter Dance and Helen Scales; and Jenny Staletovich, the reigning queen of Queen Conch reporting here in Florida.

Finally, I thank our family of Barnetts, Garrisons, and Hoovers; Jim and Claude Owens; Bruce and Sue Ellen Ritchie; Charlie Hailey and Melanie Hobson; Karen and Ken Arnold; Louise OFarrell and Larry Leshan; Mary and Charles Furman; Mike and Gracy Castine;

and Susan Cerulean and Jeff Chanton, for their immeasurable support.

My grandfather, Ovid Barnett, was determined to see his 100th birthday and read *The Sound of the Sea*, but died in fall 2020. When I went to say goodbye to his house on Marco Island before it was razed for a mini-mansion, I thought of the Calusa whose homes had stood before. I carried a Lightning Whelk from his yard to sink into mine.

NOTES

INTRODUCTION: COCKLES

1 **shells gathered in Neanderthal times:** Dirk L. Hoffmann et al., "Symbolic Use of Marine Shells and Mineral Pigments by Iberian Neandertals 115,000 Years Ago," *Science Advances* 4, no. 2 (February 22, 2018): advances.sciencemag.org.

1 **It still held a reddish pigment:** João Zilhão et al., "Symbolic Use of Marine Shells and Mineral Pigments by Iberian Neandertals," *Proceedings of the National Academy of Sciences* 107, no. 3 (January 2010): 1023–28.

3 **evidence that an enslaved African likely brought it to Virginia:** "Cowrie Shell," The Thomas Jefferson Encyclopedia, Monticello, monticello.org.

3 **clams clustered at Earth's greatest depths:** "Life Surprisingly Thrives Near Deepest Spot on Earth," *LiveScience*, February 23, 2012.

4 **If anything happens to its homemade boat:** Bec Crew, "Violet Snail an Ocean Wanderer," *Australian Geographic*, March 6, 2014: australiangeographic.com.au/blogs/creatura-blog/2014/03/violet-snail-janthina-janthina/.

4 **may have helped evolve the bigger brains:** Stephen C. Cunnane and Kathlyn M. Stewart, eds., *Human Brain Evolution: The Influence of Freshwater and Marine Resources* (John Wiley & Sons, 2010), 46–51.

5 **The decorated shells represent cognition:** Josephine C. A. Joordens et al., "Homo erectus at Trinil on Java used shells for tool production and engraving," *Nature* 518 (2015): 228–31.

5 **earliest-known keepsakes tucked into graves:** Francesco d'Errico and Lucinda Backwell, "Earliest Evidence of Personal Ornaments Associated with Burial: The Conus Shells from Border Cave," *Journal of Human Evolution* 93 (March 15, 2016): 91–108.

7 **"There is something fundamental":** Author interview with João Zilhão.

8 **mollusks don't draw the attention or research dollars:** José H. Leal, "Mollusks in Peril" forum, The Bailey-Matthews National Shell Museum, May 2016.

8 **severely underestimates loss of invertebrates:** Robert H. Cowie et al., "Measuring the Sixth Extinction: What Do Mollusks Tell Us?" *The Nautilus* 131, no 1 (March 2017): 3.

9 **Hugh Cuming described drifting over a solid mile:** Lovell Augustus Reeve,

Conchologia Iconica: Illustrations of the Shells of Molluscous Animals, Vol. XIV (Lovell Reeve & Co., 1864), 188.

9 **Oceans have also taken up a third of the carbon dioxide:** National Oceanic and Atmospheric Administration (NOAA), "Ocean Acidification," noaa.gov/education/resource-collections/ocean-coasts/ocean-acidification.

9 **has begun to limit the carbonate:** Laura Parker et al., "Predicting the Response of Mollusks to the Impact of Ocean Acidification," *Biology* (June 2, 2013).

9 **Acidic waters are also boring into some shells:** Maria Byrne and Susan Fitzer, "The Impact of Environmental Acidification on the Microstructure and Mechanical Integrity of Marine Invertebrate Skeletons," *Conservation Physiology* 7, no. 1 (November 2019).

10 **triton shells living near seeps with predicted future levels of CO2:** Ben P. Harvey et al., "Dissolution: The Achilles' Heel of the Triton Shell in an Acidifying Ocean," *Frontiers in Marine Science* 5 (October 2018): 371.

12 *90 percent* **of the visitors had no idea:** The survey was the inspired idea of Dorrie Hipschman, director of the Bailey-Matthews from 2013 to 2020, who also led the museum's transition from shells to live mollusks.

12 **Calusa left the most extensive shellworks known:** Jerald T. Milanich, *Florida's Indians from Ancient Times to the Present* (University Press of Florida, 1998), 129.

ONE: FIRST SHELLS

17 **when Shell Oil hired her:** "Carol Wagner Allison," obituary, *Fairbanks Daily News-Miner*, March 24, 1991.

17 **So-called bug hunters:** " 'Bug Men' Lead Hunt: Searching for Oil with a Microscope," *Popular Mechanics*, March 1932.

17 **"looking for her bonanza":** Interview of Richard Allison by Dan O'Neill and Bill Schneider, Eagle, Alaska, August 27, 1991, University of Alaska Fairbanks Oral History Program.

18 **had to abandon her research:** Ibid.

18 **Native American people as widely separated:** Cheryl Claassen, "Shells Below, Stars Above: Four Perspectives on Shell Beads," *Southeastern Archaeology* 38, no. 1 (October 2018): 89–94.

19 **Life loves logarithmic spirals:** Mario Divio, in *The Golden Ratio* (Broadway Books, 2002), p. 117, said that nature loves them.

19 **a shell-inspired Leonardo da Vinci designed the left-handed spiral staircase:** Thomas M. Annesley, "da Vinci's Spirals," *Clinical Chemistry* 63, no. 4 (April 2017): 931–33.

19 **Utzon credits the fierce-looking cockscomb oyster:** Budd Titlow, *Seashells: Jewels from the Ocean* (Voyageur Press, 2007), 84.

20 **what they'd been looking for—the latticed plates:** Phoebe Cohen et al., "Phosphate Biomineralization in Mid-Neoproterozoic Protists," *Geology* 39 (June 2011): 539–42.

21 **with colleagues at . . . Oxford zooming in:** Phoebe Cohen et al., "Controlled

Hydroxyapatite Biomineralization in an ~810 Million-Year-Old Unicellular Eukaryote," *Science Advances* 3, no. 6 (June 28, 2017): advances.sciencemag.org.

21 **had lived in a radical geologic era:** Nicholas J. Butterfield, "The Neoproterozoic," *Current Biology* 25, no. 19 (October 2015): 859–63.

22 **Oozing green, purple, and brown slime:** Steve Parker, ed., *Evolution: The Whole Story* (Firefly Books, 2015), 38–39.

24 **"In making the shell . . . I had also made the rest":** Italo Calvino, *The Complete Cosmicomics* (Houghton Mifflin Harcourt, 2014), 146.

25 **The coils gleam from the same stone:** Patrick T. Norton, "Fossils of the Maine State Capitol," *Maine Naturalist* 1, no. 4 (1993): 193–204.

25 **a great bloom of coccolithophores:** William M. Balch et al., "Factors Regulating the Great Calcite Belt in the Southern Ocean and Its Biogeochemical Significance," *Global Biochemical Cycles* 30 (May 2016): 1124–44.

26 **"the bit of chalk which every carpenter carries":** Thomas H. Huxley, *Discourses Biological and Geological, Essays by Thomas H. Huxley* (D. Appleton and Co., 1896), "On a Piece of Chalk," 4.

27 **second-most-consumed material on Earth:** Colin R. Gagg, "Cement and Concrete as an Engineering Material: An Historic Appraisal and Case Study Analysis," *Engineering Failure Analysis* 40 (2014): 114.

27 **rebel botanists began to chalk the names:** Alex Morss, " 'Not just weeds': How Rebel Botanists Are Using Graffiti to Name Forgotten Flora," *The Guardian*, May 1, 2020.

28 **The concept of *mollis* shows up frequently:** Carl Newell Jackson, "*Molle Atque Facetum*," *Harvard Studies in Classical Philology* 25 (Harvard University Press, 1914): 118–22.

28 **sea and terra firma must sometimes change places:** Richard John Huggett, *Fundamentals of Geomorphology* (Routledge, 2017), box 1.1, "The Origin of Geomorphology."

28 **part of the world's "vital process":** Alan Cutler, *The Seashell on the Mountaintop* (Dutton, 2003), 8–9.

28 **saw the presence of fossil oysters, clams, ammonites:** Adrienne Mayor, *Fossil Legends of the First Americans* (Princeton University Press, 2013), 116.

28 **The landlocked Zunis had keen knowledge of the seas:** Sylvester Baxter, "An Aboriginal Pilgrimage," *The Century Illustrated Monthly Magazine* 24 (1882): 526–36.

28 **Cushing brought a Zuni party:** Famously recounted by Baxter in "An Aboriginal Pilgrimage."

28 **"we often see among the rocks":** Frank Hamilton Cushing, "Zuni Fetishes," *Second Annual Report of the Bureau of Ethnology* (Secretary of the Smithsonian Institution, 1880–81), 29.

29 **Dutch legend had it:** Kenneth J. McNamara, *The Star-Crossed Stone* (University of Chicago Press, 2011), 142.

29 **"The shells of oysters and other similar creatures":** Edward McCurdy, trans., *Leonardo da Vinci's Notebooks: Arranged and Rendered into English with Introductions* (Empire State Book Co., 1923). Available at Internet Archive, archive.org.

30 **Nicolaus Steno became obsessed:** Cutler, *Seashell on the Mountaintop*, 49.

30 **He went on to show how the bivalves:** John Garrett Winter, *The Prodromus of Nicolaus Steno's Dissertation Concerning a Solid Body Enclosed by Process of Nature Within a Solid* (Macmillan, 1916), 253.

31 **"All the diversity of hues and of spines":** Steno's *De solido*, translated in Winter, *The Prodromus*, 250.

31 **a steady supply of cadavers:** Cutler, *Seashell on the Mountaintop*, 142.

32 **"are composed of Multitudes":** John Banister, *Mollusca, Fossils, and Stones*, reconstructed in Joseph Ewan and Nesta Ewan, *John Banister and His Natural History of Virginia 1678–1692* (University of Illinois Press, 1970), 323.

32 **The immense, iron-like scallop:** Martin Lister, *Historiae sive Synopsis Methodicae Conchyliorum* (London, 1685), illustration 167. Available at Biodiversity Heritage Library, biodiversitylibrary.org. A discussion of why scholars believe Banister sent it to Lister can be found in Clayton E. Ray, "Geology and Paleontology of the Lee Creek Mine, North Carolina I," *Smithsonian Contributions to Paleobiology* No. 53 (Smithsonian Institution Press, 1983): 4–5.

32 **it had lived in the Pliocene:** Thomas Say later named this fossil *Chesapecten jeffersonius* in honor of Thomas Jefferson. It is the state fossil of Virginia.

32 **Extinction was still an unacceptable concept:** Mark V. Barrow Jr., *Nature's Ghosts: Confronting Extinction from the Age of Jefferson to the Age of Ecology* (University of Chicago Press, 2009), chap. 1, "Bones of Contention," 15–45.

32 **"Now either the Earth by some kind of Salt":** Banister, *Mollusca, Fossils, and Stones*, reconstructed in Ewan and Ewan, *John Banister*, 323.

33 **squirrels . . . by the hundreds . . . and "legion" mosquitoes:** "The Animals, Plants, and Resources of the British Atlantic Colonies, Images & Commentary, 1692–1760," in *Becoming American: The British Atlantic Colonies 1690 to 1763*, National Humanities Center, nationalhumanitiescenter.org.

33 **fortress-like oyster banks running for miles:** Ernest Ingersoll, "The Oyster Industry," in Francis W. Walker, *The History and Present Condition of the Fisheries Industries, Tenth Census of the United States* (U.S. Department of Commerce, 1881), 19.

33 **"nine inches long from the joynt to the toe":** John Josselyn, *An Account of Two Voyages to New England, Made during the Years 1638, 1663* (William Veazie, 1865), 86.

33 *wampum* **was an anglicized version:** Marc Shell, *Wampum and the Origins of American Money* (University of Illinois Press, 2013), 36.

33 **the shells were closer to language than coin:** see Shell, *Wampum*, 51–52.

33 **shell beads were considered living:** Chief Irving Powless Jr., *Who Are These People, Anyway?* (Syracuse University Press, 2016), 132.

33 **wampum was a system of discourse and diplomacy:** Shell, *Wampum*, 12.

33 **said to speak the truth:** Powless, *Who Are These People*, 132.

33 **The Iroquois welcomed the Europeans:** The Onondaga Nation, "Two Row Wampum—Gäsweñta," onondaganation.org.

33 **"mercenary transactions":** Shell, *Wampum*, 36.

34 *Venus mercenaria*: Caroli Linnaei, *Systema Naturae*, accessed at World Register of Marine Species, marinespecies.org.

34 **"these parts are supposed some ages past":** John Lederer, "The Discoveries of John Lederer in three several Marches from Virginia to the West of Carolina, and other parts of the Continent: Begun in March 1669, and ended in September 1670," accessed at Archaeology at UNC Chapel Hill, rla.unc.edu, 2.

34 **He was kneeling to collect a small something:** Nesta Dunn Ewan and Joseph Ewan, "Banister, John (ca. 1650–16 May 1692)," *Dictionary of Virginia Biography*, Vol. 1, ed. John T. Kneebone et al. (Library of Virginia, 1998), 313–15.

34 **"ye Indians to take a View of their Towns":** Banister's writing as quoted in Ewan and Ewan, *John Banister and His Natural History of Virginia*, 45.

34 **some he learned about during repeated trips to America:** Robert H. Dott, "Charles Lyell's Debt to North America," in Lyell: *The Past Is the Key to the Present*, Special Publications, Vol. 143 (Geological Society of London, 1998): introduction.

34 **became an expert in the taxonomy of fossil shells:** Martina Kölbl-Ebert, "Female British Geologists in the Early Nineteenth Century," *Earth Sciences History* 21, no. 1 (2002): 7.

35 **fossil seashells regularly turned up in the cotton fields:** Eugene W. Wilgard, "Cotton Production in the United States" (U.S. Department of the Interior, 1884).

35 **Lyell would meet the enslaved land manager:** Sir Charles Lyell, *A Second Visit to the United States of North America, Volumes 1–2* (Harper Bros., 1849), 19.

35 **"The deluge had occurred to them as a cause":** Ibid.

35 **is anonymously known to the generations:** Renee Clary, "Mary Anning: She Sold (Fossil) Sea Shells by the Seashore," collected in Rock Stars, *GSA Today*, May 2019, available at Geosociety.org.

35 **Her real legacies:** The Lyme Regis Museum, "Mary Anning: The World's Greatest Fossil Hunter," lymeregismuseum.co.uk.

36 **"she understands more":** Hugh Torrens, "Mary Anning (1799–1847) of Lyme; 'the greatest fossilist the world ever knew'," *The British Journal for the History of Science* (September 1995): 257–84.

36 **Lyell enlisted her help in measuring erosion:** Thomas W. Goodhue, "The Faith of a Fossilist: Mary Anning," *Anglican and Episcopal History* (March 2001): 80–100.

36 **"the connection or analogy between the Creatures":** Ibid. For more on Anning, see also Shelley Emling, *The Fossil Hunter* (Palgrave Macmillan 2009).

TWO: EVERYTHING FROM SHELLS

39 **"the nation's fastest-growing hobby":** Horace Sutton, "Shell Collecting Gains as Hobby," *Washington Post*, January 5, 1958.

39 **Collectors were launching shell clubs:** Paul Callomon, "The Nature of Names" (Master's thesis, Drexel University, September 2016), 87.

40 **"sometimes as high as three feet":** Merrill Folsom, "The Tahitis of the Gulf," *New York Times*, March 4, 1956.

45 **The irregular protuberances:** Erasmus Darwin, "The Botanic Garden," 1791. Accessed at Project Gutenberg, gutenberg.org.

46 **"Ere Time began, from flaming Chaos":** Erasmus Darwin, "The Temple of Nature; or, the Origin of Society" (Published posthumously, T. Bensley, 1803), Accessed at Project Gutenberg, gutenberg.org.

46 **"a single living filament":** Desmond King-Hele, ed., *Charles Darwin's The Life of Erasmus Darwin* (Cambridge University Press, 2004), xiii.

46 **Canon Thomas Seward became so incensed:** Philip Ashley Fanning, *Isaac Newton and the Transmutation of Alchemy: An Alternate View of the Scientific Revolution* (North Atlantic Books, 2009), 213.

46 **"Great wizard he! by magic spells":** Ibid.

46 **kept it on his bookplate:** King-Hele, *Charles Darwin's The Life of Erasmus Darwin*, xiii.

47 **their innovation also helped destroy them:** Simon Darroch et al., "Biotic Replacement and Mass Extinction of the Ediacara Biota," *Proceedings of the Royal Society B* 282, no. 1814 (September 2015).

47 **a mass extinction that preceded the Big Five:** Ibid.

47 **"small, shelly fossils":** Aodhán Butler, "Fossil Focus: The Place of Small Shelly Fossils in the Cambrian Explosion, and the Origin of Animals," *Palaeontology Online* 5, Article 7 (January 2015): 1–14.

47 **"small smellies":** Ibid.

48 *Kimberella* **hauled a non-mineral shell several inches long:** Mikhail A. Fedonkin and Benjamin M. Waggoner, "The Late Precambrian Fossil Kimberella Is a Mollusc-Like Bilaterian Organism," *Nature* 388 (August 1997): 868–71.

48 **possibly on a creeping foot:** Ibid., 871.

50 **a shaggy slug with a combat-style shell cap:** Jakob Vinther et al., "Ancestral Morphology of Crown-Group Molluscs Revealed by A New Ordovician Stem Aculiferan," *Nature* 542 (February 2017): 471–74.

50 **hundreds of real sea monsters:** Peter Van Roy et al., "The Fezouata Fossils of Morocco: An Extraordinary Record of Marine Life in the Early Ordovician," *Journal of the Geological Society* 172 (2015): 541–49.

50 **bivalves and gastropods split into different branches:** Jakob Vinther et al., "A Molecular Palaeobiological Hypothesis for the Origin of Aplacophoran Molluscs and Their Derivation from Chiton-Like Ancestors," *Proceedings of the Royal Society B* 279, no. 2 (November 28, 2020).

51 **"superbly adapted for existence in the deep sea":** Author interview with Peter D. Ward.

51 **nautilus relies on sophisticated chemical sensors:** Peter Ward, Frederick Dooley, and Gregory Jeff Barord, "Nautilus: Biology, Systematics, and Paleobiology as Viewed from 2015," *Swiss Journal of Paleontology* 135 (February 2016): 169–85.

52 **Up to 85 percent:** Elizabeth Kolbert, *The Sixth Extinction: An Unnatural History* (Henry Holt, 2014), 96–97.

52 **scientists find increasing evidence:** David P. G. Bond and Stephen E. Grasby, "Late Ordovician Mass Extinction Caused by Volcanism, Warming, and Anoxia, Not Cooling and Glaciation," *Geology* 48, no. 8 (May 2020): 777–81.

52 **methane gas that continued to vent:** Uwe Brand et al., "Methane Hydrate: Killer Cause of Earth's Greatest Mass Extinction," *Palaeoworld* 25, no. 4 (December 2016): 496–507.

54 **"Burrowing clams go wild":** Author interview with Lydia Tackett, professor, North Dakota State University.

55 **Cretaceous bivalve fossils collected from:** Sierra V. Petersen, Andrea Dutton, and Kyger C. Lohmann, "End-Cretaceous Extinction in Antarctica Linked to Both Deccan Volcanism and Meteorite Impact via Climate Change," *Nature Communications 7*, no. 1 (2016).

56 **suggested the name *Atlantic*:** Jeffrey Goldberg, "How the Atlantic Began," TheAtlantic.com, May 5th 2017.

56 **"sometimes with the wind":** Oliver Wendell Holmes, *The Autocrat of the Breakfast-Table* (James. R. Osgood and Co., 1873; originally published in *The Atlantic* in January 1858), accessed at Project Gutenberg, gutenberg.org.

56 **"zoological catastrophe":** R. Tucker Abbott, *The Kingdom of the Seashell* (Bonanza Books, 1982), 110.

57 **"We can excuse Mr. Holmes":** Ibid., 110.

THREE: THE VOICE OF THE PAST

59 **could carry more than a mile:** Author interview with Miriam Kolar.

60 **Mirrors made of anthracite:** John W. Rick, "Context, Construction, and Ritual in the Development of Authority at Chavín de Huántar," in William J. Conklin and Jeffrey Quilter, eds., *Chavin: Art, Architecture and Culture* (Costen Institute of Archaeology at UCLA, 2008), 29.

60 **worthy of *The Wizard of Oz*:** John W. Rick, "The Evolution of Authority and Power at Chavín de Huántar," *Archeological Papers of the American Anthropological Association* 14, no. 1 (June 2008): 87.

60 **We think of ancient life as quiet:** Tom Colligan, "The Loudest Competition on Earth," *California Sunday Magazine*, September 29, 2016.

60 **Silence is a virtue for mollusks:** Geerat J. Vermeij, "Sound Reasons for Silence: Why Do Molluscs Not Communicate Acoustically?" *Biological Journal of the Linnean Society* 100, no. 3 (2010): 485–93, 487.

60 **the ideal windway:** Bradley Strauchen-Scherer, "Brass Beginnings: A Fanfare for the Conch Trumpet," The Met, *Collection Insights* blog, March 8, 2018.

61 **their surfaces often mouth-worn:** Jeremy Montagu, *Horns and Trumpets of the World: An Illustrated Guide* (Rowman & Littlefield, 2014), 29.

61 **Hearing a shell trumpet blown:** J. Arthur Thomson, *Secrets of Animal Life* (Henry Holt, 1919), 65.

61 **"What served to scare off evil spirits":** Ibid., 65.

61 **Hundreds of Triton's Trumpets:** Robin Skeates, "Triton's Trumpet: A Neolithic Symbol in Italy," *Oxford Journal of Archaeology* 10 (1991): 17–31.

62 **"the shell was the voice of the god":** J. Wilfrid Jackson, *Shells as Evidence of the Migration of Early Culture* (Manchester University Press, 1917), 23.

62 **a "sanctuary, one of the most famous":** Richard L. Burger, "Chavín de Huán-

tar and Its Sphere of Influence," in Helaine Silverman and William Isabell, eds., *Handbook of South American Archaeology* (Springer, 2008), 681.

62 **human and wild-animal parts together in monstrous jigsaw:** Peter G. Roe, "How to Build a Raptor," in Conklin and Quilter, eds., *Chavin: Art, Architecture and Culture*, 182.

63 **offered to many deities:** Alexander Herrera Wassilowsky, "Pututu and Waylla Kepa: New Data on Andean Pottery Shell Horns," translated from R. Eichmann et al., eds. and trans., *Studien zur Musikarchaeologie* 7 (2010): 17–37, 18.

64 **its size, workmanship, and iconography, suggest:** Richard L. Burger, "The Sacred Center of Chavín de Huántar," in Richard F. Townsend, ed., *The Ancient Americas: Art from Sacred Landscapes* (Prestel Verlag, 1992), 271.

65 **During ritual drug use:** Kembel S. Rodriguez and John W. Rick, "Building Authority at Chavín de Huántar: Models of Social Organization and Development in the Initial Period and Early Horizon," in Helaine Silverman, ed., *Andean Archaeology* (Blackwell, 2004), 66–67.

65 **perhaps to call or to banish what we now understand as super El Niño cycles:** Izumi Shimada, *Pampa Grande and the Mochica Culture* (University of Texas Press, 2010), 238–39; and Shimada, "Evolution of Andean Diversity (500 BCE–CE 600)," in Frank Salomon and Stuart B. Schwartz, eds., *The Cambridge History of the Native Peoples of the Americas, Volume 3: South America, Part 1* (Cambridge University Press, 1999), 350–517.

66 **The southward movement . . . might have signaled:** Rodriguez and Rick, "Building Authority at Chavín de Huántar," 57.

66 **Eastern Pacific Giant Conch:** Called *Strombus galeatus* in archaeological literature, they are now classified by the World Register of Marine Species as *Titanostrombus galeatus*.

67 **"It was beginning to make beautiful sense":** Author interview with John Rick.

68 **The original craftsmen:** The members of the *Strombus* family formerly known as *Strombus galeatus* but now classified as *Titanostrombus galeatus*.

68 **Gentle grazers that eat microalgae:** Roberto Cipriani et al., "Population Assessment of the Conch *Strombus galeatus* in Pacific Panama," *Journal of Shellfish Research* 27, no. 4 (2008): 889–96, August.

69 **Each conch is rubbed thin:** Miriam A. Kolar with John W. Rick, Perry R. Cook, and Jonathan S. Abel, "Ancient *Pututus* Contextualized: Integrative Archaeoacoustics at Chavín de Huántar," in Matthias Stockli and Arnd Adje Both, eds., *Flower World: Music Archaeology of the Americas*, Vol. 1 (Berlin: Ekho Verlag, 2012), 28.

69 **"Triton, sea-hued":** *Ovid, Metamorphoses* (Oxford World's Classics), A. D. Melville, trans. (Oxford University Press, 1986), 74.

70 **"That tumultuous uproar rent the hearts":** Annie Besant and Bhagavan Das, *The Bhagavad-Gita*, with Sanskrit text and English translation (Theosophical Publishing Society, 1905), 7–9.

71 **dived into the sea during a great flood:** R. Tucker Abbott, *Kingdom of the Seashell* (Bonanza Books, 1982), 186.

71 **"proclaim the truth":** Robert Beer, *The Handbook of Tibetan Buddhist Symbols* (Serindia Publications, 2003), 9.

71 **The white conch came to stand:** Tseten Namgyal, "Significance of 'Eight Traditional Tibetan Buddhist Auspicious Symbols/Emblems' (*bkra shis rtags brgyad*) in Day to Day Rite and Rituals," *Tibet Journal* 41, no. 2 (2016): 29–51, 30.

71 **The speech conduit appeared to Miriam Kolar:** This section is based on author interviews with Miriam A. Kolar, Visiting Scholar, Amherst College.

72 **how sounds were transmitted and transformed:** Miriam A. Kolar, "Tuned to the Senses: An Archaeoacoustic Perspective on Ancient Chavín," *The Appendix* 1, no. 3 (July 2013).

72 **They are called *pututus*, from the Quechua:** Jeremy Montagu, *The Conch Horn: Shell Trumpets of the World from Prehistory to Today* (Hataf Segol Publications, 2018), 178.

72 **who came to blow experimental toots:** Perry R. Cook et al., "Acoustic Analysis of the Chavín *Pututus*," *Journal of the Acoustical Society of America* 128, no. 4 (September 2010).

73 **as if he'd joined his breath with the wind:** Miriam A. Kolar, "Conch Calls into the Anthropocene: *Pututus* as Instruments of Human-Environmental Relations at Monumental Chavín," *Yale Journal of Music & Religion* 5, no. 2 (August 2019): 36–37.

74 **she crawled through 3,000-year-old slate-lined canals:** Miriam A. Kolar, "Tuned to the Senses: An Archaeoacoustic Perspective on Ancient Chavín," *The Appendix* 7 (July 2013).

74 **Perhaps, as Kolar speculates:** Miriam A. Kolar, "Situating Inca Sonics: Experimental Music Archaeology at Huánuco Pampa, Peru," in *Flower World: Music Archaeology of the Americas*, Vol. 6 (Ekho Verlag, 2020), 28.

75 **Atlantic conchs have been unearthed in some Aztec ruins:** Abbott, *Kingdom of the Seashell*, 176.

75 **The new human race is born:** Karl Taube, *Aztec & Maya Myths* (Austin: University of Texas Press, 1993), 37–39.

76 **hundreds of conch-shell trumpets have been unearthed:** Barbara J. Mills and T. J. Ferguson, "Animate Objects: Shell Trumpets and Ritual Networks in the Greater Southwest," *Journal of Archaeological Method Theory* 15 (2008): 338–61; see list of pueblos and gastropod species found there, 347–49, and map, 350.

76 **to give voice to the breath of life:** Ibid.

76 **The Zuni's "Great Shell Society":** Ibid.

76 **"They are demons . . .":** Charles Hudson and Robbie Ethridge, *Knights of Spain, Warriors of the Sun: Hernando de Soto and the South's Ancient Chiefdoms* (University of Georgia, 2018), 109.

76 **the louder the eerie wails became:** Ibid., 110.

76 **Still, the shell horns blew:** Wassilowsky, "Pututu and Waylla Kepa." Wassilo-wsky writes that as late as 1700 in what is now Ecuador, a conch trumpet and *Spondylus* shells were part of the evidence in the sorcery trial against the shaman Andrés Arévalo.

77 **captured the shells' command:** William Golding, *The Lord of the Flies* (Faber & Faber, 1954). Chapter 1 is "The Sound of the Shell."

77 **In his next and favorite novel:** William Golding, *The Inheritors* (Faber & Faber, 1955).

78 **The Andes have long been seen:** Rick, "The Evolution of Authority and Power at Chavín de Huántar," 71.

78 **"conscious, calculated political strategy":** Ibid., 71.

78 **"exceptionally creative in their manipulation":** Ibid.

FOUR: GREAT CITIES OF SHELL

81 **an elevated causeway:** Sarah E. Baires, "Cahokia's Rattlesnake Causeway," *Midcontinental Journal of Archaeology* 39, no. 2 (2014): 145.

81 **pole-and-thatch houses radiated 50 miles:** Timothy R. Pauketat, *Cahokia: Ancient America's Great City on the Mississippi* (Viking, 2009), 5–8.

81 **an untouched wilderness:** Alice Beck Kehoe, "Cahokia, the Great City," *OAH Magazine of History* 27, no. 4 (2013): 17.

81 **a political and religious capital:** Pauketat, *Cahokia*, 5–8.

82 **"nations of Indians":** (Emphasis added) Thomas Say; quoted in T. R. Peale, "Ancient Mounds at St. Louis, Missouri, in 1819," *Annual Report of the Board of Regents of the Smithsonian Institution, 1861*, 388.

82 **razed before the Civil War:** Pauketat, *Cahokia*, 2.

82 **A drive-in movie theater:** "Falcon Drive-In" (formerly called Mounds Drive-In), Cinema Treasures, cinematreasures.org.

83 **The leftward spiral may have had a different meaning:** William H. Marquardt and Laura Kozuch, "The Lightning Whelk: An Enduring Icon of Southeastern North American Spirituality," *Journal of Anthropological Archaeology* 42 (2016): 1–26.

83 **"more extensively used than any other shell":** William Henry Holmes, "Art in Shell of the Ancient Americans," extract, *Second Annual Report of the Bureau of Ethnology*, 1883, 191–92.

83 **the most widely traded large marine shells:** Laura Kozuch, "The Significance of Sinistral Whelks from Mississippian Archaeological Sites," PhD dissertation, the University of Florida, 1998, 136.

83 **The farther inland they were traded, the more exotic:** Author interview with William H. Marquardt, curator and director emeritus, South Florida Archaeology and Ethnography and the Randell Research Center at Pineland, Florida Museum of Natural History.

84 **found two bodies:** Thomas E. Emerson et al., "Paradigms Lost: Reconfiguring Cahokia's Mound 72 Beaded Burial," *American Antiquity* 81, no. 3 (2016).

84 **excavators would unearth the bones of a child:** Ibid.

84 **not honed from the easiest shell to drill:** Laura Kozuch, "Shark Teeth and Sea Shells from East St. Louis," Emerson et al., eds., *Revealing Greater Cahokia, North America's First Native City* (Illinois State Archaeological Survey, 2018), 413.

84 **Most of the disk beads:** Author interview with Laura Kozuch. In her analysis of all beads from the prosaically named Mound 72, Kozuch identified 23,958 disk beads as almost certainly Lightning Whelk, and 35,289 columella beads, 93 percent of which are Lightning Whelk.

85 **as many as nine small males mating with one mother:** Personal communication, José H. Leal, editor of *The Nautilus*, Science Director & Curator, the Bailey-Matthews National Shell Museum.

86 **confessed to not having a clue:** Stephen Jay Gould, "Left Snails and Right Minds," *Natural History* 104, no. 4 (1995): 10–18.

86 **Left-handedness gives Lightning Whelks a slight advantage:** Gregory P. Dietl and Jonathan R. Hendricks, "Crab Scars Reveal Survival Advantage of Left-Handed Snails," *Biological Letters* 2 (2006): 439–42.

86 **their lower risk of being eaten as embryos:** See the section "Handedness and the Notion of Constraint" in Geerat Vermeij, *A Natural History of Shells* (Princeton University Press, 1995), 21–27.

86 **reflect preference for the right and bias against the left:** Marquardt and Kozuch, "The Lightning Whelk," 3–4.

86 **The perceived superiority of right-handedness:** "Left-Handedness," *Scientific American*, May 13, 1871, 310.

87 **most of them Lightning Whelks:** Jerald T. Milanich, "Origins and Prehistoric Distributions of Black Drink and the Ceremonial Shell Drinking Cup," in Charles M. Hudson, ed., *Black Drink: A Native American Tea* (University of Georgia Press, 1979), 84–86.

88 **"Every revolution increases its diameter":** William Bartram, *Travels* (James & Johnson, 1791), 451.

88 **Kozuch found the pattern in other Native dances and rituals:** Marquardt and Kozuch, "The Lightning Whelk: An Enduring Icon," 10.

88 **suggest a skilled workforce devoted full-time:** Kozuch, "Shark Teeth and Sea Shells."

89 **proved tricky, and expensive:** Cheryl Claassen, *Shells* (Cambridge University Press, 1998), 216.

89 **the vast majority hailed from the very same region:** Laura Kozuch, Karen J. Walker, and William H. Marquardt, "Lightning Whelk Natural History and a New Sourcing Method," *Southeastern Archaeology* 36, no. 3 (2017): 226–40.

89 **spare hide coverings:** Darcie A. MacMahon and William H. Marquardt, *The Calusa and Their Legacy* (University Press of Florida, 2004), 4.

90 **the bays and estuaries were the Calusa's domain:** MacMahon and Marquardt, in *The Calusa and Their Legacy*, explore the environmental aspects of the Calusa domain.

90 **They were fishers who had no use for agriculture:** Jack E. Davis, *The Gulf: The Making of an American Sea* (Liveright: 2017), 35–39.

90 **"decks covered with awnings of hoops and matting":** Gonzalo Solís de Merás, quoted in MacMahon and Marquardt, *The Calusa and Their Legacy*, 74.

90 **wee wooden toys:** Cushing found six or seven toy canoes, well-used. "A Preliminary Report on the Exploration of Ancient Key-Dweller Remains on the Gulf Coast of Florida," *Proceedings of the American Philosophical Society* 35, no. 153 (December 1896): 364.

90 **Those at Pineland and other estuaries feasted on gastropods:** William H. Marquardt and Karen J. Walker, "The Pineland Site Complex: An Environmental and Cultural History," in Marquardt and Walker, eds., *The Archaeology of Pineland* (University Press of Florida, 2013), 878–79.

90 **Marquardt has found thousands:** William H. Marquardt, *Culture and Environment in the Domain of the Calusa*, Monograph 1, Institute of Archaeology and Paleoenvironmental Studies (University of Florida, 1992), Josslyn pits, page 19.

90 **saucers, spoons, ladles:** For the many Calusa Lightning Whelk tools, see William H. Marquardt, Chapter 5, "Shell Artifacts from the Caloosahatchee Area," in Marquardt, ed., *Culture and Environment in the Domain of the Calusa*, 191.

90 **convinced that the Calusa were also involved in long-distance trade:** William H. Marquardt, "Trade and Calusa Complexity: Achieving Resilience in a Changing Environment," in Johan Ling, Richard Chacon, and Kristian Kristiansen, eds., *Trade Before Civilization: Long Distance Exchange and the Development of Social Complexity* (Cambridge University Press, 2020), p. 16 of review chapter.

91 **"more than sixty villages of his own":** MacMahon and Marquardt, *The Calusa and Their Legacy*, 2.

91 **Among the greatest human-caused extinctions:** Cynthia Barnett, *Blue Revolution: Unmaking America's Water Crisis* (Beacon Press, 2011), 26.

91 **advocated for the Zuni against settlers:** Sascha T. Scott, "Ana-Ethnographic Representation: Early Modern Pueblo Painters, Scientific Colonialism, and Tactics of Refusal," *Arts* 9, no. 1 (2020).

92 **"a man of genius":** John Wesley Powell, remarks, memorial to Frank Hamilton Cushing, Anthropological Society of Washington, April 24, 1900; reprinted in *American Anthropologist* 2 (1900): 366.

92 **"decided the purpose and calling":** Alex F. Chamberlain, "In Memoriam: Frank Hamilton Cushing," *Journal of American Folklore* 14, no. 49 (Spring 1900): 129.

92 **for "going native":** Nancy J. Parezo, "Reassessing Anthropology's Maverick: The Archaeological Fieldwork of Frank Hamilton Cushing," *American Ethnologist* 34, no. 3 (August 2007): 575–80.

92 **pulled off his beloved Zuni project:** The "Harvard man" was Jesse Walter Fewkes. For more on Cushing's feud with Fewkes, see Curtis Hinsley, "Ethnographic Charisma and Scientific Routine: Cushing and Fewkes in the American Southwest, 1879–1893," in George W. Stocking, ed., *Observers Observed: Essays on Ethnographic Fieldwork* (University of Wisconsin Press, 1983), 53.

92 **"a beautifully shaped and highly polished ladle or cup":** Frank Hamilton Cushing, "A Preliminary Report on the Exploration of Ancient Key-Dweller Remains on the Gulf Coast of Florida," *Proceedings of the American Philosophical Society* 35, no. 153 (December 1896): 329.

92 **"the most important archaeological discovery yet made":** Ibid., 330.

92 **"to explore as many as possible of the islands":** Ibid., 331.

92 **Smith had settled his family on sheltered Pine Island:** Mary Kaye Stevens, *Pine Island* (Arcadia Publishing, 2008), 34.

93 **A noted tarpon fishing guide:** Davis, *The Gulf*, 154.

93 **The Great Freeze that winter:** Marquardt and Walker, "The Pineland Site Complex," in *The Archaeology of Pineland*, 860.

93 **"great cities of shell":** Cushing untitled manuscript 1896, Phyllis E. Kolianos and Brent R. Weisman, *The Lost Florida Manuscript of Frank Hamilton Cushing* (University Press of Florida, 2005), 69.

93 **"a mural mosaic of volutes":** Cushing, "A Preliminary Report," 339.

93 **"I was astonished beyond measure":** Cushing, *The Lost Florida Manuscript*, 62–63.

94 **"it seemed a more gigantic undertaking":** Ibid., 63.

94 **Walking its long, spiral path to the top:** Ibid., 67.

94 **stop on every island with an ancient shell structure:** Davis, *The Gulf*, 25–26.

94 **After counting more than seventy-five shell mounds:** Cushing, "A Preliminary Report," 347.

94 **the structures had extended out over the water:** Randolph J. Widmer, new introduction to Frank Hamilton Cushing, *Ancient Key-Dweller Remains on the Gulf Coast of Florida* (University Press of Florida, 2000), xviii–xix.

94 **the Gulf's bounty, and its hardships:** Kolianos and Weisman, *The Lost Florida Manuscript*, introduction, 12.

94 **one of the most significant pre-Columbian finds:** "Preserved Against the Odds, the Key Marco Cat is Returning to Marco Island," September 7, 2018, Smithsonianmag.com.

94 **pairs of clamshells, bound closed:** Cushing, "A Preliminary Report," 386–87.

95 **"an ingenious anchor":** Ibid., 366.

95 **"a Shell Age phase of human development and culture":** Ibid., 411.

95 **"this one little court of the pile-dwellers":** Ibid., 360.

95 **Everyone present when the shell was pulled from the muck:** Marion S. Gilliland, *Key Marco's Buried Treasure* (University Press of Florida, 1989), 107.

96 **He died after swallowing a fish bone in April 1900:** John Wesley Powell, "Report of the Director," in *The Twenty-First Annual Report of the Bureau of American Ethnology* (Smithsonian Institution, July 1, 1900), xxxvii.

96 **"not a single mound of any size is left":** Brian M. Fagan, *Before California: An Archaeologist Looks at Our Earliest Inhabitants* (AltaMira Press, 2004), 253.

96 **shell middens . . . into settlements and burial mounds:** The Emeryville Historical Society, *Emeryville* (Arcadia Publishing, 2005), 9.

97 **literally danced on graves:** Laura Klivans, "There were once more than 425

shell mounds in the Bay Area. Where did they go?" KQED news, December 6, 2019.

97 **black-and-white photographs show spectators:** Steam shovel photo in the collection of the Phoebe Hearst Museum of Anthropology, University of California–Berkeley.

97 **"As you undoubtedly know":** Corrina Gould and Michelle LaPena, "Buried in Shells," *GIA Reader* (Grantmakers in the Arts) 29, no. 2 (2018).

97 **The perspective reflected a fundamental disconnect:** Ibid.

97 **Native people believed all of nature:** Timothy H. Silver, "Three Worlds, Three Views: Culture and Environmental Change in the Colonial South," the National Humanities Center, nationalhumanitiescenter.org.

97 **"To be buried in shells":** Gould and LaPena, "Buried in Shells."

97 **4 inches of clamshell graded onto 7 inches of oyster shell:** William Kaszynski, *The American Highway: The History and Culture of Roads in the United States* (Jefferson, NC: McFarland & Co., 2000), 32. (The agency was the Office of Public Roads.)

97 **Mules hauling the shell-heaped carts:** Corbett McP. Torrence, "A Topographic Reconstruction of the Pineland Site Complex as it Appeared in 1896," in Marquardt and Walker, eds, *The Archaeology of Pineland*, 164.

98 **sliced right through the rubber tires:** Torrence in Marquardt and Walker, *The Archaeology of Pineland*, 161 and 863.

98 **It took mechanical rolling:** Torrence, "A Topographic Reconstruction," 161.

98 **Collier pressured Florida to carve Marco Island:** Tracy Owens, "How Barron Collier Got His County," *Gulfshore Life* magazine, July 2018.

98 **the Calusa capital with its raised chief's house:** MacMahon and Marquardt, *The Calusa and Their Legacy*, 88.

98 **Indian artifacts popping up in oyster-shell driveways:** E. D. Estevez, "Mining Tampa Bay for a Glimpse of What Used to Be," *Bay Soundings*, Spring 2010.

99 **shell dredgers began to dig into the sacred beast:** William Grant Mcintire, "Prehistoric Settlements of Coastal Louisiana," PhD dissertation, Louisiana State University, June 1954, 69.

99 **built their 20-foot lookout tower atop:** "Face of Marco Island Undergoing a Change," *Miami Herald*, August 8, 1971.

99 **more miles of canals than roads:** Frank Mackle III wrote in a company history that the development company built 95 miles of canals and 90 miles of roads on Marco. Chapter 16, "Marco Island," themacklecompany.com.

99 **a billboard appeared on a rare empty lot on Marco:** Quentin Quesnell, "Relocating Cushing's Key Marco," *Florida Anthropologist* 49, no. 1 (March 1996): 4.

99 **Across the street, trench diggers laying TV cable:** Quesnell, "Relocating Cushing's Key Marco," 4.

99 **Locals readied the site:** Randolph J. Widmer, "Recent Excavations at the Key Marco Site, 8CR48, Collier County, Florida," *Florida Anthropologist* 49, no. 1 (March 1996): 10.

99 **spire to siphon, in a concrete-like matrix:** Randolph J. Widmer, "The Key

Marco Site, A Planned Shell Mound Community on the Southwest Florida Coast," in Mirjana Roksandic et al., eds., *The Cultural Dynamics of Shell-Matrix Sites* (University of New Mexico Press, 2014), 11.

100 **the site is one of the more modest condominiums:** Eagles Retreat, a 16-unit condominium on Bald Eagle Drive.

100 **Calusa masons had covered the Lightning Whelks with a veneer:** Widmer, "Recent Excavations at the Key Marco Site," 24.

100 **Another crew showed up in 1926:** Torrence, "A Topographic Reconstruction," 162–63.

100 **They were met with a shotgun:** Ibid., 162–63. The shotgun story was recounted by Smith's son, Ted Smith, during interviews in 1992 and 1993.

100 **New York retirees Pat and Don Randell:** Marquardt and Walker, *The Archaeology of Pineland*, 868–69.

101 **archaeological research and teaching center the public can visit:** The Randell Research Center, Pineland, Florida; www.floridamuseum.ufl.edu/rrc/.

101 **Construction of the live wells coincided with a sea-level drop:** Karen J. Walker, "The Pineland Site Complex: Environmental Contexts," in Marquardt and Walker, eds., *The Archaeology of Pineland*, 42–43. The sea-level drop was associated with the Little Ice Age. For new research on the water courts, see Victor D. Thompson, et al., "Ancient Engineering of Fish Capture and Storage in Southwest Florida," *Proceedings of the National Academy of Sciences* 117, no. 15 (2020).

101 **Lightning Whelk research that finds the animals' greatest density:** Sarah P. Stephenson et al., "Abundance and distribution of Large Marine Gastropods in Nearshore Seagrass Beds along the Gulf Coast of Florida," *Journal of Shellfish Research* 32, no. 2 (August 2013).

102 **Jules Verne captured the excitement:** Jules Verne, *Twenty Thousand Leagues Under the Sea* (Brothers, 1870), 141–42.

103 **Alas, Aronnax:** The fictional character would, however, have a mollusk named for him by Philippe Bouchet and Anders Warén: *Marginella aronnax*.

103 **where they take different roles:** Sukanya Sarbadhikary, "Shankh-er Shongshar, Afterlife Everyday: Religious Experience of the Evening Conch and Goddesses in Bengali Hindu Homes," *Religions*, January 15, 2019, 7–8.

104 **massive shell-working sites:** V. H. Sonawane, "Harappan Shell Industry: An Overview," *Indian Journal of History of Science* 53, no.3 (2018): 253–62.

104 **woven into "almost every phase of Hindu life":** James Hornell, *The Sacred Chank of India: A Monograph of the Indian Conch*, Madras Fisheries Bureau, Bulletin no. 7 (1914): 1.

104 **"the men seated on the ground":** Hornell, *The Sacred Chank of India*, 11.

104 **millions of chanks:** K. Nagappan Nayar and S. Mahadevan, "Chank Fisheries and Industrial Uses of Chanks," in *The Commercial Mollusks of India*, Central Marine Fisheries Research Institute (CMFRI) bulletin no. 25, 1974, 122–40.

105 **government paid divers a thousand times more:** Nayar and Mahadevan, "Chank Fisheries," 122–40.

105 **smaller numbers of free-divers:** Aarthi Sridhar, producer, *Fishing Palk Bay*, documentary film, Dakshin Foundation, 2016.

105 **they hold their breath for 90 seconds:** Ibid.

105 **The divers collect the live animals:** Aarthi Sridhar, "A Journey with the Sacred Chank," *Frontline* of India, March 30, 2018.

105 **The free-divers have watched their incomes shrink:** S. Senthalir, "Tamil Nadu Deep Sea Divers Who Bring Conches to the World Are Losing Their Hearing," *Scroll.in*, February 10, 2019.

105 **The catch is no longer counted by individual mollusks:** R. Ravinesh et al., "Status of Marine Molluscs in Illegal Wildlife Trade in India," *G Wild Cry*, 34–39.

107 **He went broke in the Great Depression:** Stephen Hegarty, "Residents, City Can't Agree on Who Pays for Repairs to Blushing Concrete," *St. Petersburg Times*, October 29, 1984.

107 **Two men were convicted of stealing state artifacts:** George M. Luer, "Response to Looting on Josslyn Island," *Florida Anthropologist* 61 (March–June 2008): 34–35.

108 **Texas limited the daily catch to no more than two:** "Lightning Whelk, the State Shell of Texas," Texas Parks & Wildlife, tpwd.texas.gov.

108 **A decade ago, state marine researchers launched a base study:** Stephenson et al., "Abundance and Distribution of Large Marine Gastropods."

108 **Slowly, it turns out:** At least for the 20 whelks recaptured out of 280 tagged in the first 20 months of the study.

109 **There is no explanation:** There is no explanation at Pinellas Point. But for a fascinating dive into the nature of truth, memory, and historic interpretation at this and other Southeastern Indian sites, see Chapter 1, "Shell Mounds and Indianness," in Thomas Hallock, *A Road Course in Early American Literature: Travel and Teaching from Atzlán to Amherst* (University of Alabama Press, 2021).

109 **lore still taught to fourth-graders:** "The Calusa: 'The Shell Indians,'" Florida Center for Instructional Technology, Short History of Florida section, fcit.usf .edu.

FIVE: SHELL MONEY

113 **ruled the islands of the Maldives with epic command:** Fatima Mernissi, *The Forgotten Queens of Islam* (University of Minnesota Press, 2003), 107. Mernissi spells her name Khadija; I have gone with the more common spelling with two H's.

114 **but the shell money spread farther:** Bin Yang, *Cowrie Shells and Cowrie Money: A Global History* (Routledge, 2019), x.

114 **in India as early as the fourth century:** Ibid., 1.

114 **In the mountains of Yunnan:** Ibid., xi.

114 **helped settle the Maldives much earlier than other outposts:** Mirani Lister, *Cowry Shell Money and Monsoon Trade: The Maldives in Past Globalizations*, PhD dissertation, The Australian National University, 2016, 14.

114 **a cache of more than three thousand Money Cowries:** A. C. Christie and A.

Haour, "The 'Lost Caravan' of Ma'den Ijafen revisited: Re-appraising Its Cargo of Cowries, a Medieval Global Commodity," *Journal of African Archaeology* 16 (2018): 125–44.

115 **Painstakingly measuring these shells:** Anne Haour and Annalisa Christie, "Cowries in the Archaeology of West Africa: The Present Picture," *Azania: Archaeological Research in Africa* 54, no. 3 (2019): 287–321.

115 **cowries circulated longer:** Lister, *Cowry Shell Money and Monsoon Trade*, 4.

115 **"contrary to the tenets of Islam":** As quoted from the Maldives constitution, "2019 Report on International Religious Freedom: Maldives," Office of International Religious Freedom, U.S. Department of State, 2019, state.gov/reports/2019-report-on-international-religious-freedom/maldives/.

116 **They are not sinking, of course:** J. J. Robinson, *The Maldives: Islamic Republic, Tropical Autocracy* (Hurst & Company, 2015), 80–81. Robinson, a foreign journalist himself, has a fascinating section on the foreign media's obsession with the "sinking" of the Maldives.

116 **and transferring coconut palms and other large trees:** Mohamed Junayd, "#MVTreegrab: Removal of Trees Continues for Landscaping Resorts," *Maldives Independent*, November 22, 2018.

116 **More than . . . 200,000 more:** Maldives Housing Development Corporation, "Hulhumalé: The City of Hope," hdc.com.mv/hulhumale/.

117 **new laws prohibited books and papers:** Alison Flood, "Maldives Will Censor All Books to Protect Islamic Codes," *The Guardian*, September 25, 2014.

117 **"worshipped objects of nature":** Naseema Mohamed, "Note on the Early History of the Maldives," *Archipel* 70 (2005): 7.

117 **are believed to have drawn a Buddhist high culture:** Egil Mikkelsen, "Archaeological Excavations of a Monastery at Kaashidhoo: Cowrie Shells and Their Buddhist Context in the Maldives" (National Centre for Linguistic and Historical Research, Republic of Maldives, 2000).

117 **"island of women":** Lopamudra Maitra Bajpai, "Maldives-Bengal Trade Flourished under Maldivian Queen Khadeeja of Bengal Origin," *News In Asia*, October 1, 2018.

118 **"the islands of cowries":** Yang, *Cowrie Shells and Cowrie Money*, 26.

118 **"a paradise for cowries":** Ibid., 6.

118 **discovered in the rubble at Pompeii:** Peter S. Dance, *Shell Collecting: An Illustrated History* (University of California Press, 1966), 227–28..

118 **"We may in truth compare its beauteous fulvous hues":** E. Donovan, *The Naturalist's Repository* (Printed for the author and W. Simpkin and R. Marshall, 1823), Vol. 1, Plate XXXII.

118 **Scientists describe the sometimes precise shell cumulations:** David Scheel and Peter Godfrey-Smith, "*Octopus tetricus* as an Ecosystem Engineer," *Scientia Marina*, December 2014.

119 **Cowries differ from other mollusks:** This section is based on author interviews with Christopher Meyer, curator of mollusks, Smithsonian National Museum of Natural History.

120 **lost an estimated 95 percent:** Clive Wilkinson et al., "Ecological and Socio-economic Impacts of 1998 Coral Morality in the Indian Ocean," *Ambio* 28, no. 2 (March 1999): 190.

120 **no evidence that the storms:** Neville Coleman et al., *Marine Life of the Maldives Indian Ocean 2019* (Atoll Editions, 2019), viii.

121 **"The wealth of the people":** Quoted in A. Gray, "The Maldives Islands," *Journal of the Royal Asiatic Society of Great Britain and Ireland* 10, no. 2 (April 1878): 178–79.

121 **"To these the creatures attach themselves":** Quoted in Jan Hogendorn and Marion Johnson, *The Shell Money of the Slave Trade* (Cambridge University Press, 1986), 23.

121 **Ancient stories of how the Buddhist archipelago turned Muslim:** Husnu Al Suood, *Political System of the Ancient Kingdom of the Maldives* (Maldives Law Institute, 2014), 9.

122 **described cowries as sold in:** H.C.P. Bell, *The Maldive Islands: An Account of the Physical Features, Climate, History, Inhabitants, Productions and Trade* (Government Printer, Ceylon, 1883), 117.

122 **"one of the wonders of the world":** Ibn Batuta, *The Travels of Ibn Batuta*, Cambridge professor of Arabic Samuel Lee, trans. (Oriental Translation Fund, 1829), 176. (The spelling of Batuta was corrected in later literature.)

122 **"an instrument of Thy grace":** Mernissi, *The Forgotten Queens of Islam*, 108.

122 **"I was quite unable to get them covered":** Batuta, *The Travels of Ibn Batuta*, 179.

122 **"mountains of cowries":** Yang, *Cowrie Shells and Cowrie Money*, 30–31.

122 **"There is a great trade at the Maldives":** Francois Pyrard et al., *The Voyage of Francois Pyrard of Laval to the East Indies, the Maldives, the Moluccas and Brazil* (Hakluyt Society, 1887), 236.

123 **"solely to load with these shells":** Ibid., 240.

123 **slipped colonialism:** The Maldives was one of the few Asian nations spared foreign colonization. Xavier Romero-Frias, *The Maldive Islanders: A Study of the Popular Culture of an Ancient Ocean Kingdom* (Nova Ethnographia Indica, 1999), introduction.

123 **"as rife as of old":** Bell, *The Maldive Islands*, 58.

123 **home to 215,000 residents:** *Statistical Pocketbook of Maldives 2019* (National Bureau of Statistics, 2020), statisticsmaldives.gov.mv.

124 **well documented from the early years of Islam:** Mernissi, *The Forgotten Queens of Islam*, 107–109.

124 **legendary sun-worshipping ancestors:** Mohamed, "Note on the Early History of the Maldives."

125 **"the start of a climate of impunity for fundamentalism":** Robinson, *The Maldives*, 131.

125 **"What about the history of our Dhivehi Raajje?!":** Yameen Rasheed, "The Republic of Whatever!" November 11, 2008, curated and archived by Hani Amir, Hani-Amir.com.

125 **In the Legend of the Sandara Shell:** Xavier Romero-Frias, *Folk Tales of the Maldives* (Nordic Institute of Asian Studies, 2012), 250.

126 **"bear a marked resemblance":** Catherine Blackledge, *The Story of V: A Natural History of Female Sexuality* (Rutgers University Press, 2004), 50.

126 **born from a cowrie:** Joseph Heller, *Sea Snails: A Natural History* (Springer, 2015), 268.

126 **girdles decorated with cowries:** Stephen Quirke, *Exploring Religion in Ancient Egypt* (Wiley, 2015), 60.

126 **the word for vagina is *kai*—shell:** Blackledge, *The Story of V*, 49.

126 **"strong reasons for believing":** J. Arthur Thompson, *Secrets of Animal Life* (Henry Holt, 1919), 68.

126 **"The whole of the complex shell cult":** G. Elliot Smith, quoted in Thompson, *Secrets of Animal Life*, 69.

126 **cowries draw all kinds of admirers:** Yang, *Cowrie Shells and Cowrie Money*, 230.

127 **the villagers of Tell Halula had faith in cowries to heal:** Hala Alarashi et al., "Sea Shells on the Riverside: Cowrie Ornaments from the PPNB Site of Tell Halula (Euphrates, northern Syria)," *Quaternary International* 490 (2018): 111.

127 **bejeweled elephant harnesses:** Yang, *Cowrie Shells and Cowrie Money*, 230.

127 **"Early occurrences of the cowrie . . . ":** M. A. Murray, "The Meaning of the Cowrie-Shell," *Man* 39 (October 1939): 167.

128 **"He who hangs a necklace of cowries":** Quoted in Yang, *Cowrie Shells and Cowrie Money*, 232.

128 **"would not discern any human habitation":** Romero-Frias, *The Maldive Islanders*, 243.

128 **once frequented by early traders:** (Kaashidhoo was then called Kardiva Island.) Commander A. D. Taylor, *The West Coast of Hindustan Pilot Including the Gulf of Manar, the Maldive and Laccadive Islands* (Hydrographic Office, Admiralty, 1898), 334.

130 **first modern scientific excavations:** Egil Mikkelsen, *Archaeological Excavation of a Monastery at Kaashidhoo: Cowrie Shells and Their Buddhist Context in the Maldives* (National Centre for Linguistic and Historical Research, 2000), 4.

130 *A place called Kuruhinna Tharaagandu:* Mikkelsen, *Archaeological Excavations of a Monastery at Kaashidhoo*, 9.

133 **"bikinis, bacon, beer, and Bibles":** Author interview with J. J. Robinson; and see Robinson, *The Maldives*, 115.

134 **"The whole country is shaded with trees":** Batuta, *The Travels of Ibn Batuta*, 177.

134 **Islamic national government:** Kai Schultz, "Maldives, Tourist Haven, Casts Wary Eye on Growing Islamic Radicalism," *New York Times*, June 18, 2017.

136 **they now see the human signature:** Meera Subramanian, "Humans versus Earth: The Quest to Define the Anthropocene," *Nature*, August 6, 2019.

137 **the Anthropocene does not capture the reality:** For a good introduction to argument against the term, see Matthew Henry, "Are We All Living in the Anthropocene?" Oxford University Press blog post, October 29, 2017, blog.oup .com/2017/10/are-we-all-living-in-the-anthropocene/.

SIX: SHELL MADNESS

139 **2,389 "rarities of little shells" and "little horns":** Anne Goldgar, *Tulipmania: Money, Honor, and Knowledge in the Dutch Golden Age* (University of Chicago Press, 2007), 80.

139 **The shell collection was not to be broken up:** Ibid, 80–81.

139 **"his lively imagination saw nothing but shells":** Jean-Jacques Rousseau, *The Confessions of Jean Jacques Rousseau Vol. II* (Privately printed, 1904) (many later editions leave out either the word *conchylomania* or the shell story), 106.

139 **afflicted many of the same people:** Richard Conniff, *The Species Seekers: Heroes, Fools, and the Mad Pursuit of Life on Earth* (W. W. Norton, 2011), 74.

140 **at the same auction in the 1790s:** S. Peter Dance, *Out of My Shell: A Diversion for Shell Lovers* (C-Shells-3, 2005), 11–12.

140 **part of the artist's own extensive shell collection:** Paul Crenshaw, *Rembrandt's Bankruptcy: The Artist, His Patrons, and the Art Market in Seventeenth-Century Netherlands* (Cambridge University Press, 2005), 94. Use of *extensive* is based on "The inventory of Rembrandt's insolvent estate," [Fol. 33v.] (25–26 July 1656), [179]: "A great quantity of shells, coral branches, cast from life and many other curios," The Rembrandt Documents project, www.ru.nl/remdoc/.

141 **at least five other species:** Dance, *Out of My Shell*, 23–34.

141 **"may have been prized items":** S. Peter Dance, *Rare Shells* (Faber and Faber, 1969), 24.

141 **He stressed *conchas legere*:** Leopoldine Prosperetti, " 'Conchas Legere': Shells as Trophies of Repose in Northern European Humanism," *Art History* 29, no. 3 (June 2006): 395.

142 **thriving curiosity trade:** Dániel Margócsy, *Commercial Visions: Science, Trade, and Visual Culture in the Dutch Golden Age* (University of Chicago Press, 2014), 39.

142 **During one auction in the spring of 1637:** Historic value of the guilder calculated via International Institute of Social History, www.iisg.nl/hpw/calculate .php.

142 **Some wealthy collectors had entire *Wunderkammern*:** Anna Marie Roos, *Martin Lister and His Remarkable Daughters: The Art of Science in the Seventeenth Century* (Bodleian Library, 2019), 64.

143 **"the most glorious specimens from Creation":** Prosperetti, " 'Conchas Legere', 403.

143 **"white shell resembling a spirally twisted trumpet":** Peter S. Dance, *Shell Collecting: An Illustrated History* (University of California Press, 1966), 227–28.

144 **Among the credible accounts of its market value:** Ibid., 229. The 4,000 guilders described by Dance were converted to €94,987.68 via International Institute of Social History calculator and then into U.S. dollars.

144 **Chinese forgers made imitation wentletraps:** A. Hyatt Verrill, *Strange Sea Shells and Their Stories* (Farrar, Straus and Cudahy, 1936), 52.

144 **he was a Nazi fugitive:** Dance, *Out of My Shell*, 94–95.

144 **"roasted in hot ashes or gently pan-fried":** Leo Ruickbie, *The Impossible Zoo:*

An Encyclopedia of Fabulous Beasts and Mythical Monsters (Robinson, 2016); "Golden Limpet," p. 2038 on Kindle edition.

144 **Precious Wentletrap sold for 1,611 livres:** Dance, *Rare Shells*, 87.

144 **worth more than $27,000 today:** Currency conversion via HistoricalStatistics. Org, www.historicalstatistics.org/Currencyconverter.html.

144 **Auction records show the price plummeting:** Dance, *Shell Collecting: An Illustrated History*, 230–31.

145 **When the Precious Wentletrap gets hungry:** Joseph Heller, *Sea Snails: A Natural History* (Springer International, 2015), 164–65.

145 **a seventeenth-century spider man:** Peter Hogarth, "Dr. Martin Lister (1639– 1712), The Spider Man," *Yorkshire Philosophical Society*, "Yorkshire Scientists and Innovators," ypsyork.org.

145 **and a polymath:** Anna Marie Roos, ed. and trans., *The Correspondence of Dr. Martin Lister (1639–1712), Volume One: 1662–1677* (Brill, 2015), 5.

145 **who planted 3,000 bulbs:** Roos, *Martin Lister and His Remarkable Daughters*, 70.

145 **"the meanest" . . . "five or six princes":** Martin Lister, "Dr. Martin Lister's Journey to Paris," *Retrospective Review* XIII (1826): 95–108.

145 **the first scientifically rigorous encyclopedia of seashells:** Dance, *Shell Collecting*, 44.

146 **which worked well for those he had seen:** For a turtle-like mollusk with reindeer horns, see plate 230 in Filippo Buonanni, *Ricreatione dell'occhio e della mente* (Rome: 1681), *Internet Archive*, archive.org/details/ricreationedello00buon/ page/n591.

146 **when they surely knew how to correct them:** Stephen Jay Gould, *Dinosaur in a Haystack* (Three Rivers Press, 1995), 214.

147 **mastered etching and engraving:** In *Martin Lister and His Remarkable Daughters*, Roos includes a fascinating discourse on the demands of etching and engraving on copper plates; see 104–12.

147 **"formed a new era in the science":** Thomas Thomson, *History of the Royal Society: From Its Institution to the End of the Eighteenth Century* The Royal Society, 1812), 88.

147 **Banister's last letter to Lister:** Martin Lister and John Banister, "The Extracts of Four Letters from Mr. John Banister to Dr. Lister, Communicated by Him to the Publisher," *Philosophical Transactions of the Royal Society* 17, no. 198 (December 31, 1693): 672.

147 **a patron of Banister's science:** Joseph and Nesta Ewan, "John Banister, Virginia's First Naturalist," *Banisteria* no. 1 (1992): 4.

147 **Banister was accidentally shot and killed:** Nesta Dunn Ewan and Joseph Ewan, "Banister, John (1649 or 1650–1692)," *Dictionary of Virginia Biography* 1, John T. Kneebone et al., eds. (Richmond: Library of Virginia, 1998), 313–15.

147 **featured seventeen specimens sent by Banister:** Roos, *Martin Lister and His Remarkable Daughters*, 76.

147 **the first naturalist to describe and illustrate:** Lauck W. Ward and Blake W. Blackwelder, "*Chesapecten*, A New Genus of Pectinidae from the Miocene and

Pliocene of Eastern North America," U.S. Department of the Interior, Geological Survey Professional Paper 861, (U.S. Government Printing Office, 1975), 1.

147 **"the biggest Scallop I have ever seen":** Martin Lister, with credit to Susanna and Anna Lister on title plate, *Historiae sive Synopsis Methodicae Conchyliorum* (London, 1685), notes to illustration 167. Via Biodiversity Heritage Library, biodiversitylibrary.org.

147 **speculating it might be a grand Lion's Paw:** Martin Lister, *An Index to the Historia Conchyliorum of Lister* (Lister's notes to *Conchyliorum*, published in London by J. W. Dillwyn, 1823), 14. The blue-gray scallop was the first fossil described from America. Martin wrote that "it appears to me to be a variety of *Ostrea nodosa.*" This was Linnaeus's name for what is today *Nodipecten nodosus,* the much-loved Lion's Paw.

147 **The colossal bivalve:** Ward and Blackwelder, *"Chesapecten,* A New Genus of Pectinidae," 5.

148 **the concept was still anathema:** Mark V. Barrow, *Nature's Ghosts: Confronting Extinction from the Age of Jefferson to the Age of Ecology* (Chicago: University of Chicago Press, 2009). 42–46.

148 **"very much of ye Galant & honest man":** Alan Cutler, *The Seashell on the Mountaintop* (Dutton, 2003), 134.

148 **"My argument will fall":** Martin J.S. Rudwick, *The Meaning of Fossils: Episodes in the History of Palaeontology* (University of Chicago Press, first edition 1972; American edition 1985), 63.

149 **a large monogram of the king embellished with shells:** Dance, *Shell Collecting,* 54.

149 **Expectant mothers would leave an offering:** Jennifer Larson, *Greek Nymphs: Myth, Cult, Lore* (Oxford University Press, 2001), 251.

149 **transformed rustic altars into hydraulic edifices:** Brenda Longfellow, *Roman Imperialism and Civic Patronage: Form, Meaning and Ideology in Monumental Fountain Complexes* (Cambridge University Press, 2011), 113.

149 **New generations piled on:** Riccardo Cattaneo-Vietti, *Man and Shells: Molluscs in the History* (Bentham Books, 2016), 89.

150 **"swallowed vast quantities of shells":** Hazelle Jackson, *Shell Houses and Grottoes* (Shire Publications, 2001), 11.

150 **the Grotto of Tethys:** "Versailles on Paper," exhibit, Princeton University Library, rbsc.princeton.edu/versailles/grotto.

151 **including a hydraulic organ:** André Félibien, *Description de la Grotte de Versailles* (France: 1672), Princeton University Library, rbsc.princeton.edu/versailles/item/862.

151 **some of history's most famous shell art:** François Chauveau, Etching, *Masques de coquillages et de rocaille.* Plate 15 in *Description de la Grotte de Versailles,* 1675.

151 **Only the marble sculptures were preserved:** "Versailles on Paper," exhibit, Princeton University Library, rbsc.princeton.edu/versailles/grotto.

151 **encrusted with thousands of seashells:** Geri Walton, *Marie Antoinette's Confidante: The Rise and Fall of the Princesse de Lamballe* (Pen and Sword Books, 2016), 145.

151 **The duke lost Rambouillet, and the princess her shell cottage:** Ibid., 146.

152 **They say she was hiding in her grotto:** Annette Condello, *The Architecture of Luxury* (Routledge, 2016), 83–84.

152 **London society flocked to see the spectacle:** Jackson, *Shell Houses and Grottoes*, 13.

152 **"Were it to have nymphs as well":** letter from Pope to his friend Edward Blount, quoted in Tom Brigden, *The Protected Vista: An Intellectual and Cultural History* (Routledge, 2019), 30.

152 **one of the great works of early modern feminism:** Hannah Smith, "English 'Feminist' Writings and Judith Drake's *An Essay in Defence of the Female Sex* (1696)," *Historical Journal* 44, no. 3 (2001): 727.

153 **amid almost constant prosecution:** Ibid., 737–38.

153 **from "A Lady":** Judith Drake, *An Essay in Defence of the Female Sex* (London, 1696), Gutenberg.org edition, published 2018.

154 **"she only treated women and children":** Ruth Watts, *Women in Science: A Social and Cultural History* (Routledge, 2007), 65.

154 **He also refused Queen Louisa's offer:** Londa Schiebinger, "Gender and Natural History," in N. Jardine et al., eds., *Cultures of Natural History* (Cambridge University Press, 1996), 169.

154 **A "tyranny of custom":** Mark Knights, *The Devil in Disguise: Deception, Delusion and Fanaticism in the Early English Enlightenment* (Oxford University Press, 2011), 124.

155 **The first published print of a Precious Wentletrap:** Dance, *Shell Collecting*, 228.

155 **a town more famous for . . . the brothers Grimm:** George Sarton, "Rumphius, *Plinius Indicus* (1628–1702)." *Isis* 27, no. 2 (1937): 242–57.

155 **"Burning with an insatiable desire":** Georgius Everhardus Rumphius, *The Ambonese Curiosity Cabinet*, translated, edited, annotated, and with an introduction by E. M. Beekman (Yale University Press, 1999), xlii.

155 **He'd been deceived:** Georg Eberhard Rumpf, *The Poison Tree: Selected Writings of Rumphius on the Natural History of the Indies*, ed. and trans. E. M. Beekman (University of Massachusetts Press, 1981), 2.

155 **where he spent three years as a soldier:** Ibid.

156 **"begun a work that describes":** Rumphius, *The Ambonese Curiosity Cabinet*, lxvi.

156 **"my first Companion and Helpmate":** Georg Eberhard Rumpf, *Rumphius' Orchids*, editor and translator, E. M. Beekman (Yale University Press, 2003), 86.

157 **"dazzling array of lyrical metonymies":** Rumpf, *The Poison Tree*, 29.

157 **malacology's true pioneer:** Dance, *Shell Collecting*, 49.

157 **He was the first to describe a living nautilus:** Beekman in Rumphius, *The Ambonese Curiosity Cabinet*, xcviii.

157 **declaring him *Plinius indicus*:** Sarton, "Rumphius, *Plinius Indicus* (1628–1702)," 242–57.

157 **"Today's covetousness and pomp":** Rumphius, *The Ambonese Curiosity Cabinet*, 317.

157 **also came to accept its ethos:** Conniff, *The Species Seekers*, 81.

158 **"they will only be lucky for the one who found them":** Rumphius, *The Ambonese Curiosity Cabinet*, 328.

158 **He blamed his affliction:** Ibid., lxviii.

158 **"terrible misfortune":** Ibid., lxviii.

158 **"They got the name Kinkhooren":** Ibid., 134.

159 **"What vexed Rumphius":** Beekman in Rumphius, *The Ambonese Curiosity Cabinet*, civ–cv.

SEVEN: AMERICAN SHELLS

163 **agreed to a detour:** Patricia Tyson Stroud, *Thomas Say: New World Naturalist* (University of Pennsylvania Press, 1992), 163.

163 **amid a fervor of scientific patriotism:** Robert McCracken Peck and Patricia Tyson Stroud, *A Glorious Enterprise: The Academy of Natural Sciences of Philadelphia and the Making of American Science* (University of Pennsylvania Press, 2012), 9.

163 **first American research institution devoted solely to natural history:** Simon Baatz, "Philadelphia Patronage: The Institutional Structure of Natural History in the New Republic, 1800–1833," *Journal of the Early Republic* 8, no. 2 (Summer 1988): 122.

163 **INSECTS, as he once wrote in all caps:** Harry B. Weiss and Grace M. Weiss, *Thomas Say, Early American Naturalist* (C. C. Thomas, 1931), 53.

164 **Mollusks came second:** Ibid., 53.

164 **Say and Sistare were two of the few on the boat who enjoyed it:** Stroud, *Thomas Say*, 179.

164 **He also planned a library and printing press:** Ibid., 188.

164 **Revolutions created:** Bryony Onciul, Museums, *Heritage and Indigenous Voice: Decolonizing Engagement* (Routledge, 2015), 1-25.

165 **Bartram's Boxes:** Caroline Winterer, *American Enlightenments: Pursuing Happiness in the Age of Reason* (Yale University, 2016), 56.

165 **The Quaker collector explored the hazy peaks:** John Bartram, map, "Middle Atlantic states, showing rivers and mountains and location of sea shells on the tops of mountains," circa 1750, American Philosophical Society Library, Philadelphia.

166 **" 'Tis certainly the Wreck of a World":** Ibid.

166 **American scientists were trying to build their own authority:** Charlotte M. Porter, "The Concussion of Revolution: Publications and Reform at the Early Academy of Natural Sciences, Philadelphia, 1812–1842," *Journal of the History of Biology* 12, no. 2 (Autumn 1979): 273.

166 **South Carolina colonists established America's first museum:** William G. Mazyck, "History of the Museum, The Period Previous to 1798," *Bulletin of The Charleston Museum* 3, no. 6 (October, 1907): 49–51.

166 **Enslaved Africans had unearthed mammoth teeth:** Albert E. Sanders and

William D. Anderson Jr., *Natural History Investigations in South Carolina from Colonial Times to the Present* (University of South Carolina Press, 1999), 12–13.

166 **It would be more than eighty years:** Adrienne Mayor, *Fossil Legends of the First Americans* (Princeton University Press, 2005), 56.

166 **"Athens of America":** Benjamin Franklin, in "Humble Address of the Directors of the Library Company of Philadelphia" to patron Thomas Penn, May 16, 1733, *National Archives*, "Founders Online," https://founders.archives.gov.

167 **about one in ten Philadelphians:** William Souder, *Under a Wild Sky: John James Audubon and the Making of The Birds of America* (Milkweed Editions, 2014 edition), 47–48.

167 **"would be aroused by the clatter of knives and forks":** Stroud, *Thomas Say*, 22.

167 **"disdain of riches":** Ibid., 25.

167 **"The thrift of trade":** George Ord, "A Memoir of Thomas Say," in *The Complete Writings of Thomas Say*, Vol. 1 (Bailliere Bros, 1859), viii.

167 **Ord claimed that this "injudicious application to study":** Ibid., xix.

168 **They envisioned a research enterprise more democratic:** Simon Baatz, "Philadelphia Patronage: The Institutional Structure of Natural History in the New Republic, 1800–1833," *Journal of the Early Republic* 8, no. 2 (Summer 1988): 112.

168 **"dogmatic reproof, persecution, or proselytism":** Ibid., 118.

168 **tossing a blanket over a horse skeleton:** Stroud, *Thomas Say*, 40.

168 **he wished for a hole in his side:** Ibid., 40–41.

168 **Say wrote a friend of this patriotic obligation:** Porter, "The Concussion of Revolution," 274.

169 **constantly traveling between Europe and America:** John S. Doskey, *The European Journals of William Maclure* (American Philosophical Society, 1988), xviii.

169 **"I am mortified beyond measure":** Maclure, quoted in J. Percey Moore, "William Maclure—Scientist and Humanitarian," *Proceedings of the American Philosophical Society* 91, no. 3 (August 1947): 235.

169 **"the promised land":** Thomas Say, "A Letter of the Distinguished Naturalist, Thomas Say, to Rev. J. F. Melsheimer," *Journal of the Linnaean Society* (June 10, 1818): 37.

169 **They explored Georgia's barrier islands:** Ibid.

169 **"filled with multitudes of beautiful Medusae":** Titian Peale, quoted in L. Peale, "A Visit to Florida in the Early Part of the Century," manuscript, American Museum of Natural History digital repository. digitallibrary.amnh.org/handle/2246/6166

169 **Aesthetic-minded Peale found it "incongruous, almost ludicrous":** Ibid.

170 **In Florida they sailed through dense dolphin pods:** Ibid.

170 **The concept that species could die out:** Elizabeth Kolbert, *The Sixth Extinction: An Unnatural History* (Henry Holt, 2014), 36–43.

170 **"this most cruel & inhuman war":** Say, "A Letter of the Distinguished Naturalist," 38.

170 **"The animals that inhabit them should guide us":** Thomas Say, "Conchology," in William Nicholson, *The British Encyclopedia: Or, Dictionary of Arts and Sciences*, Vol. 4 (London: The Philosophical Journal, 1819), A-4.

171 **"so that their souls might easily return to Africa":** "Low Country Gullah Culture Special Resource Study," National Park Service, U.S. Department of the Interior, December 2003, 76.

171 **"The shells stand for the sea":** Ibid., 221.

171 **who saw seashells as evidence of the hand of God in nature:** Lester D. Stephens, *Science, Race, and Religion in the American South* (University of North Carolina Press, 2000), 67.

171 **Distinguishing the American Lettered Olive from Lamarck's *Oliva litterata*:** Ravenel, 1834, Malacolog Version 4.1.1, the Academy of Natural Sciences, malacolog.org.

172 **But Ravenel didn't have a daughter named Annabel:** Henry Ravenel, *Ravenel Records: A History and Genealogy of the Huguenot Family of Ravenel of South Carolina* (Franklin Printing, 1898), 157–58.

172 **Poe biographer Arthur Hobson Quinn found it "probable":** Arthur Hobson Quinn, *Edgar Allan Poe: A Critical Biography* (Johns Hopkins University Press, 1941; new paperback edition 1998), 130.

172 **He hired Poe for $50:** Stephen Jay Gould, "Poe's Greatest Hit," *Natural History* 102, no. 7 (1993).

173 **"It is grimly ironic, however":** Quinn, *Edgar Allan Poe*, 277.

173 ***The Conchologist's First Book* was one of the earliest to contrast malacology:** Gould, "Poe's Greatest Hit."

175 **as a curious detour:** Stroud, *Thomas Say*, 173.

175 **Marie Fretageot, a force:** Carol A. Kolmerten, *Women in Utopia: The Ideology of Gender in the American Owenite Communities* (Syracuse University Press, 1998), 143.

175 **one of ten children:** Harry B. Weiss and Grace M. Ziegler, "Mrs. Thomas Say," *Journal of the New York Entomological Society* 41, no. 4 (December 1933): 554.

175 **which included hands-on natural history at Peale's museum:** Leonard Warren, *Maclure of New Harmony: Scientist, Progressive Educator, Radical Philanthropist* (Indiana University Press, 2009), 97.

175 **"the singular spectacle which this place presented":** Stroud, *Thomas Say*, 183.

176 **"the trinity of the most monstrous evils":** Chris Jennings, *Paradise Now: The Story of American Utopianism* (Random House, 2016), 137.

176 **a first-rate school that enrolled four hundred children its first year:** Ibid., 140.

176 **Maclure ultimately convinced Owen to transfer ownership:** Ibid, 140.

176 **They were "shut up in their cabinet":** Stroud, *Thomas Say*, 193.

177 **a gentle treatise on nature study:** John Craig, ed., with Julia E. Rogers, assisting, "Nature Study Outlines for the Use of Teachers in the State," Iowa State Horticultural Society and State Agricultural College, Ames, Iowa, 1899. Rog-

ers wrote the introduction for teachers and the chapter on grasshoppers. In his annual report of 1899, the president of the Horticultural Society described the project as an important joint effort of the society, agricultural college, and State Superintendent of Public Instruction to bring nature study to the schools of Iowa.

177 **they would develop good character and an ethos:** Kevin Armitage, *The Nature Study Movement: The Forgotten Popularizer of America's Conservation Ethic* (University of Kansas Press, 2009), 3.

177 **"To feel an intimacy":** Rogers, quoted in Craig, "Nature Study Outlines for the Use of Teachers in the State," 1899, 4.

178 **"Let the books alone for a while":** Ibid.

178 **one of her many shell-collector fans described her:** Dorma P. Goley, "Julia Ellen Rogers," *Pacific Northwest Shell News* III, no. 4 (July 1963): 41. Rogers died in May 1958 in Long Beach, but her fans continued to mourn her for many years.

178 **Ed Ricketts twice laments in his journal:** Katharine A. Rodger, *Breaking Through: Essays, Journals, and Travelogues of Edward F. Ricketts* (University of California Press, 2006), see 147 and 163.

179 **The best-known shell man of the twentieth century:** Goley, "Julia Ellen Rogers," 41.

179 **"I spent too much time outdoors":** Richard Wolkomir, "Seeking Gifts from the Sea, Sanibel-style," *Smithsonian* 26, no. 5 (August 1995).

179 **"Pedagogue and Dress Reformer":** *The Hawkeye*, Junior Annual, Vol. 2, 1893, page 200. Rogers graduated from Iowa State in 1892, but this remembrance of seniors is included in the 1893 annual.

179 **two-story Craftsman house . . . where she lived:** City of Long Beach, Ordinance 17-0016, designating the property at 355 Junipero Avenue as a historic landmark, p. 2.

180 **Her grandfather was also known for his poetic nature writing:** Nathaniel Peabody Rogers, "Letter from the Old Man of the Mountain," in Crispin Sartwell, ed., *Herald of Freedom: Essays of Nathaniel Peabody Rogers, American Transcendentalist and Radical Abolitionist* (CreateSpace, 2016), 69.

180 **"the glory of the countryside":** *Autobiography of Daniel Farrand Rogers*, unpublished memoir, Michael J. Spinelli Jr. Center for University Archives and Special Collections, Plymouth State University, gift of Sarah Kinter of Canterbury, NH, 1998, undated.

180 **before landing as a high school principal in Minnesota:** "Julia Ellen Rogers," *The Arrow of Pi Beta Phi* 23 (November 1906): 204–207.

180 **She began her second career in double byline:** Daniel F. and Julia E. Rogers, "Camping and Climbing in the Big Horn," *Midland Monthly* 6, no. 2 (August 1896): 99–107.

180 **he optimistically envisioned America building out in three phases:** L. H. Bailey, *The Holy Earth* (Charles Scribner's Sons, 1916), 22.

181 **"At first man sweeps the earth":** Ibid.

181 **"we begin to enter the productive stage":** Ibid, 22–23.

181 **"earth righteousness":** Ibid, 23–24.

181 **Comstock collected the classic essays:** Karen Penders St. Clair, ed., and Anna Botsford Comstock, *The Comstocks of Cornell: The Definitive Autobiography* (Cornell University Press, 2020), see epilogue.

181 **completing her thesis on how to teach nature studies through winter:** Julia E. Rogers, "Materials for Winter Work in Nature Study," master's thesis, College of Agriculture, Cornell University, Ithaca, NY, 1902.

182 **a phenomenon scientists reported on for the first time more than a century later:** Eetu Puttonen et al., "Quantification of Overnight Movement of Birch (*Betula Pendula*) Branches and Foliage with Short Interval Terrestrial Laser Scanning," *Frontiers in Plant Science*, February 29, 2016. The authors describe this as the first study to report spatial changes in tree branch geometry over a daylong cycle.

182 **It was her fellow nature writer Anthony Dimock:** Rogers was researching seashells by 1904, the year she published "The Common Shells of the Seashore and the Queer Creatures that Live Inside," *Country Life Magazine* 6, July 1904.

182 **to his son's everlasting horror:** A. W. Dimock, *Wall Street and the Wilds* (Outing Publishing, 1915), 13.

182 **eschewed college for Wall Street:** Dimock's *New York Times* obituary has him taking a degree, but he wrote in his autobiography that his education ended at Andover. Dimock, *Wall Street and the Wilds*, 27.

182 **"dominated the gold market":** "A. W. Dimock Dead; Financier-Author," *New York Times*, September 13, 1918.

183 **by a fourth bankruptcy and fraud charges:** Jerald T. Milanich and Nina J. Root, *Hidden Seminoles: Julian Dimock's Historic Florida Photographs* (University Press of Florida, 2011), 12.

183 **"so near that it was wicked to shoot it":** Dimock, *Wall Street and the Wilds*, 426.

183 **"To get within camera range":** A. W. Dimock, "Camera vs. Rifle," in Edward L. Wilson, ed., *Photographic Mosaics: An Annual Record of Photographic Progress*, Vol. 26, (Edward L. Wilson, 1890), 75.

183 **a "long and leisurely summer cruise":** Rogers, *The Shell Book*, viii.

183 **built especially for him:** Anthony Weston Dimock, *The Book of the Tarpon* (Outing Publishing, 1911), 60.

183 **Rogers hooked her share of tarpon:** Julia Ellen Rogers, talk, "Minutes of the Conchological Club of Southern California," October 1944, 28.

183 **"On those tide-washed shores":** Rogers, *The Shell Book*, viii.

184 **"gaily painted shells, full of life":** Ibid., 412.

184 **his father's view that the mangroves and the oysters:** A. Dimock and J. Dimock, *Florida Enchantments* (Hodder and Stoughton, 1909), 290.

184 **ninety-two of Julian's photographs:** Jack E. Davis, *The Gulf: The Making of An American Sea* (Liveright, 2017), 166.

185 **refers to Rogers only as "the tree lady":** see A. W. Dimock, "Cruising on the Gulf Coast of Florida," *Harper's*, December 1, 1906.

185 **A Cornell researcher and nature-study historian:** Anna Botsford Comstock, *The Comstocks of Cornell: The Definitive Autobiography*, ed. Karen Penders St. Clair (Cornell University Press, 2020), editor's commentary and epilogue.

185 **The published version:** That would be Glenn W. Herrick and Ruby Green Smith, eds., Anna Botsford Comstock, author, *The Comstocks of Cornell: John Henry Comstock and Anna Botsford Comstock* (Cornell University Press, 1953).

186 **Nature study became part of public education in every state:** Armitage, *The Nature Study Movement*, 3–4.

186 **"the corporate values of mass consumption and commercialized leisure":** Ibid., 197.

186 **disputes between scientists and schoolteachers in the early twentieth century:** Nathalie op de Beeck, "Children's Ecoliterature and the New Nature Study," *Children's Literature in Education*, February 2018, 77. Op de Beeck writes that "The movement was characterized by gendered debates around whether schoolteachers, coded female, could practice the scientific method objectively."

186 **"many modern nature books suffer from what might be called effeminization":** Kim Tolley, *The Science Education of Girls* (Routledge, 2003), 183.

186 **on the Long Beach Board of Education for a decade:** City of Long Beach, Ordinance 17-0016. This historic designation for Rogers's home lists her board service from 1918 to 1928.

186 **In a nature study program:** "Talk on Nature," *The Long Beach Telegram*, November 15, 1921.

186 **Rachel Carson's mother, Maria:** Linda Lear, *Witness for Nature* (Houghton Mifflin Harcourt, 1997), 14.

187 **He raved about "bird study":** Curt Meine, *Aldo Leopold: His Life and Work* (University of Wisconsin Press, 2010), 17.

187 **very likely with Julia Ellen Rogers's lessons plans:** Personal communication with the scientist and nature study historian Stanley A. Temple, Senior Fellow and Science Advisor, The Aldo Leopold Foundation.

187 **"his old enemy, the bilious":** Stroud, *Thomas Say*, 210–11.

187 **prepared the plates from her drawings:** Patricia Tyson Stroud, "'At What Do You Think the Ladies Will Stop?'" Women at the Academy, *Proceedings of the Academy of Natural Sciences of Philadelphia* 162 (March 2013): 196.

187 **Fretageot's letters, revealing friction:** Vera Norwood, *Made From This Earth: American Women and Nature* (University of North Carolina, 1993), 65.

188 **The other two are by the naturalist Lesueur:** Stroud, *Thomas Say*, 254.

188 **"first attempt, August 1834":** Norwood, *Made From This Earth*, 65.

189 **George W. Tryon Jr. worked in the music publishing business:** W. S. W. Ruschenberger, *A Biographical Notice of George W. Tryon Jr., Conservator of Conchological Section of the Academy of Natural Sciences of Philadelphia* (H. Binder, 1888), 9.

190 **He'd planned to describe all mollusks:** Ibid.

190 **saw Pilsbry's brilliance in taxonomy:** Horace Burrington Baker, "The Pilsbry Nautilus," *The Nautilus* 71, no. 3 (January 1958): 112–15.

190 **"I fell upon his shell":** Ibid., 247.

191 **believed that natural science was a means for uniting humanity:** Hideo
Mohri, *Imperial Biologists: The Imperial Family of Japan and Their Contributions
to Biological Research* (Springer, 2019), 70.

191 **His telegraph asked:** The American Malacological Union Annual Report of
1952, 19.

EIGHT: SHELL OIL

193 **opened a small curio shop in Sailor's Town:** Robert Henriques, *Bearsted: A
Biography of Marcus Samuel, First Viscount Bearsted and Founder of 'Shell' Transport and Trading Company* (Viking Press, 1960), 16–17.

193 **Marcus and Abbie both grew up in Jewish merchant families:** Ibid., 13–14.

193 **"small Shells for Ladies' Work":** "Official Catalogue of the Educational Exhibition," St. Martin's Hall, Society for the Encouragement of Arts, Manufac-
tures, and Commerce, London, July 4, 1854, 140.

194 **a craze in Victorian England propelled by Her Majesty:** Lauren Miskin,
"The Victorian 'Cameo Craze'," *Victorian Review* 42, no. 1 (Spring 2016): 176.

194 **"that might have been designed in dreams":** Charles Dickens, *The Old Curiosity Shop* (Pollard & Moss, 1885), 5.

194 **during a beach holiday at Margate:** Isaac F. Marcosson, *The Black Golconda:
The Romance of Petroleum* (Harper & Brothers, 1923), 72. Marcosson tells the
obviously apocryphal story that the Samuel children invented the Victorian
shell box at Margate. It makes sense that Samuel hit upon the idea to sell shell
boxes as souvenirs while at Margate.

194 **crown pincushions in honor of Queen Victoria:** Carole and Richard Smyth,
Neptune's Treasures (Carole Smyth Antiques, 1998), 77. The Smyths, experts in
dating Victorian shell craft, credit Samuel with manufacturing crown pincushions popular throughout Queen Victoria's reign.

194 **"from near-poverty to relative wealth":** Henriques, *Bearsted*, 21.

195 **had dominated the region since its Charter:** The British East India Company was chartered in 1601; the Dutch East India Company—the VOC that
employed Rumphius—was chartered in 1602.

195 **its first mechanical looms:** Henriques, *Bearsted*, 32.

195 **a "curiosity dealer" in the London business directory of 1841:** Mark West-
garth, *A Biographical Dictionary of Nineteenth Century Antique and Curiosity
Dealers* (The Regional Furniture Society, 2009), 163.

195 **amounted to . . . nearly £5 million in today's:** Henriques, *Bearsted*, 37. (Henriques reported £40,000 in 1870 currency, worth an estimated £4.8 million in
2020, via the UK Consumer Price Index inflation calculator, www.officialdata/
UK-inflation.org.)

195 **A quarter of it was in seashells:** Ibid., 39.

196 **according to their father's will:** Ibid., 39.

196 **Abbie made her venturous middle son work for Joseph:** Ibid., 38.

196 **an uncommonly hot monsoon season:** Cornelius Walford et al., "The Fam-

ines of the World: Past and Present," *Journal of the Statistical Society of London* (1878): 17.

197   **India's colonial government had put up £3 million:** Peter B. Doran, *Breaking Rockefeller: The Incredible Story of the Ambitious Rivals Who Toppled an Oil Empire* (Penguin, 2017), 30. (Doran reported £3 million in 1873 currency, which is worth an estimated £300 million in 2020, via the UK CPI inflation calculator www.officialdata/UK-inflation.org.)

197   **"his father's name and his brother's credit":** Henriques, *Bearsted*, 42.

197   **almost all merchants were importing:** Doran, *Breaking Rockefeller*, 31.

197   **bought low, sold high:** Ibid., 31.

197   **The first global capitalists:** Michael C. Howard, *Transnationalism in Ancient and Medieval Societies: The Role of Cross-Border Trade and Travel* (McFarland & Co., 2012), 128.

197   **They acquired common items:** Mark Cartwright, "Trade in the Phoenician World," *Ancient History Encyclopedia*, April 2016, ancient.eu/article/881/.

197   **Jewelry from Greece, linen from Egypt:** Joe Carlin, *A Brief History of Entrepreneurs: The Pioneers, Profiteers, and Racketeers Who Shaped Our World* (Columbia University Press, 2016), 36–38.

197   **They built some of the earliest networks to enslave:** Dierk Lange, *Ancient Kingdoms of West Africa* (J. H. Röll, 2004), 278.

198   **derived from the word for purple:** Michael C. Astour, "The Origin of the Terms 'Canaan,' 'Phoenician,' and 'Purple'," *Journal of Near Eastern Studies* 24, no. 4 (October 1965): 348.

198   **"the loveliest, most exquisite":** S. Peter Dance, *Rare Shells* (Faber and Faber, 1969), 75.

198   **"the ribs of a fish picked clean":** R. Tucker Abbott, *Kingdom of the Seashell*, (Bonanza Books, 1982), 155.

199   **some type of biochemical block:** Winston F. Ponder et al., *Biology and Evolution of the Mollusca*, Volume 1 (CRC Press, 2020), 153.

199   **Scientists find the precursor to purple:** Ibid.

199   **Each is made up of a hundred or more capsules:** Mehmet Güler and Aynur Lök, "Embryonic Development and Intracapsular Feeding in *Hexaplex trunculus*," *Marine Ecology* 35, no. 2 (2014): 193–203.

199   **also survives on textiles:** Alessandro Ciccola et al., "Dyes from the Ashes: Discovering and Characterizing Natural Dyes from Mineralized Textiles," *Molecules* 25, no. 6 (March 2020).

199   **Pacific peoples used large sea snails:** Riccardo Cattaneo-Vietti, *Man and Shells: Molluscs in the History* (Bentham Books, 2016), 96.

199   **Greasy deposits burned into shells:** Kerlijne Romanus et al., "Brassicaceae Seed Oil Identified as Illuminant in Nilotic Shells from a First Millennium A.D. Coptic Church in Bawit, Egypt," *Analytical & Bioanalytical Chemistry* 290, no. 2 (January 2008): 783–93.

200   **"born from the sand":** Mao Huahe, *The Ebb and Flow of Chinese Petroleum: A Story Told by a Witness* (Brill, 2019), 4.

200   **Drake scrambled for empty whiskey barrels:** Doran, *Breaking Rockefeller*, 12.

200 **Tarbell's Tank Shops:** Daniel Yergin, *The Prize: The Epic Quest for Oil, Money & Power* (Free Press, 2008), 87.

201 **90 percent of America's oil industry:** Ibid., 56.

201 **The first bulk oils arrived:** Stephen Howarth, *SeaShell: The Story of Shell's British Tanker Fleets, 1892–1992* (Thomas Reed, 1992), 18.

201 **packing oil in 5-gallon rectangular tins:** Ibid., 20.

201 **had commissioned the world's first oil tanker, the *Zoroaster*:** Lincoln Paine, *The Sea and Civilization: A Maritime History of the World* (Vintage, 2013), 543.

201 **employing his famous strategy:** Doran, *Breaking Rockefeller*, 49–50.

201 **accepting a small slice of the market:** Ibid., 50–51.

201 **crammed to the ceiling with seashells:** Yergin, *The Prize*, 49.

202 **the leading British business concern in Japan:** Henriques, *Bearsted*, 159.

202 **half the country's annual rice exports:** Doran, *Breaking Rockefeller*, 53.

202 **Even Standard Oil could not possibly compete:** Howarth, *SeaShell*, 24.

202 **the idea man at the front of the company:** Henriques, *Bearsted*, 52.

202 **Oil soaked Baku's forest of wooden derricks:** Sarah Searight, "Region of Eternal Fire," *History Today*, August 1, 2000, 48–49.

203 **marvelously named marine engineer:** Howarth, *SeaShell*, 20.

203 **Paleolithic hunter-gatherers ate *H. trunculus*:** Catherine Perlès, *Ornaments and Other Ambiguous Artifacts from Franchthi Cave, Greece*, Vol. 1 (Indiana University Press, 2018), 70.

203 **by smashing "a great number of shells":** John Coakley Lettsom, *History of the Origin of Medicine: An Oration*, Medical Society of London (J. Phillips for E. & C. Dilly, 1778), as told by the Greek scholar Pollux, 113.

204 **The palatial beachfront complex is paved with slabs and drainage channels:** Deborah Ruscillo, "Reconstructing Murex Royal Purple and Biblical Blue in the Aegean," in Daniella E. Bar-Yosef Mayer, ed., *Archaeomalacology: Mollusks in Former Environments of Human Behavior* (Oxbow Books 2005), 100.

204 **She grew weary:** Personal communication with Deborah Ruscillo.

204 **Crushing murex and concocting dye through the summer:** Ruscillo, "Reconstructing Murex Royal Purple and Biblical Blue in the Aegean," 100–106.

205 **"divine the minds of the Canal authorities":** Howarth, *SeaShell*, 25.

205 **designed the rest of the ship's features to protect them:** Ibid.

205 **The rest of the oil world learned about the *Murex*:** Ibid.

206 **Standard Oil hired London solicitors to sow doubt:** Ron Chernow, *Titan: The Life of John D. Rockefeller, Sr.* (Vintage, 2004), 248.

206 **"purely of Hebrew inspiration":** "Petroleum in Bulk and the Suez Canal," *The Economist*, January 9, 1892, 36–37.

206 **efforts to cancel the canal authorities' permission failed:** Joost Jonker and Jan Luiten van Zanden, *From Challenger to Joint Industry Leader, 1890–1939, A History of Royal Dutch Shell*, Vol. 1 (Oxford University Press, 2007), 42.

206 **"status for himself and his family":** Henriques, *Bearsted*, 87.

206 **all named for seashells in his father's honor:** Ibid., 119.

206 **sixty-five of those were on ships named for seashells:** Howarth, *SeaShell*, 29.

207 **sold the family shell-box business to nephews:** Henriques, *Bearsted*, 347–48.

207 **changed the common meaning:** Ibid., 88. Due to what was perhaps a sublim-inal typo in his book, Henriques's sentence reads, "meaning of the *world* shell" instead of the *word* shell.

207 **female murexes were growing penises:** V. Axiak et al., "Imposex in *Hexaplex trunculus*: First Results from Biomonitoring of Tributyltin Contamination in the Mediterranean," *Marine Biology* 121 (1995): 685–91.

207 **were found sterile:** Ameer Abdulla and Olof Linden, eds., *Maritime Traffic Effects on Biodiversity in the Mediterranean Sea*, Vol. 1 (IUCN Global and Med-iterranean Marine Programme, 2008), 30.

207 **a pioneer in the study of metal–carbon bonds:** Johann T.B.H. Jastrzebski and Gerard Van Koten, "Intramolecular Coordination in Organotin Chemistry," in F.G.A. Stone and Robert West, eds., *Advances in Organometallic Chemistry, Vol. 35* (Academic Press, 1993); see the dedication to Dr. van der Kerk on the occa-sion of his eightieth birthday, 241.

207 **a 1-millimeter layer of algae can slow a ship's speed 15 percent:** Deloitte, "Tributyltin and Booster Biocides: A Socio-Economic Impact Assessment of NERC Funded Research," Natural Environmental Research Council, July 2016, 10.

208 **loss of an oyster predator was considered acceptable:** Deloitte, 10.

208 **TBT was implicated in reproductive mayhem:** Jakob Strand et al., "TBT Pollution and Effects in Molluscs at U.S. Virgin Islands, Caribbean Sea," *Envi-ronment International* 35 (2009): 707–11.

208 **Only after TBT was found to deform shells:** David Santillo, Paul Johnston, and William J. Langston, "TBT Antifoulants: A Tale of Ships, Snails and Impo-sex," in Poul Harremös et al., eds., *The Precautionary Principle in the 20th Cen-tury* (European Environment Agency, 2002), 153–54.

208 **organotins may threaten human health, too:** Vinicius Bermond Marques et al., "Overview of the Pathophysiological Implications of Oranotins on the Endocrine System," *Frontiers in Endocrinology* 9 (March 2018).

208 **are recovering thanks to the ban:** Youssef Lahbib et al., "First Assessment of the Effectiveness of the International Convention on the Control of Harmful Anti-Fouling Systems on Ships in Tunisia Using Imposex in *Hexaplex trunculus* as Biomarker," *Marine Pollution Bulletin* 128 (January 2018): 17–23.

209 **Neither the ancient exploitation nor the modern fishery:** Kamel Elhansi et al., "Harvesting and Population Status of *Hexaplex trunculus* in Intertidal Areas Along the Gulf of Gabès," *Journal of Coastal Conservation* 22 (2018): 347–60. The researchers stress that nevertheless, this important fishery needs a closed season to protect the murexes during their spawning and breeding peak.

209 **expected to mold into geologic strata:** Jan Zalasiewicz et al., "The Geological Cycle of Plastics and Their Use as a Stratigraphic Indicator of the Anthropo-cene," *Anthropocene* 13 (March 2016).

209 **now eddy more than 5 trillion:** Marcus Eriksen et al., "Plastic Pollution in the World's Oceans: More than 5 Trillion Plastic Pieces Weighing over 250,000 Tons Afloat at Sea," *PLoS ONE* 9, no. 12 (December 2014).

209 **carry our plastic waste even to far-flung tropical islands:** Jennifer L. Lavers et al., "Significant Plastic Accumulation on the Cocos (Keeling) Islands, Australia," *Scientific Reports* 9, no. 1 (2019).

209 **more than 500,000 of the crustacean boarders:** Jennifer L. Lavers et al., "Entrapment in Plastic Debris Endangers Hermit Crabs," *Journal of Hazardous Materials* 387 (November 2019).

209 **called "garbage season":** Amanda Tazkia Siddharta, "Bali Fights for its Beautiful Beaches by Rethinking Waste, Plastic Trash," *National Geographic*, October 14, 2019.

209 **albatross chicks die with stomachs full:** Kenneth R. Weiss, "Altered Oceans Part Four: Plague of Plastic Chokes the Seas," *Los Angeles Times*, August 2, 2006.

210 **average eleven bits of microplastic:** Britta Baechler et al., "Microplastic Concentrations in Two Oregon Bivalve Species: Spatial, Temporal, and Species Variability," *Limnology and Oceanography Letters* 5, no. 1 (November 2019).

210 **Blue mussels in Arctic outposts thought immune:** Alister Doyle, "Plastic Found in Mussels from Arctic to China—Enters Human Food," *Reuters*, December 20, 2017.

210 **some of the highest concentrations of plastic in the world:** Giuseppe Suaria et al., "The Mediterranean Plastic Soup: Synthetic Polymers in Mediterranean Surface Waters," *Scientific Reports* 6 (November 2016).

210 **the sixth-largest accumulator:** Ibid.

210 **scientists analyzed six economically important mollusks:** Sami Abidli et al., "Microplastics in Commercial Molluscs from the Lagoon of Bizerte," *Marine Pollution Bulletin* 142 (May 2019).

210 **"an additional exposure route":** Ibid.

210 **into water and even air:** Kieran D. Cox et al., "Human Consumption of Microplastics," *Environmental Science and Technology*, June 5, 2019.

211 **Shell's contractors have relocated state highway and interchanges:** "Shell Pennsylvania Petrochemicals Complex," Hydrocarbons-Technology.com.

211 **Beginning in the 2020s, ethane gas will flow:** Michael Corkery, "A Giant Factory Rises to Make a Product Filling Up the World: Plastic," *New York Times*, August 12, 2019.

211 **Scientists have identified ethylene as the famous vapor:** John R. Hale et al., "Questioning the Delphic Oracle," *Scientific American*, August 2003, 67–73.

211 **two types of nurdles:** Explanations of the Pennsylvania Petrochemical Complex's processing units, products, and markets are from Shell; see www.shell.com/chemicals. Two polyethylene units will manufacture high-density polyethylene (HDPE) grades of pellets, and a third will produce linear low-density polyethylene grades (LLDPE).

211 **more than three thousand freight cars:** Corkery, "A Giant Factory Rises."

212 **often led the oil industry in developing petrochemicals:** Jonker and van Zanden, *A History of Royal Dutch Shell Vol. 1*; see especially pages 5 and 381.

212 **Shell held exclusive rights to what its chemists called:** Stephen Howarth and Joost Jonker, *A History of Royal Dutch Shell, Vol. 2* (Oxford University Press), 427.

212 **tackling the public relations problem rather than the pollution:** Ibid., 433.

212 **While the company's scientists insisted:** Ibid.

212 **"Markets have to be found":** Marcus Samuel Jr., "Liquid Fuel," remarks to the Society of Arts, London, March 15, 1889. In *Journal of the Society of Arts*, XLVII (1889): 385.

213 **had never been recycled:** Roland Geyer et al., "Production, Use, and Fate of All Plastics Ever Made," *Science Advances* 3, no. 7 (July 2017): advances.sciencemag.org.

213 **"extraordinary" rogue wave:** quote from Southampton harbor Captain Steven Young, "Storm-hit Ship Docks," *Southern Daily Echo*, February 15, 1997. The story includes a photograph of the 40-foot cargo containers fallen over like dominoes, some still hanging off the side of the *Tokio Express*.

213 **Her son still remembers:** The Lego Lost at Sea section is based on personal communication with Tracey Williams.

214 **The plastic pellets:** Zoë Schlanger, "Virgin Plastic Pellets Are the Biggest Pollution Disaster You've Never Heard Of," *Quartz*, August 19, 2019.

215 **A countywide marine debris database:** The Marine Debris Database is a project of the nonprofit organization Heal the Bay, healthebay.org.

NINE: SHELL SHOCK

217 **valentines when she was young:** Daniel Mark Epstein, *What Lips My Lips Have Kissed: The Loves and Love Poems of Edna St. Vincent Millay* (Henry Holt 2001), 153.

217 **"The very thought of the words *Conus gloriamaris*":** Edna St. Vincent Millay, February 1937 foreword, in Karl Yost, *A Bibliography of the Works of Edna St. Vincent Millay* (Harper & Bros, 1937), 67.

217 **"Earth-ecstatic":** Allan Ross Macdougall and Edna St. Vincent Millay, *The Letters of Edna St. Vincent Millay* (Harpers, 1952), 38.

218 **The couple checked in at the lodge:** Epstein, *What Lips My Lips Have Kissed*, 246.

218 **an exhausting and nerve-wracking time:** Ibid., 246.

218 **"I live in the strong though ebbing hope":** Yost, *A Bibliography of the Works of Edna St. Vincent Millay*, foreword by Millay, 67.

219 **"the Floridians and the conchological aliens":** Julia Ellen Rogers, *The Shell Book* (Doubleday, 1908), 83.

219 **Only four specimens were then recorded:** S. Peter Dance, *Rare Shells* (Faber and Faber, 1969), 92.

219 **Audubon painted Ravenel's shell:** William Gaillard Mazÿck, *Contributions from the Charleston Museum: Catalog of Mollusca of South Carolina* (The Charleston Museum, 1913), 10.

220 **"innumerable millions of shells":** Fritz Haas, "Ecological Observations on the Common Mollusks of Sanibel Island, Florida," *American Midland Naturalist* 24, no. 2 (September 1940): 369.

221 **Historians consider it the likely area:** Samuel Turner, "Juan Ponce de León

and the Discovery of Florida Reconsidered," *Florida Historical Quarterly* 92, no. 1 (Summer 2013): 22.

221 **The explorers kidnapped four Calusa women:** Ibid., 24.

221 **what a Spanish mariners' guide described as** *Costa de Carocoles*: Alonso de Chaves, cosmographer to the King of Spain, issued this guide in 1527. Elinore M. Dormer, *The Sea Shell Islands: A History of Sanibel and Captiva* (Rose Printing Co., 1987), 35.

222 **were likely the products of Anglo land boosters:** André-Marcel d'Ans, "The Legend of Gasparilla: Myth and History on Florida's West Coast," *Tampa Bay History* 2, no. 2 (Fall/Winter 1980): 5–25.

222 **set up the Florida Peninsular Land Company:** E. A. Hammond, "Sanibel Island and Its Vicinity, 1833," *Florida Historical Quarterly* 48, no. 4 (April 1970): 392–93.

222 **He complained that on his first night:** Ibid., 397–98.

222 **"On the south side is a beautiful sea beach":** Ibid., 399.

223 **"may be had with very little trouble":** Ibid., 408.

223 **His Sanibel dreams would come true:** Dormer, *The Sea Shell Islands*, 131.

223 **the iron tower lighthouse . . . and the first tarpon:** Betty Anholt, *Sanibel's Story: Voices and Images Calusa to Incorporation* (City of Sanibel, 1998), 19.

223 **Sanibel still rose with hulking Calusa mounds:** Jerald T. Milanich, *Florida's Indians from Ancient Times to the Present* (University Press of Florida 1998), 129.

223 **hauling the shell away:** Charles LeBuff and Betty Anholt, *Protecting Sanibel and Captiva Islands: The Conservation Story* (History Press, 2018), 29.

223 **They turned out to be Henry Ford, Harvey Firestone, and Thomas Edison:** Anholt, *Sanibel's Story*, 36.

224 **"seems to be the meeting ground":** Rogers, *The Shell Book*, 83.

224 **"Sanibel is too popular":** Ibid., 88.

224 **"this might in part explain the paucity":** William J. Clench, "The Marine Shells of Sanibel Florida," from 1921, in R. Tucker Abbott, ed., *The Best of the Nautilus* (American Malacologists, 1976), 19–20.

225 **the best fisher and sheller:** Sam Bailey, quoted in the Esperanza Woodring display, Sanibel Historical Museum and Village.

225 **Surviving the night:** Anholt, *Sanibel's Story*, 151.

226 **"The animal is strikingly marked":** Louise M. Perry, *Marine Shells of the Southwest Coast of Florida* (Paleontological Research Institution, 1940), 155–56.

226 **"always the conversation":** Albert Field Gilmore, "The Spell of 'Shell Shock,'" *Christian Science Monitor*, March 11, 1939.

226 **"a wiry little man":** Theodore Pratt, "Shell Shock," *Saturday Evening Post*, February 22, 1941.

227 **Large shell clubs launched in New York City and Philadelphia:** Paul Callomon, "The Nature of Names: Japanese Vernacular Nomenclature in Natural Science," Master of Science thesis, Drexel University, Philadelphia, Pennsylvania, September 2016, 87.

227 **By 1940, conchologists had set up in thirty-eight states:** Pratt, "Shell Shock."

227 **Shell auctions such as those at Philadelphia's Buttonwood Farm:** Paul

Callomon, the Academy of Natural Sciences of Drexel University, "The Buttonwood Farm Auctions," presentation, 2019 Conchologists of America Convention, June 15–21, 2019, Captiva, Florida.

227 **"From the Aleutians to Africa":** Adeline Pepper, "A Field Day for Shell Collectors," *New York Times*, September 7, 1958.

227 **The police found not one dead body:** Pratt, "Shell Shock."

228 **"one she described later as a 'journey toward insight'":** Reeve Lindbergh, "Of Flight and Life," *Virginia Quarterly Review* 88, no. 4 (Fall 2012): 78.

228 **Charles described their approach on the Sanibel ferry:** Charles Augustus Lindbergh, *The Wartime Journals of Charles A. Lindbergh* (Harcourt, Brace, Jovanovich, 1970), 308–309.

228 **Nine months later, their first daughter was born:** Susan Hertog, *Anne Morrow Lindbergh: Her Life* (Nan A. Talese Doubleday, 1999), 383.

228 **Ickes called it "the bible of every American Nazi":** Lynne Olson, *Those Angry Days: Roosevelt, Lindbergh, and America's Fight Over World War II* (Random House, 2013; quote is from the 2014 paperback edition), 313.

229 **She rented a cottage at the bend:** Hertog, *Anne Morrow Lindbergh*, 425.

229 **"not primarily beautiful, but functional":** Anne Morrow Lindbergh, *Gift from the Sea* (Pantheon, 1955), 83.

229 **320,000 copies its first year:** Elsie F. Mayar, *My Window on the World: The Works of Anne Morrow Lindbergh* (Archon, 1988), 64.

229 **remained on the bestseller list for eighty weeks:** Eric Pace, "Anne Morrow Lindbergh, 94, Dies; Champion of Flight and Women's Concerns," New York Times, February 8, 2001.

230 **"The acquisitive instinct is incompatible":** Lindbergh, *Gift from the Sea*, 114.

230 **"To dig for treasures":** Ibid., 17.

231 **a hellacious stench in the Sanibel night air:** Pratt, "Shell Shock."

231 **"Beach-worn empty shells":** Pepper, "A Field Day for Shell Collectors."

231 **"Real shell collectors must take a shell alive":** Horace Sutton, "Shell Collecting Gains as Hobby," *Washington Post*, January 5, 1958.

232 **a warm mentor and peerless field leader:** R. Tucker Abbott, "A Farewell to Bill Clench," *The Nautilus* 98, no. 2 (April 1984): 57.

232 **Abbott was assigned to the Navy's Medical Research Unit in Guam:** Lynn Scheau, "Robert Tucker Abbott, Lieutenant, U.S. Navy," *Arlington National Cemetery* website, arlingtoncemetery.net.

233 **she'd idolized Abbott:** Personal communication with Carole Marshall.

233 **"The ivory-tower boys look askance":** Sam Hodges, "A Populist for Mollusks," *Orlando Sentinel*, May 31, 1989.

234 **began to suggest the unthinkable:** LeBuff and Anholt, *Protecting Sanibel and Captiva Islands*, 142–43.

234 **"certainly not from over-collecting":** Martha Riley Kinney, "Mr. Seashell," *Island Reporter*, July 22, 1977.

235 **He made the case that Sanibel had an obligation:** Cindy Chalmers, "Proposed Ordinance Prohibits Any Live Shelling," *Sanibel Captiva Islander*, December 28, 1982.

235 **implored the panel to limit live shelling:** "Live Shell Limit to Be Submitted to Council," *Sanibel Captiva Islander*, February 27, 1979.

235 **It was 1987 before Emerson and fellow conservationists:** Chalmers, "Proposed Ordinance Prohibits Any Live Shelling."

235 **By that time, Abbott's dream of a "monument to shells-for-people":** Lynn Scheu, "Robert Tucker Abbott, Lieutenant, U.S. Navy," arlingtoncemetery.net.

236 **"the most widely known malacologist":** M.G. Harasewych, "The Life and Malacological Contributions of R. Tucker Abbott (1919-1995)," *The Nautilus*, 110 (1997): 55.

236 **Herbert analyzed drill holes:** Gregory P. Dietl, Gregory S. Herbert, and Geerat J. Vermeij, "Reduced Competition and Altered Feeding Behavior among Marine Snails After a Mass Extinction," *Science* 306 (December 2004): 2229–31.

238 **The paralysis suggests a fast-acting toxin:** Jose H. Leal and Rebecca Mensch, "Swift Strike by the Gastropod *Scaphella junonia* on Its Gastropod Prey *Americoliva sayana*," *Bulletin of Marine Science*, August 2018.

238 **precipitous drop in the number of shells on the beach:** Michal Kowalewski et al., "Vanishing Clams on an Iberian Beach: Local Consequences and Global Implications of Accelerating Loss of Shells to Tourism," PLoS ONE 9(1) (January 8, 2014).

TEN: THE END OF ABUNDANCE

244 **bearing a striking resemblance:** Benjamin A. Palmer et al., "The Image-Forming Mirror in the Eye of the Scallop," *Science* 358, no. 6367 (December 2017): 1172.

245 **for every ten species of gastropods:** Elizabeth Gosling, *Marine Bivalve Molluscs* (John Wiley & Sons, 2015), 10.

245 **"truly stood among the most favored":** R. Tucker Abbott, *Kingdom of the Seashell* (Crown, 1982), 170.

245 **The words *scallop* and *shell*:** B. Wiledge, "Shell: A Word's Pedigree," in Ian Cox, ed., *The Scallop: Studies of a Shell and Its Influence on Humankind* (The Shell Transport and Trading Co., 1957), 11–12.

246 **The modern Dutch word for shell:** Ibid., 12.

246 **the Danish biologist Otto Friedrich Müller:** W. J. Rees, "The Living Scallop," in Cox, *The Scallop: Studies of a Shell*, 18.

246 **Müller is better remembered:** "Otto Müller," *Linda Hall Library*, lindahall.org.

246 **Scallops became such common architectural motifs:** Sir Mortimer Wheeler, "A Symbol in Ancient Times," in Cox, *The Scallop: Studies of a Shell*, 37.

246 **an earth mother dating to Paleolithic times:** Frances Stahl Bernstein, "This is Where I Found Her: The Goddess of the Garden," *Journal of Feminist Studies in Religion* 12, no. 2 (Fall 1996): 105.

246 **Venus appears as this older goddess:** Ibid., 113.

246 **Around her neck:** Ibid.

246 **In an ancient celebration of Demeter:** Rena Veropoulidou and Daphne Niko-

laidou, "Ritual Meals and Votive Offerings: Shells and Animal Bones at the Archaic Sanctuary of Apollo at Ancient Zone, Thrace, Greece," in Alexandra Livarda et al., eds., *The Bioarchaeology of Ritual & Religion* (Oxbow, 2018), 95.

247 **Archaeologists researching archaic feasts:** Ibid., 94.

247 **a ritual connected with a watery underworld:** Hope B. Werness, *The Continuum Encyclopedia of Animal Symbolism in Art* (Continuum International, 2003), 359.

247 **"To see hundreds of scallops":** Julia Ellen Rogers, *The Shell Book* (Doubleday, 1908), 412.

247 **as many as 50 million organisms:** *Smithsonian Ocean*, "Seagrass and Seagrass Beds," ocean.si.edu.

248 **rank close to estuaries and wetlands in all they give:** Robert Costanza et al., "The Value of the World's Ecosystem Services and Natural Capital," *Nature* 387 (1997): 253–60.

248 **the grasses contribute $1.9 trillion a year:** Michelle Waycott et al., "Accelerating Loss of Seagrasses around the Globe Threatens Coastal Ecosystems," *Proceedings of the National Academy of Sciences* 106, no. 30 (July 2009): 12377–381.

248 **President Lincoln's pointy chin:** For this and other descriptions of the Bay Scallop's life cycle and habitats, I am indebted to the research scientist Dr. Stephen P. Geiger in the Marine Fisheries Research section of the Florida Fish and Wildlife Conservation Commission.

249 **with special attention to mollusks:** Ernest Ingersoll, "Special Report on the Mollusca," in *Bulletin of the United States Geological and Geographical Survey of the Territories*, Second Series, No. 1 (1875): 127.

249 **"little snails with slippery tales":** Ernest Ingersoll, "In a Snailery," *Scribner's Monthly* 17, no. 6 (April 1879): 796.

249 **"vast multitude and variety":** Ibid.

249 **Native Americans harvested fewer scallops:** Ernest Ingersoll, "The Scallop Industry," in G. B. Goode, ed, *The Fisheries and Fish Industries of the United States*, Section V, Vol. II (U.S. Government Printing Office, 1887), 567.

249 **more common in the younger layers:** Clyde L. MacKenzie Jr., "History of the Bay Scallop: *Argopecten irradians*, in Eastern North America, Massachusetts through Northeastern Mexico," *Marine Fisheries Review* 70, no. 3–4 (2008): 16.

249 **Ingersoll visited America's commercial scalloping industry in 1879:** Clyde L. MacKenzie Jr. "Biographic Memoir of Ernest Ingersoll: Naturalist, Shellfish Scientist, and Author," *Marine Fisheries Review* 53, no. 3 (1991): 26.

250 **"with cradles behind them":** Ingersoll, "The Scallop Industry," 572–73.

250 **it spread the young scallops:** Ibid., 568.

250 **Dredgers that hauled up to a hundred bushels:** Ibid., 571.

250 **"the ruin which was being perpetrated":** Ibid., 569.

250 **"Then would come a sudden accession":** Ibid.

251 **"the yield there not does not compare":** Ibid., 570.

252 **Those not trying to scam them felt obliged to help:** Walter Starkie, *The Road to Santiago: Pilgrims of St. James* (University of California Press, 1965), 67.

252 **It's also full of stereotypes about the local people:** *Codex Calixtinus, Pilgrim's Guide*, "Chapter VII. The Lands and Peoples along the Camino de Santiago," culturedcamino.com/history/codex-calixtinus-v/.

252 **"For health is given to the sick":** *Codex Calixtinus, Pilgrim's Guide*, "Chapter IX. The City and Basilica of St. James, Apostle of Galicia," culturedcamino. com/history/codex-calixtinus-v/.

253 **a "miraculous fountain":** Ibid.

253 **"*obsiti conchis*, 'all beshelled about' ":** Thomas Fuller, *The Church History of Britain: From the Birth of Jesus Christ Until the Year* MDCXLVIII (T. Tegg and Son, 1837), 227.

254 **DNA evidence showed that he suffered:** Simon Roffey et al., "Investigation of a Medieval Pilgrim Burial Excavated from the *Leprosarium* of St. Mary Magdalen, Winchester, UK," *Neglected Tropical Diseases* 11, no. 1 (January 2017).

254 **Today, hikers set out for a month:** Rick Steves, *Rick Steves' Europe*, "A Medieval Pilgrimage in Modern Times," ricksteves.com.

254 **"An effective reply":** Ernest Ingersoll, "The Scallop and Its Fishery," *American Naturalist* XX, no. 12 (December 1886): 1004.

254 **"now become so rare as to be a prize":** Ibid. Ingersoll was writing about *Pecten tenuicostatum*. The current name is *Placopecten magellanicus*.

255 **The collapse was long blamed:** Matthew P. J. Oreska et al., "The Bay Scallop Industry Collapse in Virginia and Its Implications for the Successful Management of Scallop-Seagrass Habitats," *Marine Policy* 75 (2017): 119.

255 **commercial fishers took young scallops:** Ibid., 122.

255 **"safe from economic or biological depletion":** James F. Murdock, "Investigation of the Lee County Bay Scallop Fishery," University of Miami Marine Laboratory for the Florida State Board of Conservation, March 1955, 9.

256 **Aerial photographs from the late 1930s:** Sara-Ann F. Treat et al., eds., *Proceedings, Tampa Bay Area Scientific Information Symposium, May 1982* (Bellwether Press, 1985), 213.

256 **By the early 1970s:** Ibid., 212.

256 **averaging 300,000 bushels a year nationwide:** Clyde MacKenzie, "The Bay Scallop, *Argopecten irradians*, Massachusetts through North Carolina: Its Biology and the History of Its Habitats and Fisheries," *Marine Fisheries Review* 70, no. 3–4 (2008): 6.

256 **less than 1 percent:** *Final Report of the New York State Seagrass Task Force*, New York State Department of Environmental Conservation, December 2009, 14.

256 **"everyone treated the brown tides as alien encounters":** Brett Walton, "Ecosystems Are Dying as Long Island Contends with a Nitrogen Bomb," *Circle of Blue*, November 4, 2015.

256 **in no less than twenty-four articles:** The *New York Times* archive.

257 **The turnaround began with tough sewage regulations:** Hannah Waters, "Bringing Back Tampa Bay's Seagrass," *Smithsonian Ocean*, January 2017.

257 **the populations would again collapse:** Jay R. Leverone et al., "Increase in Bay Scallop Populations Following Releases of Competent Larvae in Two West Florida Estuaries," *Journal of Shellfish Research* 29, no. 2 (2010): 395–406.

257 **In 2019 they found fifty***: Tampa Bay Watch,* "Scallop Search Counts Since 1996," tampabaywatch.org.

258 **scientists have reared Bay Scallops:** *Cornell Cooperative Extension, Suffolk County,* "Scallop Program: Overview and Results," ccesuffolk.org.

258 **a massive die-off:** Charity Robey, "The Baymen's Nightmare: All the Scallops Are Dead," *New York Times,* November 7, 2019.

258 **this die-off seemed different:** Christopher Walsh, "Is Climate the Culprit?" *East Hampton Star,* November 7, 2019.

259 **a realized utopia:** A. W. and Julian A. Dimock, *Florida Enchantments* (Hodder and Stoughton, 1909), 3.

259 **an Alcatraz for a small group of monkeys:** "Monkey Island of Homosassa," *Atlas Obscura,* atlasobscura.com.

262 **"a man who knew the wild":** John McPhee, "Wild Man," *New York Times,* January 10, 1976.

262 **"express the joy of life":** Euell Gibbons, *Stalking the Blue-Eyed Scallop* (David McKay Co. 1964; paperback field guide, 1973), 66.

262 **"express the joy of living":** Rogers, *The Shell Book,* 412.

263 **"They have the cornea, lens, choroid coat, and optic nerve":** Ibid.

263 **"having cornea, lens, choroid coat, and optic nerve":** Gibbons, *Stalking the Blue-Eyed Scallop,* 67.

263 **told McPhee his childhood story:** John McPhee, "A Forager," *New Yorker,* April 6, 1968.

263 **"There are dozens of others":** Gibbons, *Stalking the Blue-Eyed Scallop,* vii.

263 **"that very peak of molluscan evolution":** Ibid., 308.

264 **The secret . . . is the rule of threes:** Gibbons, *Stalking the Blue-Eyed Scallop,* 88–89.

264 **It is, rather, the exponential growth:** Lijing Cheng et al., "Record-Setting Ocean Warmth Continued in 2019," *Advances in Atmospheric Sciences* 37 (February 2020): 137–42.

ELEVEN: SAVING THE QUEENS

267 *Aliger gigas***:** the accepted name was changed from *Lobatus gigas* in 2020. Originally named by Linnaeus, *Strombus gigas,* in 1758.

267 **to elude predators and break up their scent trail:** Megan Davis, Species Profile, Queen Conch, Southern Regional Aquaculture Center, USDA.

268 **knives clank and conch guts fly:** My visit to the annual Conch Fest event was in October 2019.

269 **in no less than twenty-six countries:** Martha C. Prada et al., *Regional Queen Conch Fisheries Management and Conservation Plan* (Rome: Food and Agriculture Organization of the United Nations, 2017), 39.

269 **archaeologists found imagery of Queen Conchs as melee weapons:** Ashley

E. Sharpe, "The Ancient Shell Collectors: Two Millennia of Marine Shell Exchange at Ceibal, Guatemala," *Ancient Mesoamerica* 30 (2019): 509.

269 **"The seashell would be loaded with explosives":** David W. Belin, CIA Commission Report, "Summary of Facts—Investigation of CIA Involvement in Plans to Assassinate Foreign Leaders," June 5, 1975, 60. Agents never got so far as to specify the species for the shell decoy, though before the papers were declassified it was always rumored to be a Queen Conch. The CIA considered a related plot to have General James Donovan, then negotiating with Castro for release of Bay of Pigs prisoners, give the dictator a contaminated skin-diving suit as a gift. According to the CIA Commission report of 1975: "The CIA plan was to dust the inside of the suit with a fungus producing Madura foot, a disabling and chronic skin disease, and also contaminating the suit with *tuberculosis bacilli* in the breathing apparatus." Donovan did not know about the scheme, which was dropped "because Donovan on his own volition gave Fidel Castro an uncontaminated skin-diving suit as a gesture of friendship."

269 **leap to deep-sand channels when they reach old age:** Prada et al., "Regional Queen Conch Fisheries Management and Conservation Plan," 39–41.

270 **scientists say it takes at least ninety:** Gabriel A. Delgado and Robert A. Glazer, "Demographics Influence Reproductive Output in Queen Conch: Implications for Fishery Management," *Bulletin of Marine Science* 96, no. 4 (March 2020).

270 **Fewer than 1 percent:** I have relied on several conch scientists to describe the Queen Conch's life cycle, especially Dr. Megan Davis.

270 **The atom-shaped veliger:** I have based this section on video captured in Dr. Davis's microscope at Harbor Branch.

271 **hired out a boat and some sandy-haired "Conch boys":** Charles Frederick Holder, "In Conch Land," *The Outlook*, January 5, 1895, 101. (See next note; Holder is describing an earlier time.)

271 **"a world beneath the sea":** Charles Frederick Holder, "On a Coral Reef," *The Californian*, October 1892, 611. (Holder is describing the reef he saw in Key West twenty-eight years before.)

271 **Large cowries, which the Conch boys called micramocks:** Ibid.

271 **"the conchs were feeding on the weed in such numbers":** Holder, "In Conch Land," 101.

271 **with a vision to turn the fishing archipelago into an empire of cotton:** Thelma Peters, "The American Loyalists in the Bahama Islands: Who They Were," *Florida Historical Quarterly* 40, no. 3 (January 1962): 226–40.

272 **"usually applied in derision":** "Key West and the Conchs," *New York Times*, May 11, 1884, 612.

272 **"a large, rough class of men":** "Life on the Coral Keys: Odd Customs of a People in the Gulf of Mexico," *Atlanta Constitution*, March 10, 1887.

272 **The Conchs still held Cockney accents:** *Florida: A Guide to the Southernmost State*, Works Progress Administration Federal Writers' Project (Trinity University Press, 2014), 3099.

272 **The spectacle was not as lighthearted:** Katherine Alex Beaven, "That Time

the Florida Keys Tried to Secede from the U.S. by Dropping Conch Fritter Bombs," *Vice*, April 21, 2017.

272 **Wardlow read a proclamation:** Dennis Wardlow, "The Conch Republic Proclamation of Secession," speech to U.S. federal court in Miami, Florida, April 23, 1982.

273 **strung choice meats on a stick:** Henry O'Malley, "Report of the U.S. Commissioner of Fisheries for the Fiscal Year 1923," U.S. Department of Commerce Bureau of Fisheries, 59.

273 **"The supply could easily be depleted":** Ibid., 59.

273 **Commercial shellers steered barges over reefs:** Gilbert Voss, "First Underseas Park," *Sea Frontiers* 6, no. 2 (May 1960): 90–91.

273 **if Americans did not preserve their great undersea monuments:** Ibid., 87–94.

273 **the queen was "becoming rare in the Florida Keys":** Rachel Carson, *The Edge of the Sea* (Houghton Mifflin, 1955), 231–32.

274 **Thompson compared their extirpation to the slaughter of plume birds:** Christopher Lane, "Snails Popular, Endangered," *Miami Herald*, November 18, 1980.

274 **"will enable the species to replenish itself":** United Press International, "State Okays New Limits on Fishing, Scalloping," *St. Petersburg Times*, May 27, 1985.

274 **Its 7,500 residents:** *2010 Census of Population and Housing*, The Commonwealth of The Bahamas Department of Statistics, see Andros. This was a population decline of 2.5 percent from 2000, while the Bahamas population increased 15 percent.

275 **who had lived in the Bahamas for more than a thousand years:** William F. Keegan, *The People Who Discovered Columbus: The Prehistory of the Bahamas* (University of Florida Press, 1992).

275 **widely circulated in Italy the following year:** Theodore J. Cachey Jr., "Between Humanism and New Historicism: Rewriting the New World Encounter," *Annali d'Italianistica* 10 (1992): 28–46. This is the woodcut by Giuliano Dati that accompanied Dati's Italian translation of Columbus's letter announcing "discovery."

275 **found bone built up around their ears in response to the pressure:** Andrew Todhunter, "Deep Dark Secrets: The Blue Holes of the Bahamas Yield a Scientific Trove That May Even Shed Light on Life Beyond Earth," *National Geographic* 28, no. 2 (2010). nationalgeographic.com/magazine/2010/08/bahamas-caves-underwater-blue-holes/.

275 **and roasted the conch meat on *barbacoa*:** Keegan, *The People Who Discovered Columbus*, 147.

275 **the single Lucayan canoe that survives:** William F. Keegan and Lisabeth A. Carlson, *Talking Taino: Caribbean Natural History from a Native Perspective* (Fire Ant Books, 2008), 85.

275 **They called the conchs *cobo*:** Ibid., 58.

275 **no Native people remained:** Carl Ortwin Sauer, *The Early Spanish Main* (University of California Press, 1966), 160.

275 **"an extension of slave hunting beyond the empty islands":** Ibid.

276 **when newly freed mariners relocated to Andros:** Grace Turner, *Honoring Ancestors in Sacred Space* (University of Florida Press, 2017), 35.

276 **the first adhesive postage:** Michael Craton and Gail Saunders, *Islanders in the Stream: A History of the Bahamian People*, Vol. Two (University of Georgia Press, 2000), 76.

276 **"this fish is not so plentiful as it used to be":** Augustus J. Adderley, "The Fisheries of the Bahamas," 1883, collected in *Conferences Held in Connection with The Great International Fisheries Exhibition*, Vol. V, Part II (William Clowes & Sons, 1884), 20.

277 **helped islanders weather economic downturns and disasters:** Thomas Henry Magness III, "The Conch: An Expandable Folk Food of the Bahamas," master's thesis, University of Wisconsin, Madison, 1969, 54–55. Magness was researching on behalf of the Lerner Marine Laboratory of the American Museum of Natural History, on Bimini, to explore "potential for intensifying utilization of (Queen Conchs) in the Caribbean," iv.

277 **"hurricane ham":** Dee Carstarphen, *The Conch Book* (Pen & Ink Press), 1982, 41.

277 **Conch and Lobster decide to race:** Elsie Clews Parsons, *Folk-Tales of Andros Island, Bahamas* (The American Folk-Lore Society, 1918), 102–103.

277 **"It was like after the Zombie Apocalypse":** author interview with Gabriel Delgado.

279 **FedExed their gonads:** Gabriel Delgado, Claudine T. Bartels, Robert A. Glazer, Nancy J. Brown-Peterson, and Kevin J. McCarthy, "Translocation as a Strategy to Rehabilitate the Queen Conch (*Strombus gigas*) Population in the Florida Keys," *Fishery Bulletin* 102, no. 2 (April 2004): 278–88.

279 **"I've never seen":** Author interview, Dr. Nancy Brown-Peterson.

280 **Taylor's grandfather fished for conchs:** Bertram Taylor, Conch Fest 2018, South Andros Video.

280 **some of the last commercial sailing vessels:** Carstarphen, *The Conch Book*, 26.

280 **"may boat 1,000 or more of the mollusks":** John E. Randall, "Monarch of the Grass Flats," *Sea Frontiers* 9, no. 3 (1963): 160–67.

280 **as many as sixty divers:** Prada et al., "Regional Queen Conch Fisheries Management and Conservation Plan," 3.

281 **conch densities in the historic grounds at Andros:** Allan Stoner and Martha Davis, "Queen Conch Stock Assessment: Historical Fishing Grounds, Andros Island, Bahamas," Report of Community Conch to the Nature Conversancy, Nassau, Bahamas, June 2010.

281 **The once-massive herds:** Allan W. Stoner, Martha H. Davis, and Andrew S. Kough, "Relationships between Fishing Pressure and Stock Structure in Queen Conch (*Lobatus gigas*) Populations: Synthesis of Long-Term Surveys and Evidence for Overfishing in the Bahamas," *Reviews in Fisheries Science & Aquaculture*, October 2018.

281 **"The data are showing a serial depletion":** Author interview, Andrew Kough.

282 **it's the shell's thickness that counts:** Allan W. Stoner et al., "Maturation and

Age in Queen Conch: Urgent Need for Changes in Harvest Criteria," *Fisheries Research* 131 (2012): 76–84.

282 **some fishers say they simply don't trust:** Elizabeth H. Silvy et al., "Illegal Harvest of Marine Resources on Andros Island and the Legacy of Colonial Governance," *British Journal of Criminology* 58, no. 2 (2018): 332–50.

282 **In 2017 he led a sweeping study:** Andrew S. Kough et al., "Ecological Spillover from a Marine Protected Area Replenishes an Over-Exploited Population across an Island Chain," *Conservation Science and Practice* 1, no. 3 (March 2019).

283 **its indirect death toll nearly three thousand:** "Ascertainment of the Estimated Excess Mortality from Hurricane Maria in Puerto Rico," Project report, The Milken Institute School of Public Health, The George Washington University, August 2018.

283 **can read signs:** Jordan A. Massie et al., "Going Downriver: Patterns and Cues in Hurricane-Driven Movements of Common Snook in a Subtropical Coastal River," *Estuaries and Coasts*, July 2019.

283 **With their carbonate strength and rough textures:** Rowan Jacobsen, "Beyond Seawalls," *Scientific American*, April 2019.

284 **It finally quieted down:** The description of Buck Island's displaced, groaning reef is based on author interviews with Zandy Hillis-Starr, resource management chief, Buck Island Reef National Monument.

284 **Such was the fate of hordes of juvenile Queen Conchs:** Enrique Pugibet Bobea et al., "Hurricane Matthew Impacts to Marine and Coastal Biodiversity on Southern Coast of Dominican Republic," *Journal of Agricultural Science and Technology* B, no. 7 (2017): 415–25.

285 **It smothered conchs:** Andres Viglucci, "Puerto Rico Recovery: Roofless Homes, Closed Schools, an Island Left to Fend for Itself," *Miami Herald*, September 20, 2018.

285 **"It brings out the tears":** Ibid.

286 **"It's not only the science but it's also the art of growing the conch":** Author interview with Dr. Megan Davis, research professor of aquaculture and stock enhancement at Florida Atlantic University's Harbor Branch Oceanographic Institute.

286 **"The Queen Conch seems an alert and sentient creature":** Carson, *The Edge of the Sea*, 232.

287 **they were abnormal compared with those offshore:** Gabriel A. Delgado, Robert A. Glazer, and Nancy J. Brown-Peterson, "Arrested Sexual Development in Queen Conch (*Lobatus gigas*) Linked to Abnormalities in the Cerebral Ganglia," *Biological Bulletin* 237, no. 3 (December 2019): 241.

287 **While those waters "are hardly pristine":** Ibid., 247.

287 **more than 90 percent of Earth's warming:** LuAnn Dahlman and Rebecca Lindsey, "Climate Change: Ocean Heat Content," Climate.gov, NOAA, August 17, 2020.

287 **sea-temperature records going back to the lighthouse keepers:** Ilsa B. Kuff-

ner et al., "A Century of Ocean Warming on Florida Keys Coral Reefs: Historic In Situ Observations," *Estuaries and Coasts* 38 (2015): 1085–96.

287 **Such hot spots can lower egg and sperm counts:** Delgado et al., "Arrested Sexual Development in Queen Conch," 247.

TWELVE: GLOWING FUTURE

291 **Alison Sweeney lingers at a plunging coral ledge:** This chapter is based on author interviews with Alison Sweeney and her team in Philadelphia and Palau.

293 **"the clam signifies power and it signifies persistence":** Author interview with Elsa Sugar.

293 **Many of the islands' oldest surviving tools:** Many of the archeological and cultural details in the chapter are from Belau National Museum, Koror, Palau.

293 **the earliest islanders depleted the giant clams:** S. M. Fitzpatrick and T. J. Donaldson, "Anthropogenic Impacts to Coral Reefs in Palau, Western Micronesia during the Late Holocene," *Coral Reefs* 26 (2007): 915–29.

293 **the culture's creation lore began with a giant clam:** Belau National Museum, Koror, Palau.

294 **The microalgae, called zooxanthellae:** Lincoln Rehm, "Giant Clams," in *Paradise of Nature: Understanding the Wonders of Palau* (Palau International Coral Reef Center, 2017), 160.

294 **who married in the church in 1822 with the help of a fake baptism certificate:** Graham Robb, *Victor Hugo: A Biography* (W. W. Norton, 1997), 96.

295 **"ensured that the sonorous notes of the conch shell":** John Wrightson, *Mission to Melanesia: Out of Bondage* (Janus Publishing Co., 2005), 194.

295 **"Who would not feel cosmically heartened":** Javier Senosiain, *Bio-Architecture* (Architectural Press, Elsevier, 2003), 49.

295 **"some spendthrift and gourmand":** Pliny the Elder, *The Natural History of Pliny*, Vol. VI, trans. John Bostock and H. T. Riley (H. G. Bohn, 1857), 27.

295 **In the Maori legend *Rata's Voyage*:** Roland Burrage Dixon, *Oceanic Mythology*, Vol. IX in Louis Herbert Gray, ed., *The Mythology of All Races* (Marshall Jones Co., 1916), 69.

296 **the mollusk cuts his hand clean off:** Ibid., 194–95.

296 **"caricatures of every gigantic and imaginary creature":** Jules Verne, *Twenty Thousand Leagues Under the Sea* (Butler Brothers, 1887), 5.

296 **"a beautiful bed of tulips":** Lovell Augustus Reeve, *Conchologia Iconica: Illustrations of the Shells of Molluscous Animals*, Vol. XIV (Lovell Reeve & Co., 1864), 188.

296 **"So powerful are they":** Charles Frederick Holder and Joseph Bassett Holder, *Elements of Zoology* (ND. Appleton and Company, Appletons' Science Text-Books, 1885), 56.

296 **"they serve as gigantic traps":** "Giant Clams Trap Sea Divers in Grip of Shells," *Popular Mechanics Magazine*, May 1924, 685.

296 **"has caused the deaths of many natives":** "Giant Clam Is a Man-Killer," *New York Times*, December 12, 1937.

297 **were briefed on the "man-eating clams":** James D. Hornfischer, *The Fleet at Flood Tide: America at Total War in the Pacific, 1944–1945* (Bantam Books paperback edition, 2017), 93.

297 **Clark profiled one of her guides:** Eugenie Clark, "Siakong—Spear-Fisherman Pre-Eminent," *Natural History*, May 1953, 237–34.

299 **appear to be thriving despite conditions:** Author interview with Yimnang Golbuu, CEO, Palau International Coral Reef Center, Koror.

300 **13,500 aluminum drawers:** Details on the mollusk collection at the National Academy of Sciences are based on interviews and tours with Paul Callomon, collection manager, and Gary Rosenberg, curator.

300 **By measuring the relative weights of the isotopes:** Author interviews with Michelle Gannon, Drexel University.

302 **"I was fascinated because it's self-assembly":** Author interview with University of Pennsylvania materials science and engineering professor Shu Yang, Penn Laboratory for Research on the Structure of Matter.

303 **only about 10 percent of the animal's flesh:** John S. Lucas, "Giant Clams: Mariculture for Sustainable Exploitation," in M. Bolton, ed., *Conservation and the Use of Wildlife Resources* (Chapman & Hall, 1997), 81.

303 **poachers illegally hauled in up to half a million clams a year:** Ibid., 82.

303 **During a crackdown in Palau:** South Pacific Commission, *Fisheries Newsletter*, no. 33, June 1985, 4.

303 **driven the largest species locally extinct:** Mei Lin Neo et al., "Giant Clams (Bivalvia: Cardiidae: Tridacninae): A Comprehensive Update of Species and Their Distribution, Current Threats and Conservation Status," *Oceanography and Marine Biology: An Annual Review* 55 (2017): 87–388; 132.

303 **Giant clamshells have long been venerated in China:** Lin Qiqing, "In China's Hawaii, an End to Small Fortunes from Giant Clams," *Sixth Tone*, January 19, 2017, sixthtone.com.

304 **fishers could pull in 80,000 yuan, or $12,000:** Christina Larson, "Shell Trade Pushes Giant Clams to the Brink," *Science* 351 (January 2016): 323–24.

304 **The Tanmen workshops and retail shops:** Qiqing, "In China's Hawaii, An End to Small Fortunes from Giant Clams."

306 **the first sightings of juvenile *T. gigas*:** Patrick Cabaitan and Cecilia Conaco, "Bringing Back the Giants: Juvenile *Tridacna gigas* from Natural Spawning of Restocked Giant Clams," *Coral Reefs* 36, no. 2 (June 2017): 519.

306 **hires guards to protect them:** Anne Moorhead, "Giant Clam Aquaculture in the Pacific Region: Perceptions of Value and Impact," *Journal of Development in Practice* 28, no. 5 (2018).

307 **If she could control for color and glow:** Author interview with Bernice Ngirkelau.

310 **"In the old days, clams saved us":** Author interview with F. Umiich Sengebau, Minister for Natural Resources, Environment, and Tourism, Palau.

THIRTEEN: TRUST IN NATURE

313 **Mandë Holford's mother would drop her and her four brothers and sisters:** Author interview with Mandë Holford.

313 **The museum's collection holds more than 6 million:** Dr. Estefania Rodriguez, Curator of Marine Invertebrates, American Museum of Natural History.

314 **more than eight hundred chemicals:** Alan J. Kohn, quoted in Ashley Braun, Meigan Henry, and Gord More, "The Killer Kiss of Kohn's Snails," *Hakai magazine*, September 27, 2016.

315 **The Dutch East India Company had taken the Bandas:** Melvin E. Page, ed., *Colonialism: An International Social, Cultural, and Political Encyclopedia*, Vol. One (ABC-CLIO, 2003), 711.

315 **"She had only held this little Whelk":** Georgius Everhardus Rumphius, *The Ambonese Curiosity Cabinet*, translated, edited, annotated, and with an introduction by E. M. Beekman (Yale University Press, 1999), 149.

315 **first in the official record of more than 140:** Alan J. Kohn, "Conus Envenomation of Humans: In Fact and Fiction," *Toxins* 11, no. 1 (2019): 10.

316 **"apparently a powerful neurotoxin":** A. J. Kohn, "Piscivorous Gastropods of the Genus Conus," *Proceedings of the National Academy of Sciences* 42, no. 3 (1956): 168–71.

316 **more than eight hundred species:** Ibid.

316 **About a hundred are piscivores:** Baldomero M. Olivera et al., "Prey-capture Strategies of Fish-hunting Cone Snails: Behavior, Neurobiology and Evolution," *Brain, Behavior and Evolution* 86 (September 2016): 58–74.

316 **it's very likely that the Geographer Cone:** Kohn, "Conus Envenomation of Humans."

316 **"Only local findings such as pain":** Shigeo Yoshiba, "Venom of a Stinging Snail Bekko-Imogai *Chelyconus fulmen*—Especially on Its Toxicities against Various Animals," *Japanese Journal of Medical Science and Biology* 32, no. 2 (April 1979): 112.

317 **At age seventeen, Olivera prepared the collection:** "Official Week in Review," *Official Gazette*, November 23–29, 1958.

317 **He prized the large cones especially:** Gisela Telis, "Finding Venom's Silver Lining," Howard Hughes Medical Institute, *HHMI Bulletin* 27, no. 2 (Spring 2014).

317 **"It was clear I wasn't going to be very competitive":** Rajendrani Mukhopadhya, "From DNA Enzymes to Cone Snail Venom: The Work of Baldomero M. Olivera," *Journal of Biological Chemistry* 287, no. 27 (June 2012).

317 **During the political and economic upheavals:** Ibid.

318 **"we were not dealing with just a few paralytic components":** Ibid.

318 **never actually made it all the way around the globe:** "April 27, 1521 CE: Magellan Killed in Philippine Skirmish," *National Geographic Resource Library*, "This Day in Geographic History," http://admin.nationalgeographic.org/thisday/apr27/magellan-killed-philippine-skirmish/.

319 **some of the poorest and most marginalized:** "Fish Forever in the Philippines," *Rare Conservation*, https://rare.org/program/philippines/.

319 **The exporters pack mixed tropical shells by the ton:** Adonis S. Floren, "The Philippine Shell Industry with Special Focus on Mactan, Cebu," Coastal Resource Management Project, Department of Environment and Natural Resources, Philippines, 2003.

319 **They ship them to China, Europe, and North America:** Paterno Esmaquel II, "Cebu Launches Crackdown on Illegal Seashell Trade," *GMA News Online*, May 18, 2011.

319 **The pressure has decimated giant clams, top shells, and once-thick beds:** Floren, "The Philippine Shell Industry."

319 **"may have been harvested beyond their sustainable yield":** Ibid.

319 **sold openly throughout the tourist markets:** Vincent Nijman, Denise Spaan, and K. Anne-Isola Nekaris, "Large-Scale Trade in Legally Protected Marine Mollusc Shells from Java and Bali, Indonesia," *PLoS ONE* 10, no. 12 (December 2015).

320 **Amey Bansod describes the stench:** Amey Bansod, "Re-Framing Sustainability in Livelihoods," unpublished manuscript, 2015, ameybansod.com.

320 **"Scenes of the apocalypse come early":** Amey Bansod, "The Story of Shells," *YouTube*, June 4, 2015.

322 **marine biologist Philippe Bouchet:** Author interview with Philippe Bouchet, Senior Professor, Muséum National d'Histoire Naturelle.

322 **still await description:** Philippe Bouchet et al., "How Many Species of Molluscs Are There in the World's Oceans, and Who Is Going to Describe Them?" in *Tropical Deep-Sea Benthos*, Vol. 29 (Muséum National d'Histoire Naturelle, 2016), 9–24.

322 **The 50,000 species of marine mollusks known:** Bouchet and MolluscaBase.

323 **more than half of all cancer drugs:** "Nature's Bounty: Revitalizing the Discovery of New Cancer Drugs from Natural Products," *National Cancer Institute*, March 22, 2019, cancer.gov.

323 **tarantulas seem to hold particular promise:** Emma Sargent, "From Venoms to Medicine," *Chemistry World*, December 12, 2017.

323 **Indigenous cultures on three continents:** Yann Henaut et al., "The Use of Tarantulas in Traditional Medicine on Three Continents," presented at 20th International Congress of Arachnology, Golden, Colorado, July 2–9, 2016.

323 **collect terebrids:** Cone snails, terebrids, and turrids are all part of the Conoidea superfamily of mollusks.

324 **"not an unalloyed good":** "Rembrandt's Century," *Art in Print*, May–June 2013, 12.

325 **placed Marbled Cones and Humpbacked Conchs together:** Sue-Ann Watson et al., "Ocean Acidification Alters Predator Behavior and Reduces Predation Rate," *Biology Letters* (February 2017).

325 **The cones in the more acidic water:** Ibid.

325 **increasing evidence that some are adapting their shells:** Manon Fallet et al., "Epigenetic Inheritance and Intergenerational Effects on Mollusks," *Gene* 729 (March 2020).

325 **we are altering the evolution of marine animals:** Sarah P. Otto, "Adaptation, Speciation and Extinction in the Anthropocene," *Proceedings of the Royal Society B: Biological Sciences* 285 (November 14, 2018).

325 **They shrank in modern centuries:** Stephen G. Hesterberg et al., "Prehistoric Baseline Reveals Substantial Decline of Oyster Reef Condition in a Gulf of Mexico Conservation Priority Area," *Biology Letters* (January 2020).

326 **"They have all died—there is nothing":** Mark Harrington, "They All Died: Peconic Bay Scallop Harvesting Season Appears Lost," *Newsday*, October 31, 2020.

326 **remained locked up in museums overseas:** Jim Specht and Lissant Bolton, "Pacific Islands' Artefact Collections: The UNESCO Inventory Project," *Journal of Museum Ethnography* No. 17 (2005): 58.

327 **an elaborate ritual of social connections and political trust:** Bronislaw Malinowski, "The Essentials of the Kula," chapter III in *Argonauts of the Western Pacific* (Dutton, 1922), 81.

327 **cut from large cone shells:** "The Kula Ring and Sir William MacGregor," University of Aberdeen Museums, December 22, 2015, uoamuseums.wordpress .com/2015/12/22/the-kula-ring-and-sir-william-macgregor/

327 **some are now being repatriated:** "Repatriation," Queensland Museum, South Brisbane, qm.qld.gov.au.

327 **"The Southern Hemisphere basically was our supermarket":** David Schindel, "Practicing Science Diplomacy at Museums and Science Centers," presented at "Science Diplomacy 2017," American Association for the Advancement of Science Center for Science Diplomacy, Washington, D.C., March 29, 2017 aaas .org/scidip2017.

327 **led one hundred . . . nations:** K. Divakaran et al., "When the Cure Kills—CBD Limits Biodiversity Research," *Science* 360 (June 29, 2018): 14056.

327 **the treaties did not stem:** Robert Blasiak et al., "Corporate Control and Global Governance of Marine Genetic Resources," *ScienceAdvances* 4, no. 6 (June 2018), advances.sciencemag.org/content/4/6/eaar5237

327 **Transnational corporations have now patented:** Ibid.

327 **the world's largest chemical maker:** Ibid.

327 **"has become the main obstacle to inventorying":** Author interview with Phillipe Bouchet.

CONCLUSION: THE OPEN END

331 **an estimated 12.5 million enslaved people:** The Trans-Atlantic Slave Trade Database. As of September 2020, scholars had identified 12,521,337 captive people as having embarked between 1501 and 1875; slavevoyages.org.

331 **Two million:** Ibid. As of September 2020, scholars had identified 10,702,656 captive people as having disembarked in those years.

332 **as many as a third of the enslaved people:** Bin Yang, *Cowrie Shells and Cowrie Money: A Global History* (Routledge, 2019), xii.

332 **gold rather than shell money dominated exchange:** Ibid., 169.

332 **hidden in drains and crevices:** Doig Simmonds, "A Note on the Excavations in Cape Coast Castle," *Transactions of the Historical Society of Ghana* 14, no. 2 (December 1973): 267.

332 **the "mercenary transactions":** Marc Shell, *Wampum and the Origins of American Money* (University of Illinois Press, 2013), 36.

332 **discourse and diplomacy:** Ibid.

333 **were called "slave money":** Akinwumi Ogundiran, "Of Small Things Remembered: Beads, Cowries, and Cultural Translations of the Atlantic Experience in Yorubaland," *International Journal of African Historical Studies* 35, no. 2–3 (2002): 440.

333 **The animals, themselves, were said to follow:** Historic and cultural sources recount many variations of this theme. Saidiya Hartman, *Lose Your Mother: A Journey Along the Atlantic Slave Route* (Farrar, Straus and Giroux, 2008), 209–10, covers some of the myths about cowries feeding on people and finds that they essentially come down to "the human cost of money."

333 **fear of the shells:** Author interview with Akinwumi Ogundiran.

333 **Yorùbá people had developed glassmaking technology:** Akinwumi Ogundiran, *The Yorùbá: A New History* (Indiana University Press, 2020), 252.

333 **chiefs and princes were encrusting rooms:** Ibid., 302.

333 **in the shape of a cowrie shell:** The late University of Ghana archaeologist James Anquandah, who led the first site excavations in the 1980s, provided the original description of the cavities as "cowrie motifs."

333 **some leading to hidden channels:** According to CT scanning on some of the figurines. Timothy Insoll, Benjamin Kankpeyeng, and Sharon Fraser, "Internal Meanings: Computed Tomography Scanning of Koma Figurines from Ghana," *African Arts* 49, no. 4 (Winter 2016): 24–32.

334 **what they call *kronkronbali* ("Olden Days Children"):** Benjamin W. Kankpeyeng and Christopher R. DeCorse, "Ghana's Vanishing Past: Development, Antiquities and the Destruction of the Archaeological Record," *African Archaeological Review* 21, no. 2 (June 2004): 104.

334 **between the sixth and fourteenth centuries:** Insoll et al., "Internal Meanings," 27–28.

334 **The scholar who unearthed her:** Author interview with Benjamin W. Kankpeyeng.

334 **little evidence of cowries' use before European traders sent billions:** Akinwumi Ogundiran, "Cowries and Rituals of Self-Realization in the Yoruba Region, ca. 1600–1860," in Ogundiran et al., *Materialities of Ritual in the Black Atlantic* (Indiana University Press, 2014), 70.

334 **"the harbinger of all good things":** Ogundiran, "Of Small Things Remembered," 456.

334 **became protectors in religion and rituals, etc.:** Ibid., 74.

335 **to carry the souls of the dead across the Volta:** Robin Law, "West Africa's Discovery of the Atlantic," *International Journal of African Historical Studies* 44, no. 1 (2011): 11. The idea that the souls of the dead had to cross water to reach the afterlife was widespread in West Africa before the transatlantic trade.

335 **"over the river's various arms":** H. C. Monrad, "A Description of the Guinea Coast and Its Inhabitants," collected in H. C. Monrad (1805–1809) and Johannes Rask (1708–1713), *Two Views from Christiansborg Castle*, trans. Selena Axelrod Winsnes (Sub-Saharan Publishers, 2009), 41.

335 **a nation of eight clans:** C.O.C. Amate, *The Making of Ada* (Woeli Publishing Accra, revised edition of 2017), 1; salt production, 126–27; European merchants and cowries, 136–37.

335 **absorbed most of the cowrie money:** Ogundiran, *The Yorùbá: A New History*, 253. Ogundiran found that more than 70 percent of the cowries reaching West Africa over these two hundred years came into the Bight of Benin.

335 **traders became middlemen:** Amate, *The Making of Ada*, 138–39.

335 **piled Maldivian cowries:** Yang, *Cowrie Shells and Cowrie Money*, 169, describes the Volta as a major route. See also J. Hogendorn and M. Johnson, *The Shell Money of the Slave Trade* (Cambridge University Press, 1986), who estimate that the cost of transporting a ton of cowries from Ada to Salaga in the late nineteenth century would have been $55 to $60 plus the price of the canoe.

335 **The journeys . . . took months:** Amate, *The Making of Ada*, 135.

335 **People and donkeys:** Yang, *Cowrie Shells and Cowrie Money*, 169.

335 **Kongensten is the first of the colonial forts:** Victoria A. Aryee et al., "Climate Change and the Mitigating Tool of Salvage Archaeology: The Case of the Fort Kongensten Site at Ada Foah, Ghana," *Legon Journal of the Humanities* 29, no. 2 (2018): 82–83.

335 **Cape Coast and others are also at risk:** Chris Stein, "Rising Seas Washing Away Ghana's Former Slave Forts," *Christian Science Monitor*, October 4, 2012.

336 **Glacial and ice-sheet melting are now the primary driver:** M. Oppenheimer et al., "Sea Level Rise and Implications for Low-Lying Islands, Coasts and Communities," *IPCC Special Report on the Ocean and Cryosphere in a Changing Climate (International Panel on Climate Change, 2019)*, 323.

336 **Coastlines altered by dams:** Ibid. The IPCC's special report mentions the Volta delta specifically on pages 371–72.

336 **at an average of 2 meters a year:** K. Appeaning Addo et al., "Assessment of the Dynamics of the Volta River Estuary Shorelines of Ghana," *Geoenvironmental Disasters* 7, (May 2020): 1–11.

336 **the largest reservoir by surface area in the world:** Benjamin Ghansah et al., "Mapping the Spatial Changes in Lake Volta Using Multitemporal Remote Sensing Approach," *Lakes & Reservoirs* 21, no. 3 (August 2016): 206–15.

336 **The dam forced 80,000 people:** Ben Daley, "The Impacts of the Akosombo

Dam," Unit 3.2.1 in *Understanding Sustainable Development* (University of London Centre for Development, Environment and Policy), soas.ac.uk.

336 **worth as much as the meat:** Daniel Adjei-Boateng et al., "The Current State of the Clam, *Galatea paradoxa*, Fishery at the Lower Volta River, Ghana," *IIFET Tanzania Proceedings* (2012), 9.

336 **Milled into grit, whitewash, or lime:** Ibid., 8.

336 **about two thousand people made their living:** Ibid., 1.

337 **5 million cedis a year:** Ibid., 8–9.

337 **a sustained shift to clam farming:** Personal communication with Fisheries Professor Daniel Adjei-Boateng, Kwame Nkrumah University of Science and Technology, Kumasi, Ghana.

337 **depend on a series of choices:** Ibid.

338 **ease our reliance on quarried limestone:** James P. Morris et al., "Shells from Aquaculture: A Valuable Biomaterial, Not a Nuisance Waste Product," *Reviews in Aquaculture* 11 (2019): 42–57.

338 **They hold potential for greening cement:** Ibid.

338 **best chance for saving imperiled marine life:** Cynthia Barnett, "Why It's Important to Save Our Seas' Pristine Places," *National Geographic*, February 2017 cover story.

338 **five times as much carbon as tropical forests:** *Project Drawdown*, "Coastal Wetland Protection," drawdown.org/solutions/coastal-wetland-protection.

339 **Wave and tidal power lag far behind:** Sophia Schweitzer, "Will Tidal and Wave Energy Ever Live Up to Their Potential?" *Yale Environment 360*, October 15, 2015.

339 **Ghanaian divers dropped six wave energy converters:** Author interview with Daniel Käller, project management officer, Seabased wave-energy company.

339 **the technology seems to pose less harm:** Andrea Copping et al., eds., "OES-Environmental 2020 State of the Science Report: Environmental Effects of Marine Renewable Energy Development Around the World," prepared by Pacific Northwest National Laboratory for the U.S. Department of Energy (September 2020), ii–327. Scientists are still studying possible auditory or electromagnetic effects on marine animals.

339 **Capturing tidal power may also slow:** Ibid., 133 and author interview with Daniel Käller.

339 **the first commercial wave energy plant:** "Ada Foah Wave Energy Project: TC's Energy USA and Power China Ltd. Sign Financing Agreement," *Modern Ghana*, July 26, 2020.

339 **Their reefs often beat:** The labyrinthine oyster beds and reefs that defined New York's outer harbor when the colonists arrived kept waves from washing over settlements even in the greatest tempests. Without the reefs to absorb the force of storms, wave energy reaching the coast today is up to 200 percent higher. Christine M. Brandon, "Evidence for Elevated Coastal Vulnerability Following Large-Scale Historical Oyster Bed Harvesting," *Earth Surface Processes and Landforms* 41, no. 8 (February 2016): 1136–43.

339 **traditional homes in Ada Foah are fortified with seashells:** Author interview with anthropologist Netty Carey.

340 **bright yellow numbers:** Netty Carey, "We Are in the Air: Land Claims and Liminal Space on the Volta River Estuary," presentation at 59th Annual African Studies Association annual meeting (December 2016), and author interview with Carey.

340 **green dots imposed by planners:** Calvin Hennick, "A Tale of Two Neighborhoods," *USGBC+* (magazine of the U.S. Green Building Council), September 2014.

340 **was sanctioned for its failure to work with residents:** "Trasacco Estates Development Company Faces Sanctions for Non-Compliance," *GBCGhana Online*, February 27, 2019.

340 **has finally come up with a resettlement plan:** "Trasacco EIA Presentation for Project at the Ada Estuary," February 26, 2019; www.youtube.com.

340 **"proper jobs" rather than having to fish or farm:** "Trasacco's New Project Expected to Create Jobs and Boost Tourism," *JoyNews*, March 28, 2018, www .youtube.com. The quote is from Ian Morris, managing director, Trasacco Estates Development Company.

341 **lost their power when bought or sold:** See E. M. Beekman's analysis in Georgius Everhardus Rumphius, *The Ambonese Curiosity Cabinet*, edited, annotated, and with an introduction by E. M. Beekman (Yale University Press, 1999), cv.

341 **often taking the side of the local people:** Ibid., civ–cv.

341 **warned of the rise of excess in the empire:** Andrew Wallace-Hadrill, "Pliny the Elder and Man's Unnatural History," *Greece & Rome* 37, no. 1 (April 1990): 80–96.

341 **"Never should it be forgotten":** Our guide, Ato Ashun, adds these words to those in the memorial mounted on the wall at Elmina, which says in part: "May Humanity Never Again Perpetuate Such Injustice Against Humanity." The admonition is also in his book on the slave castles of Ghana: *Ato Ashun, Elmina, the Castles, and the Slave Trade* (Nyakod Printing & Publishing, 2017 edition), 121.

341 **primary sources of wisdom rather than academic grist:** Jim Specht and Lissant Bolton, "Pacific Islands' Artefact Collections: The UNESCO Inventory Project," in *Journal of Museum Ethnography* 17 (2005): 67. Also see introduction by Anita Herle, "Pacific Ethnography, Politics and Museums," 1–7.

342 **leadership positions where they are forging a better future:** Electing women has been found to directly influence climate change action. See Astghik Maviasakalyan et al., "Gender and Climate Change: Do Female Parliamentarians Make a Difference?" *European Journal of Political Economy* 56 (January 2019): 151–64.

342 **The ancient engineers:** For new shell ring research from the Gulf, see Terry E. Barbour, Kenneth E. Sassaman, et al., "Rare pre-Columbian Settlement on the Florida Gulf Coast Revealed through High-Resolution Drone LiDAR," *Proceedings of the National Academy of Sciences USA* 16, no. 47 (November 2019): 23493–98.

342 **as the book also warned, perhaps too gently:** See this book's epigraph: "We can have a surfeit of treasures—an excess of shells, where one or two would be significant." Anne Morrow Lindbergh, *Gift from the Sea* (Pantheon, 1955), 115.

INDEX

Note: Page numbers in *italics* refer to illustrations. Page numbers after 350 refer to Notes section.